ANNALS OF
THE NEW YORK ACADEMY
OF SCIENCES

Volume 568

EDITORIAL STAFF
Executive Editor
BILL BOLAND
Managing Editor
JUSTINE CULLINAN
Associate Editor
COOK KIMBALL

The New York Academy of Sciences
2 East 63rd Street
New York, New York 10021

CALCIUM, MEMBRANES, AGING, AND ALZHEIMER'S DISEASE

ANNALS OF THE NEW YORK ACADEMY OF SCIENCES

Volume 568

CALCIUM, MEMBRANES, AGING, AND ALZHEIMER'S DISEASE

Edited by
Zaven S. Khachaturian, Carl W. Cotman, and Jay W. Pettegrew

The New York Academy of Sciences
New York, New York
1989

Library of Congress Cataloging-in-Publication Data

Calcium, membranes, aging, and Alzheimer's disease / edited by Zaven
S. Khachaturian, Carl W. Cotman, and Jay W. Pettegrew.
    p.   cm. — (Annals of the New York Academy of Sciences, ISSN
0077-8923 ; v. 568)
    Papers presented at a conference held by the John Douglas French
Foundation, the Pharmaceutical Manufacturers Association, and the
National Institute on Aging in Irvine, Calif., on Oct. 24–26, 1988.
    Includes bibliographical references.
    ISBN 0-89766-547-3 (alk. paper). — ISBN 0-89766-548-1 (pbk. :
alk. paper)
    1. Alzheimer's disease—Congresses. 2. Brain—Aging—Congresses.
3. Calcium in the body—Congresses. 4. Cell membranes—Congresses.
I. Khachaturian, Zaven S. II. Cotman, Carl W. III. Pettegrew, Jay
W. IV. John Douglas French Foundation for Alzheimer's Disease.
V. Pharmaceutical Manufacturers Association. VI. National Institute
on Aging. VII. Series.
    [DNLM: 1. Aging—congresses. 2. Alzheimer's Disease—congresses.
3. Brain—physiology—congresses. 4. Calcium—physiology—
congresses. 5. Cell Membranes—physiology—congresses. W1
AN626YL v. 568 / QV 276 C1438 1988]
Q11.N5 vol. 568
[RC523]
500 s—dc20
[618.97′831071]
DNLM/DLC
for Library of Congress   89-14029
                              CIP

B-B
Printed in the United States of America
ISBN 0-89766-547-3 (cloth)
ISBN 0-89766-548-1 (paper)
ISSN 0077-8923

ANNALS OF THE NEW YORK ACADEMY OF SCIENCES

Volume 568
December 29, 1989

CALCIUM, MEMBRANES, AGING, AND ALZHEIMER'S DISEASE[a]

Editors and Conference Chairs
ZAVEN S. KHACHATURIAN, CARL W. COTMAN, AND JAY W. PETTEGREW

CONTENTS

[a]The papers in this volume were presented at a conference entitled Calcium, Membranes, Aging, and Alzheimer's Disease, which was sponsored by the John Douglas French Foundation, the Pharmaceutical Manufacturers Association, and the National Institute on Aging and held at the Arnold and Mabel Beckman Center of the National Academies of Sciences and Engineering in Irvine, California on October 24–26, 1988.

Financial assistance was received from:

- JOHN DOUGLAS FRENCH FOUNDATION (through a grant from ALLIED-SIGNAL CORP.)
- NATIONAL INSTITUTE ON AGING/NATIONAL INSTITUTES OF HEALTH
- PHARMACEUTICAL MANUFACTURERS ASSOCIATION
- SIGMA TAU FOUNDATION

Introduction and Overview

ZAVEN S. KHACHATURIAN

Neuroscience and Neuropsychology of Aging Program
National Institute on Aging
Building 31C, Room 5C35
National Institutes of Health
Bethesda, Maryland 20892

Although the field of aging research has many theories,[1] no single theory appears to account for most age-dependent brain changes. For example, at present no one seriously believes that there is a single gene controlling aging. Similarly, the concept that aging is the result of the accumulation of random errors (error catastrophe) in genes and proteins is no longer tenable. It is becoming increasingly evident that "aging" results from a number of sequential and/or parallel biochemical events regulated by different genes. However, it is not quite clear to what extent a particular genetic change is the cause or consequence of aging, because the reciprocal relationships between aging and gene expression are not well understood.[2]

Many age-dependent changes are ordained by a precise genetic program. Truman, for example, has demonstrated that metamorphosis in insects such as the moth, *Manduca sexta,* involves the programmed death of certain cells;[3] a hormonal cue, exposure to a steroid hormone followed by its withdrawal, is a necessary prerequisite signal for cell death. This study illustrates that, at least in a simple organism, there are biochemical cues that activate different genes whose products are necessary for cell death. Such a model of temporal control of gene expression suggests that, in more highly developed organisms and/or organ systems, such as the brain, there might be positive gene regulators—*e.g.,* increasing titers of a biochemical product of a metabolic reaction—or negative regulators of genes such as decreasing concentrations of metabolites, or the appearance of other signals, such as hormones or peptides. In the future, it is possible, with a better understanding of the biochemical mechanisms regulating gene expression, that some aspects of the aging process will be manipulable (see Morgan in this volume).

It is only recently that the aging brain has become a subject of intense study. It is quite evident that the neurobiology of aging field needs to develop its own theories to account for the unique aspects of brain aging. At the same time, such theories need to link the brain to other physiological processes in the body. In spite of the difficult methodological challenges, the field has made rapid advances during the last 10 to 12 years by moving increasingly away from descriptive studies towards those exploring dynamic mechanisms of brain aging.

In the early 1980s, when the Neuroscience of Aging Program was being developed at the National Institute on Aging (NIA), there were no viable theories of brain aging, nor did the state of knowledge warrant one. However, I felt that the newly emerging field did need a hypothesis in order to direct the attention of investigators away from descriptive studies and more towards studies of cellular mechanisms of neuronal aging. In a review article entitled "Towards Theories of Brain Aging,"[4] I suggested a hypothesis on the role of calcium in brain aging and Alzheimer's disease.

The "calcium hypothesis of brain aging" proposed that cellular mechanisms which regulate the homeostasis of cytosolic free calcium ion ($[Ca^{2+}]_i$) play a critical role in brain aging, and that altered $[Ca^{2+}]_i$ might account for a number of age-related changes in neuronal function. At the time (1982), the proposal was speculative and based on cir-

1

cumstantial evidence derived from the few published papers on brain aging available by investigators such as Philip Landfield, Gary Lynch, Michel Baudry, Mary Michaelis, Dean Smith, Gary Gibson, Christine Peterson, and John Blass, along with a handful of other investigators interested in the neurobiology of aging. These investigators were the first to develop preliminary evidence suggesting that cytosolic Ca^{2+} could play an important role in age-related changes in brain cellular functions. The calcium hypothesis was intended to be an alternative candidate in the search for a unifying explanation of brain aging. Although it was based on limited direct evidence, it provided a reasonable framework to link a number of discrete age-related changes observed in the nervous system, e.g., changes in neurotransmitter synthesis and release, glucose utilization, rate of axoplasmic flow, size and number of synaptic vesicles, energy-dependent extrusion of calcium, cell dysfunction and neuronal death, and other age-related pathological changes. At a time when an effort was being made to establish the field of the neurobiology of aging as a distinctive area of study, it was less critical whether the hypothesis proved to be correct than that such proposals might stimulate further research on cellular mechanisms of brain aging.

Recently, Gibson and Peterson[5] reviewed the more current literature on calcium and aging and provided further evidence supporting the hypothesis that homeostatic mechanisms regulating cytosol Ca^{2+} concentration are intricately linked with brain aging. The Gibson and Peterson review demonstrates the rapid maturation of this field of study as a significant and discrete area within the spectrum of topics covered under the broad umbrella of neuroscience. It also indicates that in the span of a very few years, the number of investigators studying the relationship between calcium and aging has substantially increased.

The purpose of this volume, which represents the proceedings of a workshop with the same title, was to re-evaluate the "calcium hypothesis of brain aging" in light of new evidence which might support or refute the proposition that cellular mechanisms which maintain the homeostasis of cytosolic $[Ca^{2+}]_i$ play a key role in brain aging and that sustained changes in $[Ca^{2+}]_i$ homeostasis could provide the final common pathway for age-associated brain changes.

In recent years, it has become increasingly apparent that Ca^{2+} functions as a nearly universal messenger system for extracellular signals to regulate cellular function in a variety of cells.[7,8] Several of the papers in this volume present the most recent data and provide further evidence that the calcium-mediated signalling system changes in the aging nervous system.[9,10] While it was previously known that there are several alternative mechanisms through which the regulation of cytosolic $[Ca^{2+}]$ can be disrupted (such as changes in ion channels, extrusion pumps, and sequestration), this volume provides new insights on the role of membrane changes in a cascade of events which might lead to disruption of $[Ca^{2+}]_i$ homeostasis. The mechanisms by which the assembly, structure, and dynamics of membrane constituents, including proteins, change during aging and pathological conditions are important topics for understanding aging and cellular mechanisms of $[Ca^{2+}]_i$ regulation. In addition, studies of membrane structure and function are essential to a better understanding of the mechanisms by which intracellular messengers mediate neuromodulation. Critical issues for studies of brain aging concern questions of how and what cellular changes may lead to destabilization of calcium homeostasis within the cytosol and/or disruption of calcium-mediated signal transduction processes. A crucial challenge is to determine the mechanism(s) that produces such changes. At present, we do not know the precise details of the processes that regulate intracellular concentration of $[Ca^{2+}]_i$ during aging. It is not clear whether $[Ca^{2+}]_i$ changes are the result or the cause of other pathogenetic effects. But it is clear that any significant and long-lasting change in the normal functioning of $[Ca^{2+}]_i$ transport systems, pumps, buffers, or storage systems that help maintain cell homeostasis could influence cytosolic $[Ca^{2+}]_i$ concentration with serious consequences. A revision of the "calcium hypothesis of brain aging," to a large extent derived from the discussion at this conference, suggests that there is a complex interaction between the *amount* of $[Ca^{2+}]$ perturbation and the *duration* of such deregulation in Ca^{2+} homeostasis. It proposes that a small disturbance in Ca^{2+} homeostasis reflecting a sustained small increase in $[Ca^{2+}]_i$ over a long period has consequences for the cell similar to those produced by a large increase in $[Ca^{2+}]_i$ over a shorter period. Many of the age-related and age-dependent changes in brain function ultimately have to be accounted for either on the basis of altered neuronal functioning or cell death. The Ca^{2+}-mediated signaling system and regulation of Ca^{2+} homeostasis appear to be the *final common pathway* for such cellular changes. There is no doubt that many of the age-related changes in the brain are determined by a genetic program, but this may involve many genes. An exciting, but inadequately explored, area of research in the neurobiology of aging concerns the relationship between Ca^{2+}-mediated second and third messengers and the induction of genes. Still unknown are the molecular events that set these processes in motion.

The principal achievement of the workshop was the creation of a forum for discussion among investigators, some of whom had not worked in the field of aging research. It is the organizers' expectation that the workshop and this volume will provide new insights and stimulate further research on cellular mechanisms of brain aging and the neuropathological processes of Alzheimer's disease.

The original idea for a conference on calcium and brain aging started germinating in 1985. At that time Michel Baudry, Gary Lynch, and Gary Gibson provided some very helpful advice on the organization, the agenda, and potential participants for the conference. For a variety of reasons, the final execution of the conference was delayed until October 1988. The final agenda was developed essentially by Jay Pettegrew and Carl Cotman, who also took the responsibility of co-chairing the various sessions. The success of the meeting, as well as the editing of this volume, are largely due to the efforts of Drs. Cotman and Pettegrew.

I want to express my special gratitude to the John Douglas French Foundation, the Pharmaceutical Manufacturers Association, and the Sigma Tau Foundation for their generous support of the workshop. I am also greatly indebted to my colleagues at the National Institute on Aging, Drs. Andrew Monjan and Creighton Phelps, and Mrs. Ann Brierly, for their help, advice and patience. Finally I am thankful to the staff at the New York Academy of Sciences for their assistance in making the publication of this volume an easy task for us.

REFERENCES

1. WARNER, H. R., R. N. BUTLER, R. L. SPROTT & E. L. SCHNEIDER. 1987. Modern Biological Theories of Aging. 1–4. Raven Press. New York, NY.
2. FINCH, C. E. & D. G. MORGAN. 1990. RNA and protein metabolism in the aging brain. Annu. Rev. Neurosci. In press.
3. TRUMAN, J. W., S. FAHRBACH & K. I. KIMURA. 1989. Hormones and programmed cell death: insights from invertebrate studies. *In* Molecular and Cellular Mechanisms of Neuronal Plasticity in Aging and Alzheimer's Disease. P. D. Coleman, G. A. Higgins & C. H. Phelps, Eds. Elsevier. Amsterdam. In press.
4. KHACHATURIAN, Z. S. 1984. Towards theories of brain aging. *In* Handbook of Studies on Psychiatry and Old Age. D.S. Kay & G. W. Burrows, Eds. 7–30. Elsevier. Amsterdam. 1984.
5. GIBSON, G. E. & C. PETERSON. 1987. Calcium and the aging nervous system. Neurobiol. Aging **8**: 329–343.
6. CHEUNG, W. Y. 1979. Calmodulin plays a pivotal role in cellular regulation. Science **207**: 19–27.

7. CARVALHO, A. P. 1982. Calcium in the nerve cell. *In* Handbook of Neurochemistry. A. Lajtha, Ed. 69–116. Plenum Press. New York, NY.
8. McGRAW, C. F., D. A. NACHSEN & M. P. BLAUSTEIN. 1982. Calcium movement and regulation in presynaptic nerve terminals. *In* Calcium and Cell Function. W. Y. Cheung, Ed. 81–110. Academic Press. New York, NY.
9. RASMUSSEN, H. 1986. The calcium messenger system. N. Engl. J. Med. **314:** 1094–1101.
10. RASMUSSEN, H. 1986. The calcium messenger system. N. Engl. J. Med. **314:** 1164–1170.

Molecular Insights into Alzheimer's Disease

JAY W. PETTEGREW[a]

Laboratory of Neurophysics
Department of Psychiatry
University of Pittsburgh School of Medicine
Pittsburgh, Pennsylvania

INTRODUCTION

The histological hallmarks of Alzheimer's disease (AD) have long been considered to be neurofibrillary tangles (NFT) and neuritic or senile plaques (SP) distributed throughout the allo- and neocortices.[1] However, recent reports suggest that AD does not necessarily require the presence of NFT[2] and the cytopathological changes may be limited to the hippocampus.[3] In addition, NFT have been demonstrated in some twenty-four neurological disorders[4] and SP occur with normal aging without associated dementia.[5] Therefore NFT and SP are not specific to AD and could represent end-stage markers. More recent studies such as event-related evoked potentials,[6–7] positron emission tomography (PET)[8–16] and 31-phosphorus nuclear magnetic resonance (^{31}P NMR) spectroscopy[17–23] have demonstrated physiological and metabolic abnormalities in areas of association neocortex, such as the prefrontal and inferior parietal cortices, that are not always the site of numerous NFT or SP. The PET studies demonstrate alterations in glucose uptake, and the ^{31}P NMR studies demonstrate alterations of membrane phospholipid metabolism in the association cortex of AD patients. Our ^{31}P NMR studies demonstrate elevations in the levels of phosphomonoesters (PME) early in the course of AD which progress to elevations in the levels of phosphodiesters (PDE) as the disease progresses.[20–21]

CENTRAL HYPOTHESIS

The central hypothesis guiding our studies is that the primary molecular/metabolic abnormalities of AD start in the neo- and allocortices and produce secondary retrograde changes in subcortical nuclei such as the nucleus basalis of Meynert, septal nucleus, locus coeruleus and dorsal raphe nucleus.[21] The retrograde changes will be simple atrophy of neurons and astrocytosis that occurs with transsynaptic degeneration under other circumstances, *e.g.*, dorsomedian neuronal atrophy in the thalamus after frontal leukotomy.[24–25] As such, the earliest molecular/metabolic changes in AD will occur in the neo- and allocortices resulting in cellular degeneration and death in these cortical areas and retrograde transsynaptic degeneration in subcortical nuclei. A type of degeneration of cholinergic neurons in the basal nucleus secondary to cerebral cortical damage has been demonstrated in the rat.[26–27] Others have suggested that a similar retrograde degeneration may

[a]Address for correspondence: c/o University of Pittsburgh, Graduate School of Public Health, Crabtree Hall, Room A-710, 130 DeSoto Street, Pittsburgh, PA 15261.

occur in AD.[28-30] Our studies suggest that the earliest molecular/metabolic changes in AD result in elevated levels of PME in the neocortex and allocortex. Cortical and subcortical elevations of PDE occur later in the course of the disease and reflect cellular degeneration and death.[17-21]

METHODOLOGY

Information Contained in the ^{31}P NMR Spectrum

The ^{31}P NMR spectrum of mammalian brain can be conveniently separated into three regions:[31] i) orthophosphate (5 to -1.5 ppm), ii) guanidophosphate (-3.5 to -5 ppm), and iii) polyphosphate (-5 to -23 ppm). The orthophosphate region can be further subdivided into phosphomonoester (PME) (5 to 1.5 ppm) and phosphodiester (PDE) regions (1.5 to -1.5 ppm). The polyphosphate region can be further subdivided into ionized ends (-5 to -8 ppm), esterified ends (-8 to -14 ppm), and middles (-18 to -23 ppm).

Contributing to the PME region are hexose 6-phosphates, triose phosphates, pentose phosphates, phosphoethanolamine, phosphocholine, inorganic orthophosphate (Pi), anomeric sugar phosphates, and several signals that have not been characterized as to the source phosphate. Contributing to the PDE region are glycerol phosphodiesters (primarily glycerol 3-phosphoethanolamine and glycerol 3-phosphocholine), a broad resonance from phosphorylated glycolipids and glycoproteins, and several uncharacterized resonances. The guanidophosphate region contains resonances from phosphocreatine (PCr) and phosphoarginine.

In the polyphosphate part of the spectrum, the ionized ends region contains resonances from the γ-phosphate of nucleotide triphosphates and the β-phosphate of nucleotide diphosphates. The esterified ends region contains resonances for the α-phosphate of nucleotide triphosphates, the α-phosphate of nucleotide diphosphates, the nicotinamide adenine dinucleotides and the uridine diphospho-sugars (galactose, glucose, mannose). The only resonance that makes a contribution to the middles region is the β-phosphate of nucleotide triphosphates. In mammalian brain, the predominant contributors to the nucleotide triphosphate and nucleotide diphosphate resonances are ATP and ADP (FIG. 1).

From a metabolic viewpoint, the ^{31}P NMR spectrum contains information about the energy status of the brain from the resonances for PCr, ATP, ADP, and Pi. Resonances related to phospholipid metabolism are contained in the PME and PDE regions.[18-19] In mammalian brain, the PME region contains resonances predominantly from α-glycerol phosphate, phosphoethanolamine and phosphocholine with smaller contributions from sugar phosphates such as hexose 6-phosphates, fructose 1,6-diphosphate, inositol phosphates, and nucleoside phosphates such as ribose 5'-phosphate, inosine 5'-monophosphate, adenosine 5'-phosphate, and NADP 2'-phosphate. In normal mammalian brain, α-glycerol phosphate, phosphocholine, and phosphoethanolamine are found predominately in the anabolic pathway of membrane phospholipid metabolism. However, a recent study demonstrates that the Ha-ras oncogene specifically activates phospholipase C hydrolysis of phosphatidylcholine (PtdC) and phosphatidylethanolamine (PtdE) giving rise to elevated levels of diacylglycerol and the PME phosphocholine and phosphoethanolamine.[32] Other biological activators of phospholipase C also could produce elevated levels of PME. The PDE region contains predominantly the resonances of glycerol 3- phosphoethanolamine and glycerol 3-phosphocholine which, in mammalian brain, are catabolic breakdown products of phospholipids due to phospholipase activity.[33-34] Therefore, the steady-state turnover of brain phospholipids (anabolism/catabolism) can be assessed by ^{31}P NMR spectroscopy.[18-19] Since neural membrane (especially synaptosomal) struc-

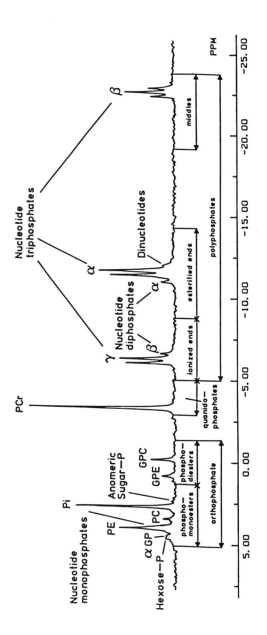

FIGURE 1. The ³¹P NMR spectrum of mammalian brain.

ture, dynamics, and function are of vital importance to normal neurochemical, neuro-physiological, and neuropharmacological function, ^{31}P NMR has the potential to provide important insights into normal and altered brain function. ^{31}P NMR spectra of post-mortem brain will contain only resonances from the PME, PDE and Pi. Therefore, ^{31}P NMR studies on postmortem brain will predominantly allow assessment of antemortem phospholipid metabolism and possible biological activation of phospholipase C and phos-pholipase B through the PME and PDE resonances, respectively.

In Vitro ^{31}P NMR Studies

The *in vitro* ^{31}P NMR studies provide chemical conditions more favorable to ^{31}P NMR analysis than occur in the living brain, and, therefore, a greater sensitivity and resolution is achieved as compared to *in vivo* analytical approaches. The enhanced sen-sitivity and resolution of *in vitro* extract studies enables the characterization and quanti-tation of many different phosphorus-containing compounds. Previous *in vitro* ^{31}P NMR studies demonstrated a remarkable correlation with more classical assay procedures and, in addition, revealed previously uncharacterized metabolites and unrecognized metabolic relationships.[35-37,17,38-40,18-19,31,41]

In order to correctly interpret ^{31}P NMR spectra, the identities of the individual res-onance signals must be carefully verified through the use of appropriate biochemical and spectroscopic procedures. The importance of this verification was recently demonstrated for a prominent ^{31}P NMR resonance at 3.84 ppm in mammalian brain which has now been definitively identified as phosphoethanolamine.[39] The identification was based on ^{1}H and ^{31}P NMR findings (including pH titrations) at 4.7 and 14.1 Tesla, as well as thin-layer chromatography studies. The ^{31}P NMR studies are in agreement with earlier studies that demonstrated a relative abundance of phosphoethanolamine in developing rabbit brain.[42] A relatively prominent PME resonance exhibiting the appropriate ^{31}P chemical shift has been reported in human neonatal brain[43,44] and childhood neuroblastoma[45] using an *in vivo* ^{31}P NMR surface coil technique.

RESULTS

In Vitro ^{31}P NMR Studies of AD Autopsy Brain

Recent ^{31}P NMR studies from this laboratory have demonstrated alterations of mem-brane phospholipid metabolism in AD brain obtained at autopsy and biopsy.[17,38,48,18-21] The alterations do not correlate with the duration of the postmortem interval and are thought to reflect antemortem metabolic changes. AD brains contain significantly ele-vated levels of PME ($p < 0.001$), which are precursors to membrane phospholipids or products of phospholipase C activity, without significant elevations of PDE, which are phospholipase A mediated breakdown products of membrane phospholipids. In contrast non-AD diseased control brains contain significant elevations of PDE ($p < 0.01$). The areas of AD brain with PME elevations are the superior and middle frontal gyri and the inferior parietal cortex. These same cortical areas of AD brain have decreased glucose utilization and abnormal electrophysiological responses to event-related evoked poten-tials.[8-16,6-7] A recent *in vivo* ^{31}P NMR spectroscopy study also demonstrates increased levels of PME in the temporo-parietal cortex of probable AD patients as compared to patients with subcortical multi-infarct dementia and demented patients with Parkinson's disease.[23] Finally, correlative ^{31}P NMR and morphological studies have been conducted on 32 samples obtained from the middle frontal and superior temporal gyri of 11 AD brains. A significant negative correlation was found between the levels of PME and the numbers of SP ($p = 0.05$; $r = 0.76$) and a significant positive correlation between the levels of PDE and the numbers of SP ($p = 0.01$; $r = 0.89$). No correlations were found between the numbers of NFT and the levels of PME or PDE. These findings suggest that elevations in the levels of PME precede the appearance of SP and that elevations in the levels of PDE coincide with the appearance of SP[21] (FIG. 2).

Recently, two other *in vitro* ^{31}P NMR spectroscopy reports have appeared which demonstrate alterations in the PDE resonances of AD autopsy brain, but no alterations in the PME resonances.[47,48] These reports deserve comment. For quantitative analyses of phosphorus metabolites by ^{31}P NMR spectroscopy, the perchloric acid extract must be carefully chemically scrubbed of divalent and trivalent cations because phosphoryl moieties chelate to varying degrees metal cations such as Al^{3+}, Ca^{2+}, Mg^{2+}, Mn^{2+}, etc. Cation chelation by phosphoryl moieties produces phosphorus resonance line-broadening due to chemical exchange and enhanced relaxation mechanisms. The net result of these interactions will be an artifactual decrease in the apparent quantities of the phosphoryl moieties present in the sample. Also, not all phosphoryl moieties chelate metal cations to the same degree; therefore, the effect of cations on phosphorus quantitation will vary with the phosphoryl moiety. For example, the chelation potential of phosphoryl moieties are in the following descending order: middle phosphates, such as β phosphate of ATP (βATP); inorganic orthophosphate (Pi); end phosphates such as PME; and diesterified phosphates, such as PDE.[49]

In our experience and as described by others[50,51] K-Chelex 100 columns are much more effective in removing cations than chelating agents such as EDTA. The use of K-Chelex 100 also avoids the NMR complications of the high salt concentration needed when using EDTA.[51] If the sample has been effectively scrubbed of cations, all phosphorus resonance peaks will appear as sharp lines with no resonance peaks appearing broader than others. Our previously reported ^{31}P NMR spectra of human autopsy brain demonstrate that all resonance lines are equally sharp.[18]

One of these reports[47] did not describe the method used to remove cations from the extract samples. In the spectra published in that report, there are narrow resonance signals for the PDE, but much broader signals for the PME and Pi. This would be consistent with the presence of cations in the extracts resulting in the broadening of the PME and Pi resonances with little or no effect on the PDE resonances. Several studies have demonstrated elevated cations, particularly Ca^{2+} and Al^{3+}, in the brains of patients with Alzheimer's disease (AD) compared to controls.[52–55] The possible presence of cations in the extract spectra reported by Barany *et al.*[47] could explain the selective broadening of the PME and Pi resonances resulting in artifactually decreased quantitative levels of PME in the AD brains.

The other *in vitro* ^{31}P NMR spectroscopy report[48] used EDTA to remove cations rather than K-Chelex 100, used 10% perchloric acid instead of 60%, and employed a total interpulse delay of 0.5 sec, which can potentially saturate some phosphorus resonance signals. The total interpulse delay used in all our studies is 1.68 sec; more than 3 times that reported by these authors. Finally, the reports by Barany *et al.*[47] and Miotto *et al.*[48] could have been on AD brain with more severe neuropathological changes than were present in the brains we have studied to date. This could explain their findings of increased levels of PDE and not PME.

In Vivo *^{31}P NMR Spectroscopy of Human Prefrontal Cortex*

In vivo ^{31}P NMR spectroscopy provides direct assessment of the high-energy phosphates phosphocreatine (PCr) and adenosine triphosphate (ATP) and their ultimate break-

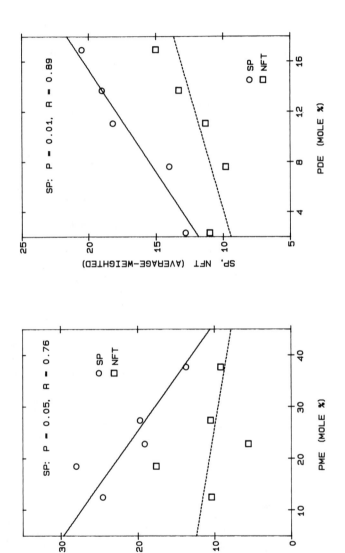

FIGURE 2. Morphological correlation of ^{31}P NMR.

down product inorganic orthophosphate (Pi).[56,31] The PCr/Pi ratio provides a convenient assessment of the energy status of the brain. In addition the ^{31}P NMR spectrum provides an assessment of brain membrane phospholipid metabolism from the PME and PDE resonances and the activity of protein and ganglioside glycosylation from the UDP-sugar resonance band.[39,18-19,20] The brain intracellular pH also can be determined by either the difference between the chemical shifts of PCr and Pi[57] or between γ and α ATP.[58] ^{31}P NMR spectroscopy is unique among the currently available techniques for assessing *in vivo* metabolism as NMR does not use ionizing radiation and is, therefore, considered safe. Because of the safety of NMR, repeated NMR spectra can be performed for longitudinal studies.

Normal Elderly Control and Probable Alzheimer's Spectra

We have obtained *in vivo* spectra under resting conditions (lying quietly, eyes open, ears unoccluded) from the dorsal prefrontal cortex of normal adults (18–44 years) and more recently from a normal elderly female and a female patient with probable AD by NINCDS-ADRDA criteria.[59] The spectra are acquired using the standard General Electric NMR spectroscopy surface coils and software. The resonances of PME, Pi, PDE, ionized ends, esterified ends, and middles are easily distinguished. The PME, Pi, PDE, and PCr resonances of brain are not completely resolved in the unprocessed *in vivo* ^{31}P NMR spectrum obtained at 1.5 Tesla. This is probably due to a combination of at least three factors: (1) The NMR natural line widths for these brain chemical species *in vivo* are probably broader than for the same chemical species in solution. (2) There is decreased resolution of these resonances at 1.5 Tesla compared to H_o fields of 4.7 Tesla or higher. (3) The H_o field homogeneity across the human head is not as good as that attainable across smaller sample diameters (5–20 mm). In order to quantitate the chemical species present, the *in vivo* spectrum needs to be deconvoluted into its component Lorentzian resonances. This is accomplished by using the General Electric spectroscopy software and data station. The integrated areas under the individual component resonances can be obtained from the deconvoluted spectrum.

The original unprocessed spectrum as well as the simulated and deconvoluted spectra from the dorsal prefrontal cortex of a 73-year-old normal female volunteer are shown in FIGURE 3. The spectrum was obtained in 5 minutes of acquisition time. Integrating the individual resonances in the deconvoluted spectrum gives the following results expressed as mole % of the total spectrum: PME = 15.09, Pi = 8.50, PDE = 35.71, PCr = 14.81, ionized ends = 8.35, esterified ends = 8.68, and middles = 8.73. From these values the following ratios were obtained: PME/PDE = 0.42, PCr/Pi = 1.74, and PCr/ATP = 1.69. The intracellular pH = 7.05. These values compare favorably with values in the literature obtained on extracted freeze clamped adult mammalian brain using classical assay methods (PCr/ATP = 1.65)[60,61] and ^{31}P NMR spectroscopy (PCr/Pi = 1.59; PCr/ATP = 1.77) (Pettegrew and Panchalingam, unpublished observations).

The values obtained from the dorsal prefrontal cortex of a 60-year-old female with probable AD and moderate dementia (Mattis Dementia Rating Scale = 74) are: PME = 23.92, Pi = 12.02, PDE = 27.96, PCr = 8.65, ionized ends = 5.49, esterified ends = 12.17, middles = 9.73, PME/PDE = 0.85, PCr/Pi = 0.72, and PCr/ATP = 0.88. The intracellular pH = 7.08. This probable AD patient, therefore, has increased levels of PME and Pi which replicates our findings in AD autopsy brain. In addition this patient has decreased levels of the high-energy phosphates PCr and ATP which cannot be determined in autopsy brain and, therefore, underscores the importance of doing *in vivo* ^{31}P NMR studies. The findings of decreased PCr and ATP are in keeping with the PET findings of decreased glucose utilization in probable AD patients. It is gratifying that these results replicate our previous PME results on autopsy brain and are also consistent with PET results from other investigators.

Two preliminary reports have also appeared on the *in vivo* ^{31}P NMR spectroscopy of AD, Parkinson disease (PD), multi-infarct dementia (MID), and controls.[62,63] These two studies utilized a 4-cm surface coil at a field strength of 1.9 Tesla. The study of AD and PD[62] noted slightly reduced ratios of PCr/βATP, PME/βATP, and PDE/βATP in PD and AD. The study of MID[63] also suggested a trend for reduced PCr/βATP, PME/βATP, and PDE/βATP in MID compared to controls. Both of these *in vivo* ^{31}P NMR studies have now been extended to 15 patients with probable AD by NINCDS-ADRDA criteria,[59] 14 age-matched controls, 9 patients with multiple subcortical infarcts and dementia, and 11 patients with PD and dementia. The results on a larger number of subjects now demonstrate increased levels of PME in the temporal-parietal cortex of the probable AD patients ($p < 0.05$) and an increased PCr/Pi ratio in the frontal and temporal-parietal cortex of the

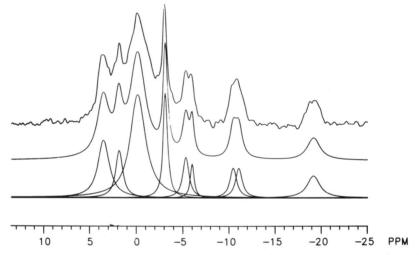

FIGURE 3. The original unprocessed spectrum (*upper*), the simulated spectrum (*middle*), and the deconvoluted spectrum (*lower*) from the dorsal prefrontal cortex of a 73-year-old normal female volunteer.

patients with multiple subcortical infarcts and dementia ($p < 0.01$). The demented PD patients generally did not differ from nondemented controls. These *in vivo* PME findings replicate our previous *in vitro* findings[17,18,20] as well as our recent *in vivo* ^{31}P NMR findings (Pettegrew and Panchalingam, unpublished observations).

Solid State ^{31}P NMR

Evidence exists for increased levels of brain aluminum (Al^{3+}) in Alzheimer's disease.[64] Al^{3+} deposition is predominantly in neurofibrillary tangles, nuclear heterochromatin, and the amyloid core of neuritic plaques, all of which are phosphorylated or contain potential phosphorylation sites. Since Al^{3+} will have to cross the plasma membrane upon entry into cells, a solid state ^{31}P NMR study investigated the possibility of Al^{3+} interaction with model membranes and its effect on the physical properties of the membranes. It is observed that Al^{3+} does not alter the rate and mode of the rotational motions of the phospholipid headgroup in 1-palmitoyl-2-oleoyl phosphatidylcholine (POPC) and bovine brain phosphatidylinositol (PtdI). However, significant alterations in the motion of the headgroup are observed in 1-palmitoyl-2-oleoyl phosphatidylethanolamine (POPE), bovine brain phosphatidylserine (PtdS), and cardiolipin lipids (FIG. 4). The ^{31}P NMR spectrum shows an Al^{3+}-induced alteration in the otherwise multi-lamellar structure similar to that induced by Ca^{2+} in model membranes.[65] Al^{3+} appears to induce membrane conformational changes from a bilayer to a micellar and hexagonal II phase in these phospholipids. These results demonstrate that Al^{3+} can induce significant alterations in the physical properties of model membranes.[66]

In an effort to study the consequences of the elevated levels of PME and PDE on plasma membranes, we performed solid state ^{31}P NMR spectroscopy on model membranes in the presence of 1.0 mM PME or PDE. It was observed that the PME do not affect the powder pattern spectrum of the phospholipid POPC, whereas significant alterations were observed in POPE and bovine brain PtdS. These results demonstrate alterations in the head group orientation and motion in POPE and PtdS but not in POPC. PDE had no effect on the line shape of any of the lipids (FIG. 5). It should be noted that POPC is predominantly on the external face and POPE and PtdS are the major components of the cytoplasmic face of cell membranes. Should similar changes occur in plasma membranes in vivo, significant functional changes can be expected. These studies taken together strongly suggest that in the presence of elevated levels of Ca^{2+}, Al^{3+}, or PME, the phospholipids located on the cytoplasmic face of membranes can be transformed from a normal bilayer phase to micellar and hexagonal II phases. The micellar phase would facilitate the formation of vesicles providing a packaging of the PME. The formation of the hexagonal II phase would facilitate the fusion of the vesicles to the cytoplasmic face of the membrane for transmembrane transport to the extracellular face of the membrane and the release of the PME into the extracellular space. These functional alterations could affect membrane receptors, ion channels, and structural and enzymatically active proteins such as those involved in the second messenger molecular cascade. Such alterations could have significant clinical implications.

Computer Molecular Modeling

Some consideration should be given to the possibility that the elevated PME could have a neurophysiological role in AD beyond merely reflecting membrane abnormalities. Available evidence now suggests that the elevated PME (phosphoethanolamine, phosphocholine, and L-phosphoserine) have other functions, e.g., as false neurotransmitters. This possibility was stimulated by our observation that the PME have chemical structural similarities to the known neurotransmitters L-glutamate and N-methyl-D-aspartate (NMDA). To investigate the possibility of conformational similarity, minimum energy conformations were computed by molecular mechanics calculations for L-glutamate, NMDA, and the PME phosphocholine, phosphoethanolamine, and L-phosphoserine as well as for inositol 1,4,5-triphosphate (IP3) and 2-amino-5-phosphonopentanoic acid (AP5). The molecular mechanics calculations are based on the AMBER force fields of Killman[67] using the Macro Model computer program. The calculated minimum potential energies (in K Joules/mole) are: L-glutamate (-233.40), NMDA (-203.05), phosphocholine (-78.18), phosphoethanolamine (-262.90), L-phosphoserine (-292.60), IP3 (axial $+199.95$), and AP5 (-59.47). The conformations corresponding to these minimum potential energies reveal definite similarities for NMDA, L-glutamate, phosphoethanolamine, and L-phosphoserine when shown in stereo projections.[20] Phosphocholine and

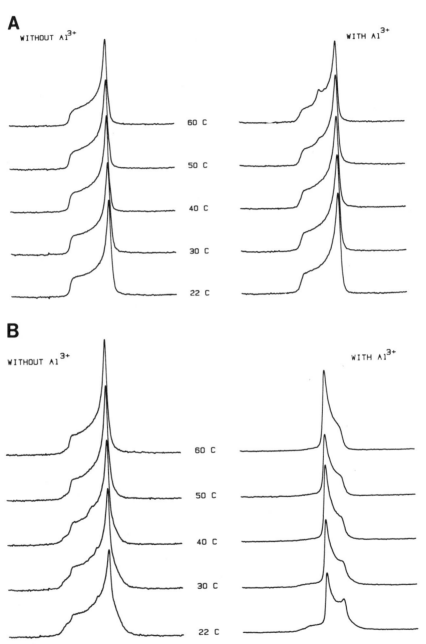

FIGURE 4. (A) 1-palmitoyl-2-oleoyl phosphatidylcholine (POPC) and (B) 1-palmitoyl-2-oleoyl phosphatidylethanolamine (POPE) without and with Al^{3+}.

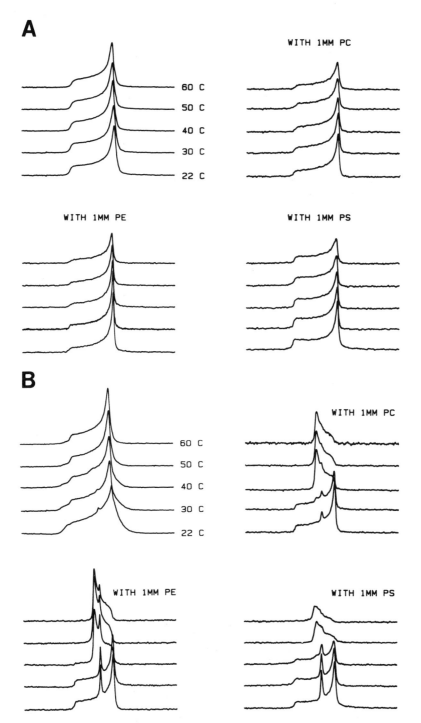

FIGURE 5. **(A)** 1-palmitoyl-2-oleoyl phosphatydylcholine (POPC) and **(B)** 1-palmitoyl-2-oleoyl phosphatidylethanolamine (POPE) without and with 1 mM of the phosphomonoesters phosphocholine (PC), phosphoethanolamine (PE), and L-phosphoserine (PS).

AP5 have conformational similarities, and IP3 has a different conformation from all the others. Phosphocholine has a conformation intermediate between L-glutamate and the L-glutamate antagonist 2-amino-5-phosphonopentanoic acid (AP5) (FIG. 6). These conformational analyses provide significant insights into the biological action of these molecules.

Electrophysiology

Recently we found that PME have profound effects on extracellular population excitatory postsynaptic potential (EPSP) responses evoked by stimulating the Schaffer collateral/commissural input to field CA1.[68] These studies of superfused rat hippocampal brain slices demonstrate that phosphoethanolamine and phosphocholine both depress the amplitude of the EPSPs of CA1 neurons in a dose-dependent fashion at 10 μm and 100 μm concentrations. However, at 1 mM concentration phosphocholine greatly enhances the amplitude of the population EPSP of CA1 neurons while phosphoethanolamine and L-phosphoserine continue to depress the amplitude of the EPSP (FIG. 7). ^{31}P NMR spectroscopy studies conducted on hippocampal slices in the presence of varying con-

A

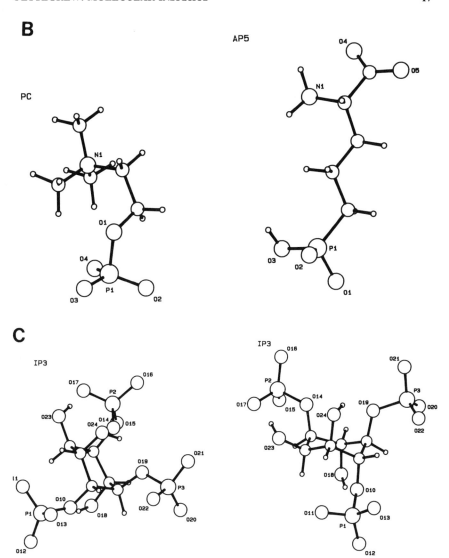

FIGURE 6. Minimum energy conformations computed by molecular mechanics calculations for **(A)** N-methyl-D-aspartate (NMDA), L-glutamate, and the phosphomonoesters phosphoethanolamine (PE) and L-phosphoserine (PS), **(B)** the phosphomonoester phosphocholine (PC) and 2-amino-5-phosphonopentanoic acid (AP5), and **(C)** inositol 1,4,5-triphosphate (IP3).

centrations (10 μM–1.0 mM) of PME demonstrate normal high-energy phosphate metabolism.[69] These results suggest that the effects of PME on the CA1 population EPSP represent a specific interaction between the PME and the L-glutamate receptor. These findings lead us to suggest that low concentrations of PME could block L-glutamate receptors in the hippocampus and might be important in the memory impairment characteristic of AD. However, at higher concentrations, phosphocholine may have the characteristics of an excitatory neurotoxin resulting in cellular degeneration and death. The levels of PME that we detect in AD brain could be in the 5–7 mM range in individual affected cells.

Recent studies, a combination of extracellular recordings and current- and voltage-clamp techniques, were conducted to examine the effects of phosphoethanolamine on the

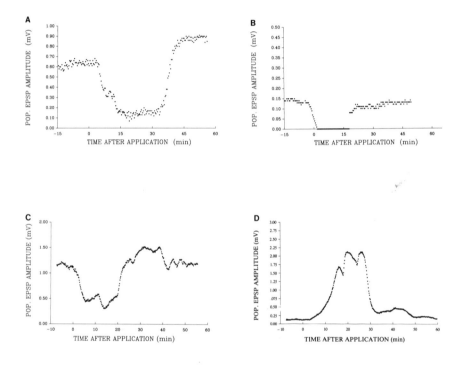

FIGURE 7. The effect on the amplitude of population excitatory postsynaptic potential (EPSP) of CA1 neurons of **(A)** 100 μM and **(B)** 1 mM o-phosphorylethanolamine, and of **(C)** 100 μM and **(D)** 1 mM phosphorylcholine chloride.

biophysical properties of CA1 neurons. As in previous studies, recordings were made from rat hippocampal slices (3-month-old Fischer 344) prepared in the conventional manner. Extracellular and intracellular EPSP as well as EPSC amplitudes were monitored prior to and during a 30-minute bath application of 1 mM phosphoethanolamine. Based on these measurements, two groups of CA1 cells were defined. In one group (N = 6), synaptic responses decreased in amplitude with a peak effect within 10–15 minutes of exposure to phosphoethanolamine (mean depression = 63%, range = 25 to 98%). In only 1 out of 5 cells in which recordings were obtained during the wash out period was a recovery observed. In 3 cells, amplitudes of synaptic responses rebounded during the wash, exceeding preexposure baseline. In the remaining cell, no recovery was detected. In 3 cells of this group, action potential duration (both orthodromic and current-evoked) decreased in the presence of phosphoethanolamine by 15, 19 and 7%. In the other group (N = 7), synaptic responses increased within the first 5 minutes of phosphoethanolamine exposure and continued to increase throughout the exposure (mean increase = 124%, range = 11 to 345%). In only 1 out of 5 cells in which recordings were obtained during wash out was a recovery observed. In 3 cells there was no recovery and in 1 cell, the amplitude of synaptic responses rebounded below original baseline levels. In 6 cells of this group, action potentials increased by 13.8% with a range of 8 to 18%. In 6 of the previously mentioned 13 cells, the after hyperpolarization measured in hybrid clamp was reduced or abolished. No recovery was seen in any case, even in recordings obtained after 40 minutes of wash.

A potential neurotoxic role for L-glutamate in AD has been hypothesized by Maragos et al.[70] In addition, a recent report demonstrated a normal density and distribution of NMDA receptors in AD hippocampus,[71] which contrasts with previous reports that claimed decreased NMDA receptors in AD cortex[72] and hippocampus.[73] The contrasting results of these two laboratories could be due to differences of the severity of the disease in the two studies, or could reflect a difference in the methods used. If the density and distribution of NMDA receptors are normal in AD brain until cell loss occurs, then cells with NMDA receptors could be vulnerable to the potential neurotoxic effects of elevated levels of phosphocholine. In addition, the lower levels of PME which would occur in the earlier stages of the disease could produce memory problems by blocking NMDA receptors.

DISCUSSION

Possible Metabolic Source(s) of Elevated Levels of PME

The findings of increased PME in AD brain cannot be explained simply by degeneration of brain tissue; with degeneration, the PDE should be elevated and not PME. Increased turnover of membrane phospholipids should result in elevations of both PME and PDE as observed in normal aging in Fischer 344 rats.[38] The PDE elevations in non-AD diseased brain could be consistent with increased phospholipid turnover or degeneration of brain tissue. The finding of increased PME in AD brain suggests one or more of the following mechanisms: 1) increased synthesis of membrane phospholipids, [18-21] 2) a relative metabolic block in the synthetic pathway,[18-21] 3) a decreased breakdown of PME secondary to decreased phospholipase D activity in AD brain,[74] or 4) stimulation of phospholipase C by neuromodulators and growth factors[75] or oncogenes[32] (FIGS. 8,9).

Elevated levels of PME are observed normally in the immature, developing brain,[39-40,18-19,20-21] particularly during the period of elaboration of dendritic processes. Our findings of elevated PME in AD brain could, therefore, suggest an increase in membrane phospholipid synthesis as occurs during the elaboration of complex membranous structures such as dendritic processes. Cytological evidence for the elaboration of dendritic processes has been previously reported in AD[76-77] and normal aged brain.[78-79] In addition, similar regenerative attempts have been observed in Huntington's disease brain,[80] another "degenerative" neurological disorder in the behavioral and neuropathological senses. However, other studies have been interpreted as evidence of failed com-

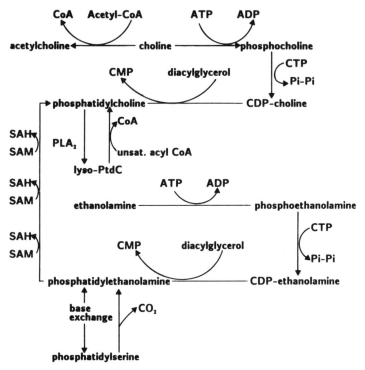

FIGURE 8. Synthesis. CoA: coenzyme A; ATP: adenosine triphosphate; ADP: adenosine diphosphate; CTP: cytidine triphosphate; CDP: cytidine diphosphate; CMP: cytidine monophosphate; Unsat: unsaturated; PLA_2: phospholipase A_2; SAM: S-adenosylmethionine; SAH: S-adenosylhomocysteine; P_i-P_i: inorganic pyrophosphate.

pensatory dendritic growth in the parahippocampal gyrus and the hippocampal dentate gyrus of AD brain.[81-82] Neither of these structures has been studied to date by ^{31}P NMR, and the ^{31}P NMR findings in cerebral cortex may not be the same as for the parahippocampal gyrus and the hippocampal dentate gyrus. In fact, studies in the rat have demonstrated regional variability for age-related dendritic changes.[83-87] Finally, a recent study of AD platelets has demonstrated increased proliferation of platelet internal membranes[88] suggesting a generalized increase in the synthesis of membranes in AD.

An elevation of the PME phosphoethanolamine and phosphocholine also could occur if there existed a relative metabolic block at the enzyme cytidine triphosphate (CTP): phosphoethanolamine (or phosphocholine) cytidyltransferase (EC 2.7.7.15), which is the rate-limiting enzyme in phospholipid synthesis.[33] An enzymatic block at this step could result in elevated levels of PME as assayed by ^{31}P NMR and decreased levels of the phospholipids, phosphatidylcholine, and phosphatidylethanolamine. In fact, HPLC studies of AD brain phospholipids do demonstrate a reduction in AD brain phospholipids as expressed as micromoles of phospholipid per gram wet weight of brain.[89] As compared to non-AD diseased controls (60 samples, 9 brains), AD brain (131 samples, 17 brains) has decreased levels of phosphatidylcholine (PtdC; $p = 0.001$), phosphatidylethanolamine (PtdE; $p = 0.003$), and cholesterol ($p = 0.01$) and small but nonsignificant decreases in the levels of phosphatidylserine (PtdS). The levels of PtdC, PtdE, PtdS, and cholesterol were correlated with the number of SP per $\times$ 200 magnification between cortical layers II and IV in the same brain regions. This revealed a nonlinear correlation for PtdC which peaked around 10 SP/$\times$200 magnification and declined with increasing SP numbers ($R^2 = 0.9, p = 0.0001$). Linear negative correlations were observed for PtdE ($R^2 = 0.6, p = 0.02$) and PtdS ($R^2 = 0.8, p = 0.003$), and no correlation for cholesterol ($R^2 = 0.08, p = 0.5$). These findings suggest an increase in PtdC synthesis early in the pathogenesis of AD at a time when PME levels are high. In theory, the augmented synthesis of PtdC could shunt available choline from acetylcholine synthesis producing a functional cholinergic deficit which is observed clinically.[90] As the disease progresses membrane degradation assumes prominence.

The increased levels of PME early in the pathogenesis of AD also could reflect altered levels of, or responses to neuromodulators, growth factors, or oncogenes. It is, therefore, of interest that CTP: phosphocholine cytidyltransferase activity has been demonstrated to be decreased under conditions that favor phosphorylation of proteins, such as stimulation of protein kinase systems.[91] There is other evidence for possible elevated protein kinase activity in AD brain. The microtubule associated tau proteins have been shown to be hyperphosphorylated in paired helical filaments found in AD brain.[92] The hyperphosphorylation of the tau proteins could secondarily result in defective microtubule assembly and altered axoplasmic flow in AD brain.[93] The hyperphosphorylated tau proteins also might be abnormally metabolized leading to the formation of paired helical filaments found in NFT. A recent report demonstrates that the AD amyloid precursor protein (ADAP) is phosphorylated by protein kinase C and Ca^{2+}/calmodulin-dependent protein kinase II.[94] Phosphorylation of ADAP potentially could interfere with membrane insertion resulting in cleavage to β-amyloid protein which is deposited intracellularly and in SP in

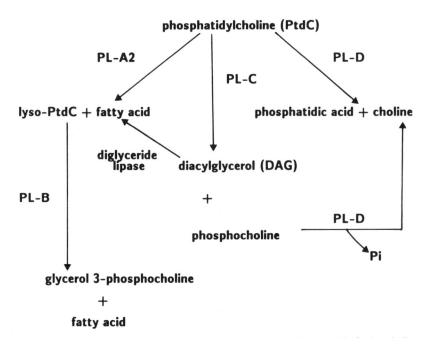

FIGURE 9. Catabolism. PL-A$_2$: phospholipase A$_2$; PL-B: phospholipase B; PL-C: phospholipase C; PL-D: phospholipase D; P$_i$: inorganic orthophosphate.

AD brain. It is conceivable that the elevated levels of PME early in the course of AD reflect increased synthesis of membrane phospholipids, but that later in the course of the disease a metabolic block develops at the rate-limiting enzyme step which further increases PME levels. The metabolic block could arise secondary to activation of receptors linked to phospholipase C and protein kinase C activity such as the L-glutamate and NMDA receptors. A recent study of protein kinase C levels in AD autopsy brain[95] is consistent with a compensatory down-regulating of protein kinase C after overactivation early in the disease process.

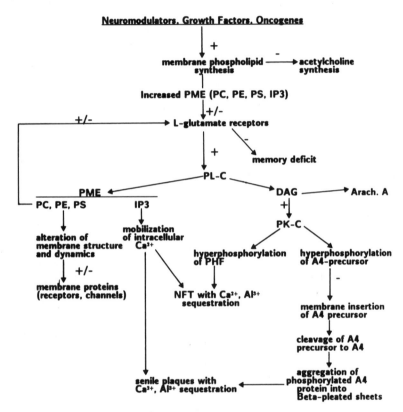

FIGURE 10. Summary of mechanisms. PME: phosphomonoesters; PC: phosphocholine; PE: phosphoethanolamine; PS: L-phosphoserine; IP3: inositol trisphosphate; DAG: diacylglycerol; Arach A: arachidonic acid; PK-C: protein kinase C; PL-C: phospholipase C; A4-Precursor: β amyloid precursor protein; A4: β amyloid protein; NFT: neurofibrillary tangles; PHF: paired helical filaments.

CONCLUSIONS

Recent studies provide substantial evidence for elevated levels of phosphomonoesters (PME) early in the course of Alzheimer's disease; perhaps even prior to the onset of clinical symptoms although this remains to be demonstrated. Similar high levels of PME are observed normally in the developing brain, especially during the period of dendritic proliferation. Along with the elevations of PME in AD brain are transient increases in the levels of phosphatidylcholine (PtdC) suggesting increased synthesis of PtdC. In theory increased synthesis of PtdC could shunt available choline away from acetylcholine synthesis and produce a functional cholinergic deficit. As the disease progresses, the elevated levels of PME decline and are replaced by elevated levels of phosphodiesters (PDE) and decreasing levels of brain phospholipids.

The elevated levels of PME also could reflect enhanced phospholipase C (PL-C) activity which would stimulate protein kinase C (PK-C) activity. Enhanced PK-C activity could lead to many diverse biological effects including the hyperphosphorylation of proteins such as the AD amyloid precursor protein (ADAP) and microtubule-associated tau proteins. The hyperphosphorylation of these proteins could alter their metabolism including membrane insertion of ADAP leading eventually to β-amyloid and paired helical filament deposition. The hyperphosphorylated β-amyloid and tau proteins would provide chelation sites for Ca^{2+} and Al^{3+} leading to further cross-linking of the individual polymers and sequestration of these cations. The enhanced PL-C activity could also elevate cytoplasmic levels of inositol-1,45-triphosphate leading to mobilization of intracellular Ca^{2+} and further exaggerating the above processes.

The PME phosphocholine (PC), phosphoethanolamine (PE), and L-phosphoserine (PS) have now been demonstrated to share striking conformational similarities with the neurotransmitters N-methyl-D-aspartate and L-glutamate. Hippocampal brain slice extra- and intracellular recordings demonstrate that these PME are neurophysiologically active in CA1 cells which contain L-glutamate receptors and appear to inhibit the slow Ca $^{2+}$-activated K^+ channel. In addition, the PME have been demonstrated to alter the structure and dynamics of the phospholipid head group region in synthetic model membranes, which could have *in vivo* biological significance for the packaging and release of the PME. These mechanisms are summarized in FIGURE 10.

An important unanswered question is what turns this whole molecular process on in AD? Since the findings early in the course of AD appear to resemble events normally occurring in the developing brain, abnormal levels of, or responses to growth factors, neuromodulators, or oncogenes should be pursued.

REFERENCES

1. TOMLINSON, B. E. & J. A. N. CORSELLIS. 1984. Aging and the dementias. *In* Greenfield's Neuropathology. J. H. Adams, J. A. N. Corsellis, and L. W. Duchen, Eds. 951–1025. Wiley. New York, NY.
2. TERRY, R. D., L. A. HANSEN, R. DETERESA, P. DAVIES, H. TOBIAS & R. KATZMAN. 1987. Senile dementia of the Alzheimer type without neocortical neurofibrillary tangles: J. Neuropathol. Exp. Neurol. **46:** 262–268.
3. BALL, M. J., V. HACHINSKI, A. FOX, A. J. KIRSHEN, M. FISHMAN, M. BLUME, V. A. KRAL, H. FOX & H. MERSKEY. 1985. A new definition of Alzheimer's disease: A hippocampal dementia. Lancet **1:** 14–16.
4. HALPER, J., B. SCHEITHAUER, H. OKAZAKI & E. LAWS. 1986. Meningioan giomatosis: A report of six cases with special reference to the occurrence of neurofibrillary tangles. J. Neuropathol. Exp. Neurol. **45(4):** 426–446.
5. TOMLINSON, B. E. 1972. Morphological brain changes in non-demented old people. *In* Aging of the Central Nervous System. H. M. Von Praag & A. F. Kalverboer, Eds. 37–57. DeErvon F. Bohn. New York, NY.
6. FLETCHER, W. A. & J. A. SHARPE. 1986. Saccadic eye movements dysfunction in Alzheimer's disease. Ann. Neurol. **20:** 464–471.
7. FLETCHER, W. A. & J. A. SHARPE. 1988. Smooth pursuit dysfunction in Alzheimer's disease. Neurology (Cleveland) **38(2):** 272–276.
8. CUTLER, N. R., J. V. HAXBY, R. DUARA, C. L. GRADY, A. M. MOORE, J. E. PARISI, J. WHITE, L. HESTON, R. MARGOLIN & S. RAPOPORT. 1985. Brain metabolism as measured

with positron emission tomography: Serial assessment in a patient with familial Alzheimer's disease. Neurology (Cleveland) **35:** 1556–1561.

9. DUARA, R. C. GRADY, J. HAXBY, M. SUNDARAM, N. R. CUTLER, L. HESTON, A. MOORE, N. SCHLAGETER, S. LARSON & S. I. RAPOPORT. 1986. Positron emission tomography in Alzheimer's disease [^{18}F] 2-fluoro-2-deoxy-D-glucose study in the resting state. Neurology (Cleveland) **36:** 879–887.

10. JAGUST, W. J., R. P. FRIEDLAND, E. KOSS, B. A. OBER, C. A. MATHIS, R. H. HUESMAN & T. F. BUDINGER. 1987. Progression of regional cerebral glucose metabolic abnormalities in Alzheimer's disease. (Abstract) Neurology (Cleveland) **37**(1): 156.

11. HAXBY, J. V., C. L. GRADY, E. ROSS, R. P. FRIEDLAND & S. I. RAPOPORT. 1987. Heterogenous metabolic and neuropsychological patterns in dementia of the Alzheimer type: Cross-sectional and longitudinal studies. (Abstract) Neurology (Cleveland) **37**(1): 159.

12. BERADI, A., J. V. HAXBY, C. L. GRADY & S. I. RAPOPORT. 1987. Asymmetrics of brain glucose consumption and memory performance in mild dementia of the Alzheimer type and in healthy aging. (Abstract) Neurology (Cleveland) **37**(1): 160.

13. BERENT, S., N. L. FOSTER, S. GILMAN, R. HICHWA & S. LEHTINEN. 1987. Patterns of cortical ^{18}F-FDG metabolism in Alzheimer's and progressive supranuclear palsy patients are related to the types of cognitive impairments. (Abstract) Neurology (Cleveland) **37**(1): 172.

14. FRIEDLAND, R. P., W. J. JAGUST, T. F. BUDINGER, E. KOSS, & B. A. OBER. 1987. Consistency of temporal parietal cortex hypometabolism in probable Alzheimer's disease (AD): Relationships to cognitive decline. (Abstract) Neurology (Cleveland) **37**(1): 224.

15. HORWITZ, B., C. L. GRADY, N. L. SCHLAGETER, R. DUARA & S. I. RAPOPORT. 1987. Intercorrelations of regional cerebral glucose metabolic rates in Alzheimer's disease. Brain Res. **407:** 294–306.

16. RAPOPORT, S. I., B. HORWITZ, J. HAXBY & C. L. GRADY. 1986. Alzheimer's disease: Metabolic uncoupling of associative brain regions. Can. J. Neurol. Sci. **13:** 540–545.

17. PETTEGREW, J. W., N. J. MINSHEW, M. M. COHEN, S. J. KOPP & T. GLONEK. 1984. P-31 NMR changes in Alzheimer's and Huntington's disease brain. (Abstract) Neurology (Minneapolis) **34**(1): 281.

18. PETTEGREW, J. W., S. J. KOPP, N. J. MINSHEW, T. GLONEK, J. M. FELIKSIK, J. P. TOW & M. M. COHEN. 1987a. ^{31}P nuclear magnetic resonance studies of phosphoglyceride metabolism in developing and degenerating brain: Preliminary observations. J. Neuropathol. Exp. Neurol. **46:** 419–430.

19. PETTEGREW, J. W., G. WITHERS, K. PANCHALINGAM & J. F. M. POST. 1987b. ^{31}P nuclear magnetic resonance (NMR) spectroscopy of brain in aging and Alzheimer's disease. J. Neurol. Trans. **24:** 261–268.

20. PETTEGREW, J. W., J. MOOSSY, G. WITHERS, D. MCKEAG & K. PANCHALINGAM. 1988a. ^{31}P nuclear magnetic resonance study of the brain in Alzheimer's disease. J. Neuropathol. Exp. Neurol. **47**(3): 235–248.

21. PETTEGREW, J. W., K. PANCHALINGAM, J. MOOSSY, T. A. MARTINEZ, G. RAO & F. BOLLER. 1988b. Correlation of ^{31}P NMR and morphological findings in Alzheimer's disease. Arch. Neurol. **45:** 1093–1096.

22. GORELL, J. M., J. A. BUERI, G. G. BROWN, S. R. LEVEINE, K. M. A. WELCH, J. W. GDOWSKI, J. A. HELPERN & M. B. SMITH. 1988. Parietal and frontal high-energy cerebral phosphate metabolism in Alzheimer and Parkinson dementia. Neurology (Cleveland) **38:** 227.

23. GDOWSKI, J. W., G. G. BROWN, S. R. LEVINE, M. SMITH, J. HELPERN, J. BUERI, J. GORELL & K. M. A. WELCH. 1988. Patterns of phospholipid metabolism differ between Alzheimer and multi-infarct dementia. (Abstract) Neurology (Cleveland) **38**(1): 268.

24. COWAN, W. M. 1970. Anterograde and retrograde transneuronal degeneration in the central and peripheral nervous system. *In* Contemporary Research Methods in Neuroanatomy. W. J. H. Nauta & S. O. E. Ebbesson, Eds. 217–251. Springer-Verlag. Berlin.

25. DUCHEN, L. W. 1984. General pathology of neurons and neuroglia. *In* Greenfields Neuropathology. J. H. Adams & J. A. N. Corsellis, Eds. 18–20. John Wiley & Sons. New York, NY.

26. SOFRONIEW, M. V., R. C. A. PEARSON, F. ECKENSTEIN, A. C. CUELLO & T. P. S. POWELL. 1983. Retrograde changes in cholinergic neurons in the basal nucleus of the forebrain of the rat following cortical damage. Brain Res. **289:** 370–374.

27. SOFRONIEW, M. V. & R. C. A. PEARSON. 1985. Degeneration of cholinergic neurons in the basal nucleus following kainic or N-methyl-D-aspartic acid application to the cerebral cortex in the rat. Brain Res. **339:** 186–190.
28. PEARSON, R. C. A., M. V. SOFRONIEW, A. C. CUELLO, T. P. S. POWELL, F. ECKENSTEIN, M. M. ESIRI & G. K. WILCOCK. 1983a. Persistence of cholinergic neurons in the basal nucleus in a brain with senile dementia of the Alzheimer's type demonstrated by immuno-histochemical staining for choline acetyltransferase. Brain Res. **289:** 375–379.
29. PEARSON, R. C. A., K. C. GATTER & T. P. S. POWELL. 1983b. Retrograde cell degeneration in the basal nucleus in monkey and man. Brain Res. **261:** 321–326.
30. PEARSON, R. C. A. & T. P. S. POWELL. 1987. Anterograde vs. retrograde degeneration of the nucleus basalis medialis in Alzheimer's disease. J. Neurol. Trans. **24:** 139–146.
31. GLONEK, T., S. J. KOPP, E. KOT, J. W. PETTEGREW, W. H. HARRISON & M. M. COHEN. 1982. P-31 nuclear magnetic resonance analysis of brain: The perchloric acid extract spectrum. J. Neurochem. **39:** 1210–1219.
32. LACAL, J. C., J. MOSCAT & S. AARONSON. 1987. Novel source of 1,2-diacylglycerol elevated in cells transformed by Ha-ras oncogene. Nature *330*(19): 269–271.
33. DAWSON, R. M. C. 1985. Enzymatic pathways of phospholipid metabolism in the nervous system. *In* Phospholipids in Nervous Tissues. J. Eichberg, Ed. 45–78. Wiley. New York, NY.
34. PORCELLATI, G. & G. ARIENTI. 1983. Metabolism of phosphoglycerides. *In* Handbook of Neurochemistry. Volume 3. Metabolism in the Nervous System. A. Lajtha, Ed. 133–161. Plenum Press. New York, NY.
35. PETTEGREW, J. W., N. J. MINSHEW, T. GLONEK, S. J. KOPP & M. M. COHEN. 1982a. Phosphorus-31 nuclear magnetic resonance analysis of Huntington and control brain. (Abstract) Ann. Neurol. **12:** 91.
36. PETTEGREW, J. W., N. J. MINSHEW, T. GLONEK, S. J. KOPP & M. M. COHEN. 1982b. Phosphorus NMR study of gerbil stroke model. (Abstract) Neurology (Cleveland) **32:** 196.
37. PETTEGREW, J. W., S. J. KOPP, T. GLONEK, N. J. MINSHEW & M. M. COHEN. 1983. Phosphorus-31 NMR analysis of normoxic and anoxic brain slices. (Abstract) Neurology (Cleveland) **33**(2): 152.
38. PETTEGREW, J. W., S. J. KOPP, N. J. MINSHEW, T. GLONEK, J. M. FELIKSIK, J. P. TOW & M. M. COHEN. 1985. ^{31}P NMR studies of phospholipid metabolism in developing and degenerating brain. Neurology (Cleveland) **35**(1): 257.
39. PETTEGREW, J. W., S. J. KOPP, J. DADOK, N. J. MINSHEW, J. M. FELIKSIK, T. GLONEK & M. M. COHEN. 1986a. Chemical characterization of a prominent phosphomonoester resonance from mammalian brain: ^{32}P and ^{1}H NMR analysis at 4.7 and 14.1 Tesla. J. Magn. Res. **67:** 443–450.
40. PETEGREW, J. W., J. F. M. POST, G. WITHERS & K. PANCHALINGAM. 1986b. ^{31}P NMR studies of brain development. (Abstract) Ann. Neurol. **20:** 400–401.
41. COHEN, M. M., J. W. PETTEGREW, S. J. KOPP, N. MINSHEW & T. GLONEK. 1984. P-31 nuclear magnetic resonance analysis of brain. Normoxic and anoxic brain slices. Neurochem. Res. **9:** 785–801.
42. COHEN, M. M. & S. LIN. 1962. Acid soluble phosphates in the developing rabbit brain. J. Neurochem. **9:** 345–352.
43. CADY, E. B., M. J. DAWSON, P. L. HOPE, P. S. TOFTS, A. M. DE L. COSTELLO, D. T. DELPY, E. O. R. REYNOLDS & D. R. WILKIE. 1983. Non-invasive investigation of cerebral metabolism in newborn infants by phosphorus nuclear magnetic resonance spectroscopy. Lancet **1:** 1059–1062.
44. YOUNKIN, D. P., M. DELIVORIA-PAPADOPOULOS, J. C. LEONARD, V. HARIHARA SUBRAMANIAN, S. ELEFF, J. S. LEIGH, Jr. & B. CHANCE. 1984. Unique aspects of human newborn cerebral metabolism evaluated with phosphorus nuclear magnetic resonance spectroscopy. Ann. Neurol. **16:** 581–586.
45. MARIS, J. M., A. E. EVANS, A. C. MCLAUGHLIN, G. J. D'ANGIO, L. BOLINGER, H. MANOS & B. CHANCE. 1985. ^{31}P nuclear magnetic resonance spectroscopic investigation of human neuroblastoma in situ. N. Engl. J. Med **312:** 1500–1505.
46. WITHERS, G., A. J. MARTINEZ, G. R. RAO, J. MOOSSY & J. W. PETTEGREW. 1987. Applications of ^{31}P NMR studies on Alzheimer's disease. Neurology (Cleveland) **37**(1): 331.

47. BARANY, M., C. YEN-CHUNG, C. ARUS, T. RISTAM & W. FREY. 1985. Increased glycerol-3-phosphorylcholine in post-mortem Alzheimer's brain. (Letter) Lancet 1: 517.

48. MIOTTO, O., R. G. GONZALEZ, F. BUONANNO & J. GROWDON. 1986. In Vitro ^{31}P NMR spectroscopy detects altered phospholipid metabolism in Alzheimer's disease. (Suppl.) Can. J. Neurol. Sci. 13: 535–539.

49. PANCHALINGAM, K., J. F. M. POST & J. W. PETTEGREW. 1987a. Evidence for increased aluminum binding ligands in Alzheimer's disease. A ^{31}P-^{27} Al NMR study (Abstract) Neurology (Cleveland 37(1): 224.

50. BARANY, M. & T. GLONEK. 1982. Phosphorus-31 nuclear magnetic resonance of contractile systems. In Methods of Enzymology. Structural and Contractile Proteins, Vol. 85: Part B. D. W. Frederiksen & L. W. Cunningham, Eds. 624–676. Academic Press. New York, NY.

51. GLONEK, T., C. T. R. BURT & M. BARANY. 1981. ^{31}P NMR analysis of intact tissue including several examples of normal and diseased human muscle determinations. In: NMR Basic Principles and Progress. P. Diehl, E. Fluck & R. Kosfeld, Eds. Vol. 19: 122–159. Springer-Verlag. Berlin & Heidelberg.

52. CRAPPER, D. R., S. S. KRISHNAN & A. J. DALTON. 1973. Brain aluminum distribution in Alzheimer's disease and experimental neurofibrillary degeneration. Science 180: 511–513.

53. CRAPPER, D. R., S. S. KRISHNAN & S. QUITTKAT. 1976. Aluminum neurofibrillary degeneration and Alzheimer's disease. Brain 99: 67–80.

54. PERL, D. & A. BRODY. 1980. Alzheimer's disease: X ray spectrometric evidence of aluminum accumulation in neurofibrillary tangle-bearing neurons. Science 208: 297.

55. CANDY, J. M., J. KLINOWSKI, R. H. PERRY, E. K. PERRY, A. FAIRBAIRN, A. OAKLEY, T. CARPENTER, J. ATACK, G. BLESSED & J. EDWARDSON. 1986. Aluminosilicates and senile plaque formation in Alzheimer's disease. Lancet 1: 354–356.

56. BARANY, M. & T. GLONEK. 1984. Identification of diseased states by phosphorus-31 NMR. In Phosphorus-31 NMR. Principles and Applications. D. G. Gorenstein, Ed. 511–515. Academic Press. New York, NY.

57. PETROFF, O. A. C., J. W. PRICHARD, K. L. BEHAR, J. R. ALGER & R. G. SHULMAN. 1985. In vivo phosphorus nuclear magnetic resonance spectroscopy in status epilepticus. Ann. Neurol. 16: 169–177.

58. PETTEGREW, J. W., G. WITHERS, K. PANCHALINGAM & J. F. M. POST. 1988c. Considerations for brain pH assessment by ^{31}P NMR. Magn. Res. Imag. 6: 135–142.

59. MCKHANN, G., D. DRACHMAN, M. FOLSTEIN, R. KATZMAN, D. PRICE & E. STADLAN. 1984. Clinical diagnosis of Alzheimer's disease: Report of the NINCDS-ADRDA work group under the auspices of Department of Health and Human Services Task Force on Alzheimer's Disease. Neurology (Cleveland) 34: 939–944.

60. PONTEN, U., R. A. RATCHENSON, L. G. SALFORD & B. K. SIESJO. 1973. Optimal freezing conditions for cerebral metabolites in rats. J. Neurochem. 21: 1127–1138.

61. VEECH, R., R. L. HARRIS, D. VELOSO & E. H. VEECH. 1973. Freeze-blowing: A new technique for the study of the brain in vivo. J. Neurochem. 20: 183–188.

62. BUERI, J. A., J. M. GORELL, S. R. LEVINE, K. WELCH, M. SMITH, J. EWING, J. HELPERN, R. BRUCE & T. KENSORA. 1987. Cerebral phosphate metabolism in Alzheimer and Parkinson dementia measured by ^{31}P NMR spectroscopy. (Abstract) Neurology (Cleveland 37(1): 160.

63. LEVINE, S. R., K. M. A. WELCH, J. A. HELPERN, J. R. EWING, R. BRUCE & M. SMITH. 1987. ^{31}P NMR spectroscopic study of multi-infarct dementia. (Abstract) Neurology (Cleveland) 37(1): 251.

64. CRAPPER, D. R., S. QUITTKAT, S. S. KRISHNAN, A. J. DALTON & U. DE BONI. 1980. Intranuclear aluminum content in Alzheimer's disease, dialysis encephalopathy and experimental aluminum encephalopathy. Acta Neuropathol. (Berlin) 50: 19–24.

65. CULLIS, P. R. & A. J. VERKLEIJ. 1979. Modulation of membrane structure by Ca^{2+} and dibucaine as detected by ^{31}P NMR. Biochim. Biophys. Acta 552: 546–551.

66. PANCHALINGAM, K., J. F. M. POST, R. JIMENEZ & J. W. PETTEGREW. 1987b. Studies of aluminum interaction with membranes using ^{13}C, ^{31}P, ^{27}Al NMR. Society of Biological Psychiatry, Forty-Second Annual Scientific Program, pg. 228.

67. WEINER, S. J., P. A. KOLLMAN, D. A. CASE, U. C. SINGH & C. GHIO. 1984. Alagona systems in mammalian brain identified with antibodies against choline acetyltransferase. Neurosci. Int. 6: 163–182.

68. BARRIONUEVO, G., J. E. BRADLER, J. W. PETTEGREW. 1988. Electrophysiological effects of phosphomonoesters on hippocampal brain slices. (Abstract) Neurology (Cleveland) 38(1): 336.

69. PETTEGREW, J. W., K. PANCHALINGAM, D. MCKEAG & G. BARRIONUEVO. 1988e. Metabolic effects of phosphomonoesters on hippocampal brain slices. (Abstract) Neurology (Cleveland) 38(1): 323.

70. MARAGOS, W. F., J. T. GREENAMYRE, J. B. PENNEY, JR. & D. A. B. YOUNG. 1987. Glutamate dysfunction in Alzheimer's disease: An hypothesis. Trends Neurosci. 10: 65–68.5.

71. GEDDES, J. W., H. CHANG-CHUI, S. M. COOPER, I. T. LOTT & C. W. COIMAN. 1986. Density and distribution of NMDA receptors in the human hippocampus in Alzheimer's disease. Brain Res. 399: 156–161.

72. GREENAMYRE, J. T., J. B. PENNY, JR., A. B. YOUNG, C. J. AMATO, S. P. HICKS & I. SHOULSON. 1985. Alterations in L-glutamate binding in Alzheimer's and Huntington's diseases. Science 227: 1496–1499.

73. YOUNG, A. B. & J. T. GREENAMYRE. 1986. Autoradiographic analysis of L-3H-glutamate receptors in mammalian brain. In Quantitative Receptor Autoradiography. C. A. Boast, E. W. Snowhill & C. A. Altar, Eds. 79–101. Alan R. Liss. New York, NY.

74. KANFER, J. N. & D. G. MCCARTNEY. 1987. Phosphatase and phospholipase activity in Alzheimer brain tissues. In Topics in the Basic and Clinical Science of Dementia. R. J. Wurtman, S. Corkin & J. H. Growden, Eds. J. Neurol. Trans. Suppl. 24. 183–188. Springer-Verlag. Vienna & New York.

75. NISHIZUKA, Y. 1986. Studies and perspectives of protein kinase C. Science 233: 305–312.

76. SCHEIBEL, A. B. 1977. Dendritic changes in senile and presenile dementias. In R. Katzman, Ed. Congenital and Acquired Cognitive Disorders. Based on the proceedings of the 57th Annual Meeting of the Association for Research in Nervous and Mental Diseases held December 2–3, 1977. (Research publication Vol. 57.) 107–124. Raven Press. New York, NY.

77. GEDDES, J. W., D. T. MONAGHAN, C. W. COTMAN, I. T. LOTT, R. C. KIM & H. C. CHUI. 1985. Plasticity of hippocampal circuitry in Alzheimer's disease. Science 230: 1179–1181.

78. BUELL, S. J. & P. D. COLEMAN. 1979. Dendritic growth in the aged human brain and failure of growth in senile dementia. Science 206: 854–856.

79. BUELL, S. J. & P. D. COLEMAN. 1981. Quantitative evidence for selective dendritic growth in normal human aging but not in senile dementia. Brain Res. 241: 23–41.

80. GRAVELAND, G. A., R. S. WILLIAMS & M. DEFIGLIA. 1985. Evidence for degenerative and regenerative changes in neostriatal spiny neurons in Huntington's disease. Science 227: 770–773.

81. FLOOD, D. G., S. J. BUELL, G. J. HORWITZ & P. COLEMAN. 1986a. Dendritic extent in human dentate gyrus granule cells in normal aging and senile dementia. Brain Res. 402(2):205–216.

82. FLOOD, G. & P. D. COLEMAN. 1986b. Failed compensatory dendritic growth as a pathophysiological process in Alzheimer's disease. Can. J. Neurol. Sci. 13: 475–479.

83. HINDS, J. W. & N. A. MCNELLY. 1977. Aging of the rat olfactory bulb: Growth atrophy of constituent layers and changes in size and number of mitral cells. J. Comp. Neurol. 171: 345–368.

84. CONNOR, J. R., M. C. DIAMOND & R. E. JOHNSON. 1980. Occipital cortical morphology of the rat: Alterations with age and environment. Exp. Neurol. 68: 158–170.

85. CONNOR, J. R., S. E. BEBAN, P. A. HOPPER, B. HANSEN & M. C. DIAMOND. 1982. A Golgi study of the superficial pyramidal cells in the somatosensory cortex of socially reared old adult rats. Exp. Neurol. 76: 35–45.

86. HINDS, J. W. & N. A. MCNELLY. 1981. Aging in the rat olfactory system: Correlation of changes in the olfactory epithelium and olfactory bulb. J. Comp. Neurol. 203: 441–453.

87. ROGERS, J., S. F. ZORNETZER, F. E. BLOOM & R. E. MERVIS. 1984. Senescent microstructural changes in rat cerebellum. Brain Res. 292: 23–32.

88. ZUBENKO, G. S., I. MALINAKOVA & B. CHOJNACKI. 1987. Proliferation of internal membranes in platelets from patients with Alzheimer's disease. J. Neuropathol. Exp. Neurol. 407–418.

89. PETTEGREW, J. W., J. MOOSSY, S. STRYCHOR, D. MCKEAG & F. BOLLER. 1988d. Membrane phospholipid alterations in Alzheimer's brain. (Abstract) Neurology (Cleveland) 38(1): 267.

90. BARTUS, R. T., R. L. DEAN, B. BEER & A. S. LIPPA. 1982. The cholinergic hypothesis of geriatric memory dysfunction. Science 217: 408–417.

91. PELECH, S. L., F. AUDUBERT & D. E. VANCE. 1985. Regulation of phosphatidylcholine biosynthesis in mammalian systems. *In* Phospholipids in the nervous system. L. A. Horrocks, J. N. Kanfer & G. Porcellati, Eds. Vol. **2**: 247–258. Raven Press. New York, NY.
92. GRUNDKE-IQBAL, I., K. IQBAL, Y. C. TUNG, M. QUINLAN, H. M. WISNIEWSKI & L. I. BINDER. 1986. Abnormal phosphorylation of the microtubule associated protein tau in Alzheimer cytoskeletal pathology. Proc. Natl. Acad. Sci. USA **83**: 4913–4917.
93. IQBAL, K., T. ZAIDI, G. Y. WEN, I. GRUNDKE-IQBAL, P. A. MERZ, S. S. SHAIKH, H. M. WISNIEWSKI, I. ALAFUZOFF & B. WINBLAD. 1986. Defective brain microtubule assembly in Alzheimer's disease. Lancet **2**: 421–426.
94. GANDY, S., A. J. CZERNIK & P. GREENGARD. 1988. Phosphorylation of Alzheimer disease amyloid precursor peptide by protein kinase C and Ca^{2+}/calmodulin-dependent protein kinase II. Proc. Natl. Acad. Sci. USA **85**: 6218–6221.
95. COLE, G., K. R. DOBKINS, L. A. HANSEN, R. D. TERRY & T. SAITOH. 1988. Decreased levels of protein kinase C in Alzheimer brain. Brain Res. **452**: 165–174.

Regulation of Membrane Function Through Composition, Structure, and Dynamics

PHILIP L. YEAGLE

Department of Biochemistry
University at Buffalo School of Medicine
State University of New York
Buffalo, New York 14214

Through recent progress in the study of cell membrane structure and function, factors involved in the regulation of membrane proteins have been uncovered. In this discussion, overall guiding principles (as best as they can be discerned at this time) will be illustrated with a few specific examples.

Lipid Regulation of Membrane Protein Function

Since membrane proteins function in a medium of membrane lipids, it is reasonable to examine whether those membrane lipids have any effect on the function of the membrane proteins imbedded in the lipid bilayer. The first example to be considered is the calcium pump protein of the fast twitch muscle.

Calcium Pump Protein

Most studies on the calcium pump protein have been utilized the Ca ATPase from the rabbit muscle sarcoplasmic reticulum. This ion pump is involved in the sequestration of calcium after muscle contraction. Sequestration of calcium in the lumen of the sarcoplasmic reticulum permits the relaxation of the muscle.

Large transmembrane ion gradients are maintained by this calcium pump protein. The difference in calcium concentrations across the sarcoplasmic reticulum can be 3 or 4 orders of magnitude. The establishment of such large ion gradients is enabled by the coupled hydrolysis of ATP.

The calcium pump protein from the rabbit muscle sarcoplasmic reticulum can be purified and reconstituted. Employing that technology it has been possible to examine the effects of various membrane lipid compositions on the activity of the calcium pump. One focus of that research has been on the role of phosphatidylethanolamine (PE) in regulating calcium transport catalyzed by this pump protein.

By reconstitution of the calcium pump protein in membranes of varying PE content, the effects of PE on this pump could be examined. FIGURE 1 shows an example of the effects of PE. As the PE content of the membrane increases, the ability to transport calcium also increases.[1] Similar effects have been observed from other laboratories, both in reconstituted systems[2] and in the biological membrane.[3]

This stimulation by PE persists at high PE contents, as long as the membrane remains in the lamellar (or bilayer) configuration. PE is known to adopt alternate phase structures, including the hexagonal (II) phase.[4] When the lipids are in nonlamellar phases, no calcium transport can occur, because closed membrane vesicles (necessary for the calcium transport assay) do not exist under such conditions.

After reconstitution procedures have been performed, it is essential to re-isolate the reconstituted membranes by density gradient centrifugation. Reconstitution is not a 100% efficient process. After density gradient centrifugation, one can usually isolate a reconstituted membrane of reasonable homogeneity with respect to lipid-protein ratio and lipid composition. In the process lipid that did not recombine with the protein during the reconstitution process is separated from the preparation one is using to study. In most reconstitution procedures, one does not harvest a reconstituted membrane of the same composition as the initial reconstitution mixture. Therefore, the separation on a sucrose gradient is of utmost importance when performing these experiments.

Another interesting effect by PE has been observed in the reconstitution process. As the PE content increased, the efficiency of incorporation of the calcium pump protein improved. That is, the lipid-protein ratio decreased with increasing PE content of the reconstitution mixture of lipids (Cheng and Yeagle, unpublished observations).

Therefore, PE both stimulates the function of the calcium pump protein and alters the membrane structure to enable the membrane to accommodate more of the calcium pump proteins in the lipid bilayer.

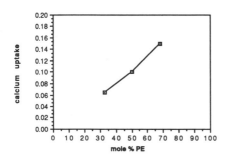

FIGURE 1. Plot of calcium uptake at 23°C as a function of mole % PE in the membrane reconstituted with the calcium pump protein from rabbit muscle sarcoplasmic reticulum (data taken from REF. 1).

Cholesterol is another lipid whose effects on the calcium pump protein have been documented. The cholesterol content of membranes can be readily modified by incubation of the biological membrane vesicles or cells with small unilamellar phosphatidylcholine vesicles (PC SUV).[5] If the PC SUV contain higher levels of cholesterol than the biological membranes, one can often achieve a net flux of cholesterol into the biological membranes, thereby increasing their cholesterol content (this can be dependent upon the lipid content).[6] If the PC SUV contain no cholesterol, then the biological membranes can be depleted of cholesterol.

Using this technology, we altered the cholesterol content of the sarcoplasmic reticulum membranes and measured the ATPase activity. The data demonstrated that in the biological membrane, no effect of cholesterol can be seen.[7] This has been observed by other groups.[8]

However, in reconstituted systems with high PE content, cholesterol stimulates the calcium pump protein.[1] The reasons for these differences have not yet been adequately established.

$Na^+K^+ATPase$

The Na^+K^+ ATPase is an important plasma membrane cation pump protein responsible for maintaining the sodium and potassium gradients across the plasma membrane that are essential for cellular viability. Furthermore, since this pumping is electrogenic, the Na^+K^+ATPase is involved in the transmembrane electrical potential characteristic of the plasma membrane.

By modulation of cholesterol levels in the membrane as described above, cholesterol has been shown to modulate the activity of this crucial plasma membrane enzyme.[9] At high levels, cholesterol inhibits the activity in the human erythrocyte[10,11] and in kidney epithelia.[9] Maximal activity is observed at the native membrane cholesterol levels for the kidney epithelia. The erythrocyte Na^+K^+ATPase is operating at substantially less than maximal activity in the human erythrocyte, with respect to cholesterol levels.

Even more interesting is the effect of cholesterol at levels less than the native cholesterol levels of the kidney epithelia. At low levels, cholesterol is *stimulatory*. In fact the data suggest that in the absence of cholesterol, the Na^+K^+ATPase will exhibit little or no activity.[9] Apparently this enzyme reuqires cholesterol for expression of its enzymatic function.

More clues to the nature of this cholesterol requirement were revealed by studying the effect of sterols with structures different from that of cholesterol on the Na^+K^+ATPase activity. Ergosterol is a yeast sterol, with a structure different from that of cholesterol. Ergosterol showed little or no ability to activate the Na^+K^+ATPase, in dramatic contrast to cholesterol.[9] Lanosterol, a precurser of cholesterol in the biosynthetic pathway for cholesterol biosynthesis,[12] exhibited about one half the potency of cholesterol in activating the Na^+K^+ATPase. Therefore, this sterol requirement for the Na^+K^+ATPase is structurally sensitive.

A viable hypothesis to explain these results is that the Na^+K^+ATPase may have a binding site for cholesterol. Such a binding site would allow the sterol to modify the activity of the enzyme and would confer the structural specificity for sterols that is observed. The inability of cholesterol to modify the activity of the calcium pump protein may be due to the lack of specific cholesterol binding to that enzyme, at least in the absence of high PE content.[8] This hypothesis must be subject to further testing, however.

What is clear from these results is a new view on the essential role of cholesterol in mammalian cells. All mammalian cells require cholesterol for function, for growth, and for differentiation.[5] Until recently, it was not known just what that essential role was. Data from membrane biochemistry[9] and from cell biology[13] demonstrate that cholesterol is required for crucial enzyme activities in mammalian cells, and that other sterols cannot substitute for the essential cholesterol content. In an analogous manner, the essential sterol in yeast is ergosterol, and cholesterol cannot substitute in yeast for that essential sterol requirement.[13]

Other Examples of Lipid Regulation of Membrane Proteins

Although this is not meant to be a comprehensive review, several other notable examples of lipid regulation of membrane proteins should be mentioned. Cholesterol regulation of the acetylcholine receptor is one such example. Cholesterol is apparently required for proper function.[14] Cholesterol is also required for the activity of the cardiac sarcolemmal Na^+–Ca^{+2} exchange transporter.[15]

In the case of the D-β-hydroxybutyrate apodehydrogenase, a specific phospholipid requirement was observed.[16] The choline headgroup of phosphatidylcholine is required for function.

Mechanisms for Regulation of Membrane Enzymes

Given that there is now considerable evidence that particular lipids are required for expression of function by membrane proteins, the question that naturally arises is by what mechanism(s) do these lipids influence the membrane proteins.

Three possible mechanisms for lipid regulation of membrane protein function will be briefly reviewed here. One possible mechanism involves the thickness of the lipid bilayer. Another possible mechanism is by binding of the lipids to the membrane protein in question, a mechanism already introduced in the case of cholesterol. The last possible mechanism to be examined involves surface-surface interactions between the extramembranous portion of the membrane proteins and the lipid bilayer surface.

Bilayer Thickness

The topology of membrane proteins makes them sensitive to the thickness of the lipid bilayer. Transmembrane proteins usually contain linear sequences of hydrophobic amino acids that can span the hydrophobic portion of the membrane lipid bilayer in an α helical structure.[17] Clustered at the ends of these hydrophobic helices are charged amino acids. Thus there is a distinct delineation between the hydrophobic and the hydrophilic regions of the membrane protein.

These hydrophobic and hydrophilic regions must match the corresponding hydrophobic and hydrophilic regions of the lipid bilayer. In the event of a mismatch, the membrane protein and/or the lipid bilayer must distort to accommodate the hydrophobic effect. For example, if the lipid bilayer is too thin, then the hydrophobic transmembrane helices of the membrane protein will be partially exposed to the aqueous media. This exposure is thermodynamically unfavorable, primarily from an entropy standpoint. Then the protein is likely to distort its conformation to try to bury those hydrophobic regions within the hydrophobic milieu of the lipid bilayer interior. Perhaps, in addition, the lipid bilayer may distort to cover up those hydrophobic regions on the membrane protein. Such distortions can be expected to cause conformational changes in the membrane protein, consequently altering the activity of the membrane protein.

A specific example of this effect was noted in the case of the calcium pump protein from rabbit muscle sarcoplasmic reticulum.[18] With lipid bilayers that were too thick or too thin, the calcium pump protein exhibited reduced activity.

Binding of Phospholipids to Membranes

Recent progress in this area has led to a new understanding of the details of interactions between membrane proteins and membrane phospholipids. The specific example which will be employed involves the glycoprotein, glycophorin, from the human erythrocyte membrane.

Human erythrocyte glycophorin containing 4 molecules of phospholipid tightly bound to the protein was isolated from human red cell ghosts. This protein preparation was reconstituted into a digalactosyldiglyceride bilayer. By so doing, the only contribution to the ^{31}P NMR spectrum would be from the phospholipids originally tightly bound to the glycophorin. The ^{31}P NMR spectrum of this reconstituted membrane produced an axially symmetric powder pattern arising exclusively from the phospholipids bound to glycophorin. The width of the powder pattern, 90 ppm, is about twice that normally exhibited by a phospholipid bilayer. The chemical shift tensor and the spin lattice relaxation rate of these tightly bound phospholipids are perturbed relative to phospholipids in a bilayer. The results are consistent with phospholipids tightly bound to the membrane protein and undergoing rotational diffusion, perhaps as a complex of phospholipid and protein.[19] Other examples of phospholipids binding to membrane proteins are available, also from NMR studies.[20–24]

These data indicate that phospholipids do, in some cases, bind tightly to membrane proteins, and act as a lipid-protein complex in the membrane. Therefore, by binding to membrane proteins, particular phospholipids may be able, as effectors or modulators, to alter the activity of the membrane protein. This requires specificity in the lipid binding which has been observed in some cases (25–27).

Surface-Surface Interactions.

In the case of transmembrane proteins with large extramembranous portions, the likelihood exists of contact between the extramembranous portion of the membrane protein and the surface of the lipid bilayer. Such contact might be facilitated if the lipid bilayer surface was easily dehydrated. That is because for contact, or even close approach, water must be removed from between the surfaces.

Recent experiments with PE have revealed that water is readily removed from the PE bilayer surface.[28] Therefore, it may be possible for PE to modulate membrane protein activity by facilitating surface-surface interactions between the surface of the extramembranous portion of the protein and the lipid bilayer surface. This hypothesis would be amenable to testing.

SUMMARY

Recent data reveal that membrane proteins can be regulated by the lipid environment of the membrane in which the proteins are located. Phospholipids and sterols are capable of regulating membrane protein activity. In some cases, the regulation is specific and may be crucial to cell function. The mechanisms by which this regulation is carried out have not been firmly established. However, the likely mechanisms include bilayer thickness, specific binding of lipids to sites on membrane proteins, and interactions between the surface of the protein and the surface of the lipid bilayer.

REFERENCES

1. CHENG, K.-H., J. R. LEPOCK, S. W. HUI & P. L. YEAGLE. 1986. The role of cholesterol in the activity of reconstituted Ca ATPase vesicle containing unsaturated phosphatidylethanolamine. J. Biol. Chem. **261:** 5081–5087.
2. NAVARRO, J., M. TOIVIO-KINNUCAN & E. RACKER. 1984. Effect of lipid composition on the calcium/adenosine 5'-triphosphate coupling ration of the Ca^{2+}-ATPase of sarcoplasmic reticulum. Biochemistry **23:**130–135.
3. HIDALGO, C., D. A. PETRUCCI & C. VERGARA. 1982. Uncoupling of calcium transport in sarcoplasmic reticulum as a result of labelling lipid amino groups. J. Biol. Chem. **257:** 208–216.
4. HUI, S. W., T. P. STEWART, P. L. YEAGLE & A. D. ALBERT. 1981. Bilayer to nonbilayer transition in mixtures of phosphatidylethanolamine and phosphatidylcholine: Implications for membrane properties. Arch. Biochem. Biophys. **207:** 227–240.
5. YEAGLE, P. L. 1985. Cholesterol and the cell membrane. Biochim. Biophys. Acta **822:** 267–287.
6. YEAGLE, P. L. & J. YOUNG. 1986. Factors contributing to the distribution of cholesterol among phospholipid vesicles. J. Biol. chem. **261:** 8175–8181.
7. SELINSKY, B. S. 1984. Ph.D. thesis, State University of New York at Buffalo, Buffalo, NY.
8. WARREN, G. B., M. D. HOUSLAY, J. C. METCALFE & N. J. M. BIRDSALL. 1975. Cholesterol is exluded from the phospholipid annulus surrounding an active calcium transport protein. Nature (London) **255:** 684–687.

9. YEAGLE, P. L., J. YOUNG & D. RICE. 1988. Effects of cholesterol on (Na^+,K^+)-ATPase ATP hydrolyzing activity in bovine kidney. Biochemistry 27: 6449–6452.
10. YEAGLE, P. L. 1983. Cholesterol modulation of $Na^+ + K^+)$ ATPase ATP hydrolyzing activity in the human erythrocyte. Biochim. Biophys. Acta 727: 39–44.
11. GIRAUD, F., M. CLARET, K. R. BRUCKDORFER & B. CHAILLEY. 1981. Effects of lipid order and cholesterol on the internal and external cation sites of the Na^+-K^+ pump in erythrocytes. Biochim. Biophys. Acta 647: 249–258.
12. YEAGLE, P. L., ED. 1988. Biology of Cholesterol. CRC Press. Boca Raton, FL.
13. DAHL, C. & J. DAHL. 1988. Cholesterol and cell function. In Biology of Cholesterol. P. L. Yeagle, Ed. CRC Press. Boca Raton, FL.
14. JONES, O. T. & M. G. MCNAMEE. 1988. Annular and nonannular binding sites for cholesterol associated with the nicotinic acetylcholine receptor. Biochemistry 27: 2364–2374.
15. VERMURI, R. & K. D. PHILIPSOLN. 1988. Biochim. Biophys. Acta 937: 258–268.
16. ISAACSON, Y. A., P. W. DEROO, A. F. ROSENTHAL, R. BITTMAN, J. O. MCINTYRE, H.-G. BOCK, P. GAZZOTTI & S. FLEISCHER. 1979. The structural specificity of lecithin for activation of purified D-β-hydroxybutyrate apodehydrogenase. J. Biol. Chem. 254: 117–126.
17. YEAGLE, P. L. 1987. The Membranes of Cells. Academic Press. Orlando, FL.
18. CAFFREY, M. & F. W. FEIGENSON. 1981. Fluorescence quenching in model membranes. 3. Relationship between calcium adenosinetriphosphatase enzyme activity and the affinity of the protein for phsophatidylcholines with different acyl chain characteristics. Biochemistry 20: 1949–1961.
19. YEAGLE, P. L. & D. KELSEY. 1989. Phosphorus NMR studies of lipid-protein interactions: Human erythrocyte glycophorin and phospholipids. Biochemistry 28: 2210–2215.
20. ALBERT, A. D., S. A. LANE & P. L. YEAGLE. 1985. ^2H and ^{31}P nuclear magnetic resonance studies of membranes containing bovine rhodopsin. J. Membr. Biol. 87: 211–215.
21. SELINSKY, B. S. & P. L. YEAGLE. 1984. Two populations of phospholipids exist in sarcoplasmic reticulum and in recombined membranes containing Ca ATPase. Biochemistry 23: 2281–2288.
22. SEELIG, A. & J. SEELIG. 1985. Phospholipid composition and organization of cytochrome c oxidase preparations as determined by ^{31}P NMR. Biochim. Biophys. Acta 815: 153–158.
23. BEYER, K. & M. KLINGENBERG. 1985. ADP/ATP carrier protein from beef heart mitochondria has high amounts of tightly bound cardiolipin, as revealed by ^{31}P NMR. Biochemistry 24: 3821–3826.
24. SELINSKY, B. S. & P. L. YEAGLE. 1985. Phospholipid exchange between restricted and nonrestricted domains in sarcoplasmic reticulum vesicles. Biochim. Biophys. Acta 813: 33–40.
25. ARMITAGE, I. M., D. L. SHAPIRO, H. FURTHMAYR & V. T. MARCHESI. 1977. ^{31}P nuclear magnetic resonance evidence for polyphosphoinositide associated with the hydrophobic segment of glycophorin A. Biochemistry 16: 1317–1320.
26. POWELL, G. L., P. F. KNOWLES & D. MARSH. 1987. Spin label studies on the specificity of interaction of cardiolipin with beef heart cytochrome oxidase. Biochemistry 26: 8138–8145.
27. BROTHERUS, J. R., P. C. JOST, O. H. GRIFFITH, J. F. W. KEANA & L. E. HOKIN. 1980. Charge selectivity at the lipid-protein interface of membraneous Na^+K^+ ATPase. Proc. Natl. Acad. Sci. USA 77: 272–276.
28. YEAGLE, P. L. & A. SEN. 1986. Hydration and the lamellar to hexagonal II phase transition of phosphatidylethanolamine. Biochemistry 25: 7518–7522.

Platelet-Activating Factor and Related Ether Lipid Mediators

Biological Activities, Metabolism, and Regulation[a]

FRED SNYDER, TEN-CHING LEE, AND MERLE L. BLANK

Medical Sciences Division
Oak Ridge Associated Universities
P.O. Box 117
Oak Ridge, Tennessee 37831-0117

INTRODUCTION

Virtually all animal cells contain ether-linked aliphatic moieties in membrane lipid constituents. With few exceptions the 0-alkyl moieties are most prominent in choline glycerolipids, whereas the 0-alk-1-enyl chains are found primarily as ethanolamine plasmalogens (FIG. 1). Both types of ether linkages also occur in the neutral lipid fraction where they exist as an ether lipid analog that is structurally similar to triacylglycerols. Brain is one of the richest sources of ethanolamine plasmalogens, but most other tissues, except liver, also contain significant quantities of this type of ether lipid.[1] The alkyl ether lipids are especially high in cells that produce biologically active phospholipids in response to inflammatory agents.[2] An exhaustive listing of the tissue distribution of ether-linked phospholipids in comparison to the diacyl types of lipids has been compiled by Sugiura and Waku[2] and Horrocks.[1]

Although ether lipids were once thought to serve strictly as structural components of membranes, it is now apparent that some types of ether lipids are highly biologically active and that the membrane forms are usually enriched in arachidonic acid and other polyunsaturated fatty acids that can be converted to bioactive lipid mediators under appropriate conditions.[3] Perhaps the most potent lipid mediator ever discovered is a novel ether-linked phospholipid class possessing the chemical structure 1-alkyl-2-acetyl-*sn*-glycero-3-phosphocholine (FIG. 1).[4-6] This phospholipid is commonly referred to as platelet-activating factor (PAF), but the term is a misnomer in view of the diverse biological activities PAF possesses. An acetylated ether-linked counterpart of diglycerides (1-alkyl-2-acetyl-*sn*-glycerol) also exhibits properties similar to PAF, but it is less active than PAF.[7,8]

This report highlights work from our laboratory and others that has contributed to our understanding of the metabolism and regulation of PAF since the discovery of its chemical structure. A brief account of the biological responses elicited by PAF and its proposed mode of action are also discussed.

Biological Activities of PAF and Its Putative Role in Cellular Signal Transduction Events

PAF elicits a wide variety of biological responses that involve many different types of cells (see reviews[9-13] of this subject). Some examples of the PAF-induced responses

[a]This work was supported by the Office of Energy Research, U.S. Department of Energy (Contract No. DE-AC05-76OR00033), the American Cancer Society (Grant BC-70T), and the National Heart, Lung, and Blood Institute, National Institutes of Health (Grant 27109-08).

35

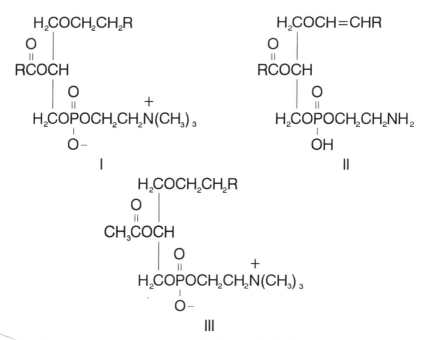

FIG. 1. Examples of several types of ether-linked phospholipids. The formulae designate: I plasmanylcholine (1-alkyl-2-acyl-*sn*-glycero-3-phosphocholine), II plasmenylethanolamine or ethanolamine plasmalogen (1-alk-1'-enyl-2-acyl-*sn*-glycero-3-phosphoethanolamine), and platelet-activating factor (1-alkyl-2-acetyl-*sn*-glycero-3-phosphocholine).

include a) hypotension, b) aggregation and degranulation of platelets and neutrophils, c) chemotaxis, d) increased Ca^{2+} influx, IP_3 production, and glycogenolysis, and e) bronchoconstriction.

The mechanisms of PAF actions are not yet understood, but some of the responses are clearly receptor-mediated since they can be blocked by specific PAF receptor antagonists.[9] Attempts to isolate PAF receptors for in-depth characterization have so far been unsuccessful because of the small number of specific receptor binding sites that occur on cell surfaces. For example, the number of receptors on the plasma membrane has been estimated to be 150–400 for rabbit and human platelets[14-16] and 1100 for human neutrophils.[17]

It has been suggested that the activated form of the PAF receptor interacts with a GTP-binding protein.[18-22] However, in platelets the GTP regulatory protein might differ from N_i or N_s since pertussis or cholera toxins exert little effect on the GTPase activity stimulated by PAF.[22] On the other hand, exposure of human neutrophils to pertussis toxin causes an inhibition of PAF-induced chemotaxis and thus implicates the involvement of an N_i regulatory protein in these cells.[18] A possibility of a connection between adenylate cyclase and the PAF receptor has also been reported.[19,21,23] Furthermore, a chymotryptic serine protease appears to contribute to PAF activation of platelets.[24]

There is some evidence that different types of receptors[25] might account for the diversity of PAF responses. The likelihood of different mechanisms being responsible for the hypotensive and platelet responses of PAF comes from a study of a PAF analog that exhibits preferential hypotensive activity;[26] however, for a wide variety of PAF analogs selectivity in the magnitudes of the hypotensive and platelet responses does not occur.[27] It is obvious that much more experimental work is required to pinpoint the precise mode(s) of action of PAF.

Biosynthesis of Ether Lipids

Literature describing the enzymatic reactions and complete pathways for the biosynthesis and catabolism of the ether-linked glycerolipids has been extensively reviewed by Hajra[28] and Snyder *et al.*[29] Alkyl ether linkages[5] in glycerolipids are formed from acyldihydroxyacetone-P and long chain fatty alcohols in a reaction catalyzed by alkyldihydroxyacetone-P synthase. Alkyldihydroxyacetone-P, a product of this reaction, is subsequently reduced by an NADPH-oxidoreductase to alkylglycero-P, the starting point for the biosynthesis of more complex ether-linked lipids, including PAF. The choline and ethanolamine types of alkyl ether lipids are formed from alkylglycero-P by a sequence of reactions analogous to those involving the biosynthesis of the diacyl phospholipid counterparts (phosphatidylcholine and phosphatidylethanolamine). The ethanolamine plasmalogens are produced from alkylacylglycerophosphoethanolamines by Δ1-alkyl desaturase, a mixed function oxidase[30,31] that involves the same cytochrome b_5 system utilized by acyl-CoA desaturase. Surprisingly, the cellular function of plasmalogens is still obscure, although it has been proposed that they could play a direct role in protecting animal cell membranes against oxidative stress.[32]

Biosynthesis of PAF by the Remodeling Pathway

The multifaceted nature of certain structural lipid components in membranes is probably most apparent for ether-linked phospholipids. For example, the alkyl lipids are not only significant structural components of cellular organelles, but the choline-containing species can also function as precursors of PAF and eicosanoid mediators[3,11,28] (FIG. 2); ethanolamine plasmalogen analogs of PAF (alk-1-enyl-2-acetyl-*sn*-glycero-3-phosphoethanolamine) are also produced by some cells[33] and this analog too might be bioactive like the choline plasmalogen analog of PAF (1-alk-1-enyl-2-acetyl-*sn*-glycero-3-phosphocholine).[34] Once PAF has served as a cellular messenger to activate a specific biochemical process in a stimulated cell, the mediator is inactivated (FIG. 2, Reactions III

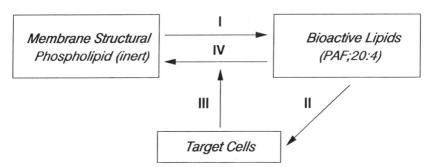

FIG. 2. Biosynthetic relationship of PAF and membrane structural lipids in the remodeling pathway of PAF biosynthesis. *Roman numerals:* I = PAF activation, II = PAF receptor interaction, and III and IV = inactivation of PAF.

and/or IV) which can then be directly converted back to a membrane structural component where it remains an inactive precursor of PAF until the cycle is repeated after appropriate cellular stimulation.[11,35]

FIG. 2 illustrates the reactions often referred to as the remodeling pathway of PAF synthesis or simply the PAF cycle. In this cycle the inert membrane structural phospholipid is a pool of alkylacylglycerophosphocholines (FIG. 1, Structure I). Cellular activation by appropriate stimuli activates a phospholipase A_2 to form lyso-PAF (alkyllysoglycerophosphocholine), the substrate for the acetyltransferase that forms PAF in the final step of this cycle. Inactivation (FIG. 2, Reactions III and/or IV) of PAF is catalyzed by an acetylhydrolase to produce the lyso-PAF intermediate which can then be converted back to the inactive PAF precursor pool in membranes by a transacylase that preferentially

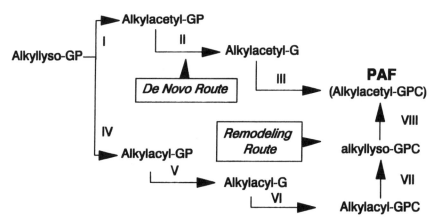

FIG. 3. Remodeling and de novo pathways for the biosynthesis of PAF. The *Roman numerals* identify the following enzymes: I. acetyl-CoA: alkyllysoglycero-P acetyltransferase, II. alkylacetylglycero-P phosphohydrolase, III. CDP-choline:alkylacetylglycerol DTT-insensitive cholinephosphotransferase, IV. acyl-CoA:alkyllysoglycero-P acyltransferase, V. alkylacylglycero-P phosphohydrolase, VI. CDP-choline:alkylacylglycerol DTT-sensitive cholinephosphotransferase, VII. Phospholipase A_2, and VIII. acetyl-CoA:alkyllysoglycerophosphocholine acetyltransferase. In this scheme the *capital letters* correspond to the following groupings: G = glycerol, P = phosphorus, and C = choline.

transfers arachidonate or other polyunsaturates into the *sn*-2 position.[36-42] Some recent evidence suggests protein kinase C might be responsible for initiating this cycle (FIG. 2, Reaction I) when cells are activated to produce PAF and eicosanoid metabolites.[43] The remodeling route is thought to be of pathophysiological relevance, since it is readily stimulated by inflammatory agents and produces both PAF arachidonic acid metabolites such as prostaglandins, leukotrienes, and hydroxy eicosanoid derivatives (see review[11]).

Biosynthesis of PAF by the De Novo Pathway

The de novo pathway of PAF biosynthesis beginning with alkylglycero-P, consists of three sequential steps that involve acetylation, dephosphorylation, and the addition of phosphocholine (FIG. 3). The enzymes responsible for catalyzing these reactions are an acetyltransferase,[44] a phosphohydrolase,[45] and a DTT-insensitive cholinephosphotransferase.[46,47] Properties of these enzymes are distinctly different from those of the acetyltransferase in the remodeling route, phosphatidate phosphohydrolase, and the DTT-sensitive cholinephosphotransferase responsible for the synthesis of phosphatidylcholine and alkylacylglycerophosphocholines.[44–47] Bioactive eicosanoid metabolites are not involved in this route of PAF biosynthesis.

Catabolism of Ether Lipids

Phospholipases, except phospholipase A_1, attack ether-linked lipids in the same manner as their diacyl counterparts. Thus, phospholipases A_2, C, and D and other lipases are capable of using ether-linked lipids as substrates, but generally at lower reaction rates than the corresponding diacyl analogs.[29] However, our knowledge of intracellular phospholipases is still quite limited and it is quite possible that highly specific phospholipases for ether-linked lipids exist in mammalian cells.

Lyso-phospholipase D is unique in its high specificity for ether-linked phospholipids; it does not utilize any acyl analogs as substrates.[48] The quantitative significance of lyso-phospholipase D in physiological situations is not known, but it appears to have a role in the metabolism of PAF.[49]

Acetylhydrolase is undoubtedly the most important enzyme activity involved in PAF catabolism, since it is capable of inactivating the biological activity of PAF. This enzyme activity is present intracellularly in most tissues[50] as well as blood.[50–54]

The ether linkages in glycerolipids can also be cleaved enzymatically. O-Alkyl and O-alk-1-enyl groups are converted to aldehydes by a Pte·H_4-dependent alkyl monooxygenase or a plasmalogenase, respectively.[29] Removal of the ether-linked aliphatic moieties by these enzymes essentially eliminates them from the lipid pool.

Regulatory Controls in Ether Lipid Metabolism

In general, our knowledge of the regulation of ether lipid metabolism still remains pretty much a "black box," although some progress has been made in our understanding of the regulatory controls involved in PAF metabolism. Branch points in the pathways such as ether versus ester biosynthesis from acyldihydroxyacetone-P or the formation of ether lipid classes in neutral lipid versus phospholipid classes from diglyceride-type analogs (alkylacylglycerols) are obvious candidates as potentially important regulatory steps.[28,29]

Regulatory control of enzymatic reactions in both the de novo and remodeling pathways of PAF biosynthesis has received considerable attention. The acetyltransferase in the remodeling pathway is activated by phosphorylation and inactivated by dephosphorylation.[55–57] A cAMP-dependent protein kinase is thought to be responsible for the phosphorylation of the serine residue of acetyltransferase in homogenates from human polymorphonuclear leukocytes.[55,56] However, experiments with microsomal preparations from the guinea pig parotid gland indicate the phosphorylation in these cells is catalyzed by a calcium/calmodulin-dependent protein kinase.[57] Little is known about the phospholipase A_2 activity in the remodeling pathway except it has been reported that activation of this step might involve protein kinase C.[43] Also, the higher levels of PAF produced in ionophore-stimulated cells supplemented with arachidonic acid[58,59] and the use of phospholipase A_2 and eicosanoid metabolite inhibitors have provided evidence to indicate a specific arachidonoyl (polyenoyl) phospholipase A_2 is required for PAF formation in the remodeling route and that this step can be rate-limiting.[60]

In the de novo pathway for PAF biosynthesis the rate-limiting steps are the acetyltransferase that catalyzes the first step in the reaction sequence forming the lipid backbone of PAF[44] and cytidylyltransferase which provides CDP-choline, the precursor of the polar head group moiety of PAF.[61] When cytidylyltransferase is activated via translocation from the cytosol to the membrane fraction, PAF formation can be stimulated at least 5-fold in cells that normally synthesize minimal quantities of PAF by the de novo route.[61] Obviously, de novo synthesis of PAF must be under very rigid regulatory control to prevent the severe consequences that would otherwise occur if physiological levels of PAF are exceeded.

Another consideration of a regulatory nature is the role of endogenous lipid inhibitors of PAF responses.[62,63] The fact that some of the lipid inhibitors described have chromatographic mobilities similar to PAF often makes it difficult to assess PAF responses, since such a combination masks the true picture of the net *in vivo* response. The importance of calcium on PAF metabolism can be appreciated from investigations that have shown this divalent ion inhibits the acetyltransferase,[44] phosphohydrolase,[45] and DTT-insensitive cholinephosphotransferase[47] in the de novo pathway and it effectively inhibits the acetylhydrolase and transacylation steps in the remodeling cycle in intact cells.[64] Lyso-phospholipase D also requires Ca^{2+} in some cells.[49] The regulatory aspects involving PAF catabolism are probably the least understood, but they undoubtedly are of equal importance (especially acetylhydrolase) as those in the biosynthetic steps in the control of PAF levels.

REFERENCES

1. HORROCKS, L. A. 1972. Content, composition, and metabolism of mammalian and avian lipids that contain ether groups. *In* Ether Lipids: Chemistry and Biology. F. Snyder, Ed.: 177–272. Academic Press. New York, NY.
2. SUGIURA, T. & K. WAKU. 1987. Composition of alkyl ether-linked phospholipids in mammalian tissue. *In* Platelet Activating Factor and Related Lipid Mediators. F. Snyder, Ed.: 55–85. Plenum Press. New York, NY.
3. SNYDER, F. 1989. Biochemistry of platelet-activating factor: A unique class of biologically active phospholipids. Proc. Soc. Exp. Biol. Med. **190:** 125–135.
4. DEMOPOULOS, C. A., R. N. PINCKARD & D. J. HANAHAN. 1979. Platelet-activating factor. Evidence for 1-0-alkyl-2-acetyl-*sn*-glycerol-3-phosphorylcholine as the active component (a new class of lipid chemical mediators). J. Biol. Chem. **254:** 9355–9358.
5. BLANK, M. L., F. SNYDER, L. W. BYERS, B. BROOKS & E. E. MUIRHEAD. 1979. Antihypertensive activity of an alkyl ether analog of phosphatidylcholine. Biochem. Biophys. Res. Commun. **90:** 1194–1200.
6. BENVENISTE, J., M. TENCE, P. VARENNE, J. BIDAULT, C. BOULLET & J. POLONSKY. 1979. Semi-synthese et structure proposee du facteur activant les plaquettes (PAF); PAF-acether, un alkyl ether analogue de la lysophosphatidylcholine. C. R. Acad. Sci. [D] (Paris) **289:** 1037–1040.
7. BLANK, M. L., E. A. CRESS & F. SNYDER. 1984. A new class of antihypertensive neutral lipids: 1-alkyl-2-acetyl-*sn*-glycerols. Biochem. Biophys. Res. Commun. **118:** 344–350.
8. SATOUCHI, K., M. ODA, K. SAITO & D. J. HANAHAN. 1984. Metabolism of 1-0-alkyl-2-acetyl-*sn*-glycerol by washed rabbit platelets: Formation of platelet activating factor. Arch. Biochem. Biophys. **234:** 318–321.
9. BRAQUET, P., L. TOUQUI, T. Y. SHEN & B. B. VARGAFTIG. 1987. Perspectives in platelet-activating factor research. Pharmacol. Rev. **39:** 97–145.
10. HANAHAN, D. J. 1986. Platelet activating factor: A biologically active phosphoglyceride. Annu. Rev. Biochem. **55:** 483–509.
11. LEE, T-C. & F. SNYDER. 1985. Function, metabolism, and regulation of platelet activating factor and related ether lipids. *In* Phospholipids and Cellular Regulation. J. F. Kuo, Ed.: 1–39. CRC Press. Boca Raton, FL.

12. O'FLAHERTY, J. T. & R. L. WYKLE. 1983. Mediators of anaphylaxis. Clin. Lab. Med. **3**: 619–643.
13. PINCKARD, R. N., L. M. McMANUS, C. A. DEMOPOULOS, M. HALONEN, P. O. CLARK, J. O. SHAW, W. T. KNIKER & D. J. HANAHAN. 1980. Molecular pathobiology of acetyl glyceryl ether phosphorylcholine: Evidence for the structural and functional identity with platelet-activating factor. J. Reticuloendothel. Soc. **28**: 95s–103s.
14. SHAW, J. O. & P. M. HENSON. 1980. The binding of rabbit basophil-derived platelet-activating factor to rabbit platelets. Am. J. Pathol. **98**: 791–810.
15. HWANG, S. B., C. L. LEE, M. J. CHEAH & T. Y. SHEN. 1983. Specific receptor sites for 1-0-alkyl-2-0-acetyl-*sn*-glycero-3-phosphocholine (platelet activating factor) on rabbit platelet and guinea pig smooth muscle membranes. Biochemistry **22**: 4756–4763.
16. HOMMA, H., A. TOKUMURA & D. J. HANAHAN. 1987. Binding and internalization of platelet-activating factor 1-0-alkyl-2-acetyl-*sn*-glycero-3-phosphocholine in washed rabbit platelets. J. Biol. Chem. **262**: 10582–10587.
17. JANERO, D. R., B. BURGHARDT & C. BURGHARDT. 1988. Specific binding of 1-0-alkyl-2-acetyl-*sn*-glycero-3-phosphocholine (platelet-activating factor) to the intact canine platelet. Thromb. Res. **50**: 789–802.
18. LAD, P. M. & C. V. OLSON. 1985. Platelet-activating factor mediated effects on human neutrophil function are inhibited by pertussis toxin. Biochem. Biophys. Res. Commun. **129**: 632–638.
19. HWANG, S-B., M-H. LAM & S-S. PONG. 1986. Ionic and GTP regulation of binding of platelet-activating factor to receptors and platelet-activating factor-induced activation of GTPase in rabbit platelet membranes. J. Biol. Chem. **361**: 532–537.
20. HOMMA, H. & D. J. HANAHAN. 1988. Attenuation of platelet activating factor (PAF)-induced stimulation of rabbit platelet GTPase by phorbol ester, dibutyryl cAMP, and desensitization: Concomitant effects on PAF receptor binding characteristics. Arch. Biochem. Biophys. **262**: 32–39.
21. AVDONIN, P. V., I. V. SVITINA-ULITINA & V. I. KULIKOV. 1985. Stimulation of high-affinity hormone-sensitive GTPase of human platelets by 1-0-alkyl-2-acetyl-*sn*-glyceryl-3-phosphocholine (platelet activating factor). Biochem. Biophys. Res. Commun. **131**: 307–313.
22. HOUSLAY, M. D., D. BOJANIC & A. WILSON. 1986. Platelet activating factor and U44069 stimulate a GTPase activity in human platelets which is distinct from the guanine nucleotide regulatory proteins, N_s and N_1. Biochem. J. **234**: 737–740.
23. HASLAM, R. J. & M. VANDERWEL. 1982. Inhibition of platelet adenylate cyclase by 1-0-alkyl-2-0-acetyl-*sn*-glyceryl-3-phosphorylcholine (platelet-activating factor). J. Biol. Chem. **257**: 6879–6885.
24. SUGATANI, J., M. MIWA & D. J. HANAHAN. 1987. Platelet-activating factor stimulation of rabbit platelets is blocked by serine protease inhibitor (chymotryptic protease inhibitor). J. Biol. Chem. **262**: 5740–5747.
25. HWANG, S. B. 1988. Identification of a second putative receptor of platelet-activating factor from human polymorphonuclear leukocytes. J. Biol. Chem. **263**:3225–3233.
26. OHNO, M., K. FUJITA, M. SHIRAIWA, A. IZUMI, S. KOBAYASHI, H. YOSHIWARA, I. KUDO, K. INOUE & S. NOJIMA. 1986. Molecular design toward biologically significant compounds based on platelet activating factor. A highly selective agent as a potential antihypertensive agent. J. Med. Chem. **29**: 1812–1814.
27. BLANK, M. L., E. A. CRESS, T-C. LEE, B. MALONE, J. R. SURLES, C. PIANTADOSI, J. HAJDU & F. SNYDER. 1982. Structural features of platelet activating factor (1-alkyl-2-acetyl-*sn*-glycero-3-phosphocholine) required for hypotensive and platelet serotonin responses. Res. Commun. Chem. Pathol. Pharmacol. **38**: 3–20.
28. HAJRA, A. K. 1983. Biosynthesis of 0-alkylglycerol ether lipids. *In* Ether Lipids. Biochemical and Biomedical Aspects. H. K. Mangold & F. Paltauf, Eds.: 85–106. Academic Press. New York, NY.
29. SNYDER, F., T-C. LEE & R. L. WYKLE. 1985. Ether-linked glycerolipids and their bioactive species: Enzymes and metabolic regulation. *In* The Enzymes of Biological Membranes. A. N. Martonosi, Ed. Vol. **2**: 1–58. Plenum, New York, NY.
30. WYKLE, R. L., M. L. BLANK, B. MALONE & F. SNYDER. 1972. Evidence for a mixed function

oxidase in the biosynthesis of ethanolamine plasmalogens from 1-alkyl-2-acyl-*sn*-glycero-3-phosphorylethanolamine. J. Biol. Chem. **247:**5442–5447.

31. PALTAUF, F., R. A. PROUGH, B. S. S. MASTERS & J. M. JOHNSON. 1974. Evidence for the participation of cytochrome b_5 in plasmalogen biosynthesis. J. Biol. Chem. **249:** 2661–2662.

32. ZOELLER, R. A., O. H. MORAND & C. R. H. RAETZ. 1988. A possible role for plasmalogens in protecting animal cells against photosensitized killing. J. Biol. Chem. **263:** 11590–11596.

33. TESSNER, T. G. & R. L. WYKLE. 1987. Stimulated neutrophils produce an ethanolamine plasmalogen analog of platelet-activating factor. J. Biol. Chem. **262:** 12660–12664.

34. NAKAYAMA, R., K. YASUDA, K. SATOUCHI & K. SAITO. 1988. 1-0-Hexadec-1′-enyl-2-acetyl-*sn*-glycero-3-phosphocholine and its biological activity. Biochem. Biophys. Res. Commun. **151:** 1256–1261.

35. SNYDER, F. 1989. Biochemistry of platelet activating factor: A unique class of biologically active phospholipids. Proc. Exp. Biol. Med. **190:** 125–135.

36. SUGIURA, T., Y. MASUZAWA, Y. NAKAGAWA & K. WAKU. 1987. Transacylation of lyso-platelet-activating factor and other lysophospholipids by macrophage microsomes. J. Biol. Chem. **262:**1199–1205.

37. CHILTON, F. H., J. S. HADELY & R. S. MURPHY. 1987. Incorporation of arachidonic acid into 1-acyl-2-lyso-*sn*-glycero-3-phosphocholine of the human neutrophil. Biochim. Biophys. Acta **917:** 48–56.

38. KRAMER, R. M., G. M. PATTON, C. R. PRITZKER & D. DEYKIN. 1984. Metabolism of platelet-activating factor in human platelets. Transacylase-mediated synthesis of 1-0-alkyl-2-arachidonoyl-*sn*-glycero-3-phosphocholine. J. Biol. Chem. **259:** 13316–13320.

39. CORNIC, M., C. BRETON & O. COLARD. 1986. Acylation of 1-alkyl- and 1-acyl-lysophospholipids by rat platelets. Pharmacol. Res. Commun. **18:** 43–49.

40. ROBINSON, M., M. L. BLANK & F. SNYDER. 1985. Acylation of lysophospholipids by rabbit alveolar macrophages: Specificities of CoA-dependent, and CoA-independent reactions. J. Biol. Chem. **260:** 7889–7895.

41. SUGIURA, T. & K. WAKU. 1985. CoA-independent transfer of arachidonic acid from 1,2-diacyl-*sn*-glycero-3-phosphocholine to 1-0-alkyl-*sn*-glycero-3-phosphocholine (lyso platelet-activating factor) by macrophage microsomes. Biochem. Biophys. Res. Commun. **127:** 384–390.

42. SUGIURA, T., Y. MASUZAWA & K. WAKU. 1985. Transacylation of 1-0-alkyl-*sn*-glycero-3-phosphocholine (lyso platelet-activating factor) and 1-0-alkenyl-*sn*-glycero-3-phosphoethanolamine with docosahexaenoic acid (22:6 ω3). Biochem. Biophys. Res. Commun. **133:** 574–580.

43. McINTYRE, T. M., S. L. REINHOLD, S. M. PRESCOTT & G. ZIMMERMAN. 1987. Protein kinase C activity appears to be required for the synthesis of platelet-activating factor and leukotriene B_4 by human neutrophils. J. Biol. Chem. **263:** 15370–15376.

44. LEE, T-C., B. MALONE & F. SNYDER. 1986. A new *de novo* pathway for the formation of 1-alkyl-2-acetyl-*sn*-glycerols, precursors of platelet activating factor. Biochemical characterization of 1-alkyl-2-lyso-*sn*-glycero-3-P:acetyl-CoA acetyltransferase in rat spleen. J. Biol. Chem. **261:** 5373–5377.

45. LEE, T-C., B. MALONE & F. SNYDER. 1988. Formation of 1-alkyl-2-acetyl-*sn*-glycerols via *de novo* biosynthetic pathway for platelet activating factor. Characterization of 1-alkyl-2-acetyl-*sn*-glycerol-3-phosphate phosphohydrolase in rat spleens. J. Biol. Chem. **263:** 1755–1760.

46. RENOOIJ, W. & F. SNYDER. 1981. Biosynthesis of 1-alkyl-2-acetyl-*sn*-glycero-3-phosphocholine (platelet activating factor and a hypotensive lipid) by cholinephosphotransferase in various rat tissues. Biochim. Biophys. Acta **663:** 545–556.

47. WOODARD, D. S., T-C. LEE & F. SNYDER. 1987. The final step in the *de novo* biosynthesis of platelet-activating factor. Properties of a unique CDP-choline:1-alkyl-2-acetyl-*sn*-glycerol cholinephosphotransferase in microsomes from the renal inner medulla of rats. J. Biol. Chem. **262:** 2520–2527.

48. WYKLE, R. L., W. F. KRAEMER & J. M. SCHREMMER. 1980. Specificity of lysophospholipase D. Biochim. Biophys. Acta **619:** 58–67.

49. KAWASAKI, T. & F. SNYDER. 1987. The metabolism of lyso-platelet-activating factor (1-0-alkyl-2-lyso-*sn*-glycero-3-phosphocholine) by a calcium-dependent lysophospholipase D in rabbit kidney medulla. Biochim. Biophys. Acta **920:** 85–93.

50. BLANK, M. L., T-C. LEE, V. FITZGERALD & F. SNYDER. 1981. A specific acetylhydrolase for 1-alkyl-2-acetyl-*sn*-glycero-3-phosphocholine (a hypotensive platelet-activating lipid). J. Biol. Chem. **256:** 175–178.
51. BLANK, M. L., M. N. HALL, E. A. CRESS & F. SNYDER. 1983. Inactivation of 1-alkyl-2-acetyl-*sn*-glycero-3-phosphocholine by a plasma acetylhydrolase: Higher activities in hypertensive rats. Biochem. Biophys. Res. Commun. **113:** 666–671.
52. FARR, R. S., M. L. WARDLOW, C. P. COX, K. E. MENG & D. E. GREENE. 1983. Human serum acid-labile factor in an acylhydrolase that inactivates platelet-activating factor. Fed. Proc. **42:** 3120–3122.
53. STAFFORINI, D. M., S. M. PRESCOTT & T. M. MCINTYRE. 1987. Human plasma platelet-activating factor acetylhydrolase. Purification and properties. J. Biol. Chem. **262:** 4223–4230.
54. STAFFORINI, D. M., T. J. MCINTYRE, M. E. CARTER & S. M. PRESCOTT. 1987. Human plasma platelet-activating factor acetylhydrolase: Association with lipoprotein particles and role in the degradation of platelet-activating factor. J. Biol. Chem. **262:** 4215–4222.
55. NIETO, M. L., S. VELASCO & M. SANCHEZ-CRESPO. 1988. Modulation of acetyl-CoA:1-alkyl-2-lyso-*sn*-glycero-3-phosphocholine (lyso-PAF acetyltransferase in human polymorphonuclears. The role of cyclic AMP-dependent and phospholipid sensitive calcium-dependent protein kinases. J. Biol. Chem. **263:** 4607–4611.
56. GOMEZ-CAMBRONERO, J., J. M. MATO, F. VIVANCO & M. SANCHEZ-CRESPO. 1987. Phosphorylation of partially purified 1-0-alkyl-2-lyos-*sn*-glycero-3-phosphocholine:acetyl-CoA acetyltransferase from rat spleen. Biochem. J. **245:** 893–898.
57. DOMENECH, C., E. MACHADO-DE DOMENECH & H. D. SOLING. 1987. Regulation of acetyl-CoA:1-alkyl-*sn*-glycero-3- phosphocholine O^2-acetyltransferase (lyso-PAF-acetyltransferase) in exocrine glands. Evidence for an activation via phosphorylation by calcium/calmodulin-dependent protein kinase. J. Biol. Chem. **262:** 5671–5676.
58. RAMESHA, C. S. & W. C. PICKETT. 1987. Platelet-activating factor and leukotriene biosynthesis is inhibited in polymorphonuclear leukocytes depleted of arachidonic acid. J. Biol. Chem. **261:** 7592–7596.
59. BILLAH, M. M., R. W. BRYANT & M. I. SIEGEL. 1985. Lipoxygenase products of arachidonic acid modulate biosynthesis of platelet-activating factor (1-0-alkyl-2-acetyl-*sn*-glycero-3-phosphocholine) by human neutrophils via phospholipase A_2. J. Biol. Chem. **260:** 6899–6906.
60. SUGA, K., T. KAWASAKI, M. L. BLANK & F. SNYDER. 1989. Phospholipase A_2 regulates the increase of platelet activating factor (PAF) synthesis in arachidonic acid (AA)-supplemented differentiated HL-60 cells. FASEB J. **3:** A1224.
61. BLANK, M. L., Y. J. LEE, E. A. CRESS & F. SNYDER. 1988. Stimulation of the *de novo* pathway for the biosynthesis of platelet activating factor (PAF) via cytidylyltransferase activation in cells with minimal endogenous PAF production. J. Biol. Chem. **263:** 5656–5661.
62. MIWA, M., C. HILL, R. KUMAR, J. SUGATANI, M. S. OLSON & D. J. HANAHAN. 1987. Occurrence of an endogenous inhibitor of platelet-activating factor in rat liver. J. Biol. Chem. **262:** 527–530.
63. NAKAYAMA, R., K. YASUDA & K. SAITO. 1987. Existence of endogenous inhibitors of platelet-activating factor (PAF) with PAF in rat uterus. J. Biol. Chem. **262:** 13174–13179.
64. TOQUI, L., A. M. SHAW, C. DUMAREY, C. JACQUEMIN & B. B. VARGAFTIG. 1987. The role of Ca^{2+} in regulating the catabolism of PAF-acether (1-0-alkyl-2-acetyl-*sn*-glycero-3-phosphocholine) in rabbit platelets. Biochem. J. **241:** 555–560.

Structure and Dynamics of the Glycolipid Components of Membrane Receptors: ^{2}H NMR Provides a Route to *In Vivo* Observation

IAN C. P. SMITH, MICHÈLE AUGER, AND
HAROLD C. JARRELL

Laboratory for Medical Biosciences
Division of Biological Sciences
Building M54
National Research Council of Canada
Montreal Road
Ottawa, Ontario, Canada K1A 0R6

INTRODUCTION

Glycolipids occur everywhere in nature—they are present in goodly number in brain. After a considerable degree of neglect in favour of the more popular phosphatidylcholines,—ethanolamines, and -serines, it has been realized that the glycolipids very likely have a greater functional significance than any of the other lipids. They are associated with the immune response, cell-cell recognition, and hormone reception, to name a few vital processes. The receptors for the signal peptides and regulatory proteins in the brain no doubt involve such glycolipids or their glycoprotein equivalents.

The chemical composition of glycolipids can be quite complex—more than a dozen carbohydrate residues can be attached to the lipid anchor. This allows a multitude of conformational possibilities for the carbohydrate region. It is extremely likely that the conformation (intramolecular spatial orientation) of this region, as well as its spatial relationships with the membrane surface, is intimately involved in its specificity of function. The masking or expression of antigenic sites could occur by simple conformational transition induced by agents such as Ca^{2+}, pH, change in ionic strength, or proximity of a bound foreign species.

How could the conformation of such a complex carbohydrate species be determined? There are various methods for small molecules, such as NMR in nonviscous solvent, or X-ray crystallography of solid samples. However, given the complexity and interconvertibility of carbohydrate conformations, it is extremely unlikely that data from these states have much relevance to an oligosaccharide residue on a cell surface. Furthermore, neither method can probe the changes involved at cell surfaces during the important processes of biological recognition. We have, therefore, developed a method which can answer the questions for these systems in a biologically relevant environment. The method involves the labelling of the sugar residues with deuterium and observation and analysis of the behaviour of these residues by ^{2}H NMR. The method has enjoyed excellent success in the study of the acyl chains of microbial membranes,[1] including the observation of intact, live cells in reasonable time.[2]

What Is Order?

The concept of order derives from the possibility that a variety of molecular conformations or orientations are possible, and that rapid interconversion between these states takes place. The consequence of this is that the average molecular area is greater, and the molecules cannot pack as closely together. Note that this can derive from conformational disorder, or from overall molecular disorder. A simplified scheme is shown in FIGURE 1, with representative values of the order parameter. For this simple description to hold, the system must have axial symmetry. This is usually achieved via rapid molecular reorientation (3-fold or higher symmetry) about a preferred axis.

The Method

Deuterium can be introduced into particular carbohydrate species by well-known chemical techniques. Often more than one site can be labelled at a time, leading to improved efficiency in both synthesis and acquisition of the NMR-derived structural

HIGH ORDER   LOW ORDER

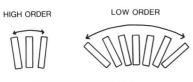

POSSIBLE ORIENTATIONS

FIGURE 1. A simple view of membrane order for a rigid molecule.

ALLOWED VOLUME

parameters. These carbohydrates must then be attached to lipids, and to each other. In certain cases the intact glycolipid may be labelled directly. In the simplest experiment these glycolipids are then studied as dispersions in aqueous buffer. They may exist as rigid lamellar (gel), fluid lamellar (so-called liquid crystalline bilayers), hexagonal phases, or phases with higher symmetry. These phases are discussed in detail by Pieter Cullis in this volume. The labelled carbohydrates may be introduced into natural membranes by biosynthesis or by intercalation.

The ^{2}H NMR spectra are obtained on instruments which are somewhat special. They must be able to excite wide spectral ranges, and obtain data very rapidly. The usual high resolution NMR spectrometers do not do well at this, but there are commercial, so-called solid state or broad line spectrometers, capable of performing these procedures satisfactorily.

A last complication in the method arises in the analysis of the NMR data. The richness of the data leads to a corresponding richness of information, but only after some difficult analysis. Simple transformations of the observed quadrupolar splittings in the ^{2}H NMR spectra to order parameters are not possible—it is necessary to obtain five or six splittings for each sugar residue, and then manipulate the data in terms of the known geometry of

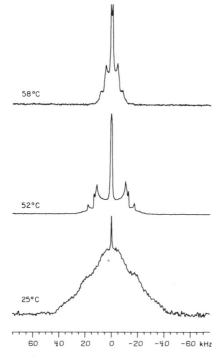

FIGURE 2. Deuterium NMR spectra of 1,2-di-0-tetradecyl-3-0-β-D-glucopyranosylglycerol, labelled at positions 2,3,4, and 6 of the glucosyl moiety. At the three temperatures shown, the lipid exists in the rigid gel, the fluid lamellar, or the fluid hexagonal phase, from bottom to top of the figure, respectively.

the sugar residues to obtain the order parameter for the sugar ring.[3] The use of the dipolar splittings and modelling calculations allows estimation of the order parameters for the more flexible CH_2 groups of the glycerol and other sugar residues.[4]

The Spectra

A set of typical ^{2}H NMR spectra of glycolipids is shown in FIGURE 2. In this case the glucose moiety, with a β-attachment to the anchoring lipid, is labelled at several positions on the glucosyl moiety and the lipid can exist in the rigid gel, the fluid lamellar, or the inverted hexagonal phase. Note that the breadth of the spectrum in the hexagonal phase (58°C) is roughly half of that in the fluid lamellar phase (52°C). This allows an easy distinction between phases. The quadrupolar splittings for lipids with α- or β-glucosyl species are very different, demonstrating the very strong dependence of the sugar surface orientation on the sense of its attachment to the lipid species. In the low temperature phase (25°C), the very broad spectrum is indicative of a high degree of rigidity for the glucosyl head group.

Glycolipid Systems and Their Interactions

We have labelled the glucosyl and mannosyl residues of monoglycolipids, and the glucosyl and galactosyl residues of a lactolipid.[5] We are just completing an NMR study of lipids containing ^{2}H-labelled sialic acid. FIGURE 3 shows a comparison of the orientations and order parameters of the lactolipid and its constituent β-glucolipid. Note that the lactolipid is extended from the surface of the bilayer, in contrast to the laid down conformation usually shown in text books. FIGURE 4 compares the conformation of the lactosyl moiety of the lactolipid in the fluid lamellar phase with that of lactose derived from X-ray studies of the crystalline solid. The orientation of the galactosyl moiety with respect to the glucosyl moiety is completely different, presumably due to the different intermolecular forces experienced by the lactosyl moiety in the two systems.

Our present research is directed towards gathering information in more complex oligosaccharide structure. We intend to synthesize the sequences corresponding to a variety of membrane-bound antigens and receptors, and to study the conformations and dynamics of these species in the absence and presence of their biologically significant interactions.

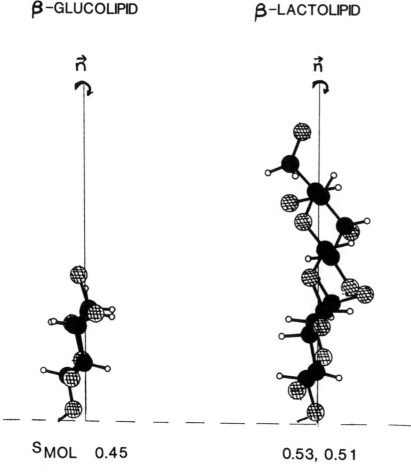

β-GLUCOLIPID   β-LACTOLIPID

S_{MOL} 0.45   0.53, 0.51

FIGURE 3. A comparison of the molecular locations of the axes about which the sugar residues rotate in the β-glucolipid and in the lactolipid derived from it by addition of a galactosyl moiety. The order parameters for the individual sugar residues are shown below the structures.

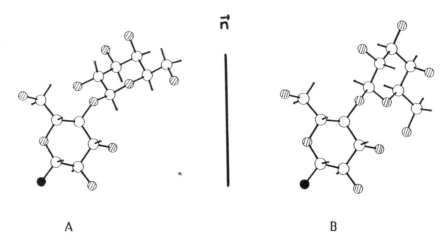

A   B

FIGURE 4. Several views of the lactolipid structure. That on the *left* is derived from the NMR data for the lamellar phase, as shown in a different perspective in FIGURE 3. The axis about which the carbohydrate headgroup rotates to achieve effectively axial symmetry is shown as the *vertical line* marked n. On the *right* is shown the structure of lactose derived from X-ray diffraction studies of single crystals.

At What Rates Do Glycolipids Move?

Measurement of the relaxation properties of lipids has been found to yield a wealth of information on the rates and natures of the various motions they can undergo.[6] For example, it is known that the upper regions of the acyl chains in bilayers move much more slowly than do the lower regions. The gradient in motion roughly parallels the gradient in order, with the less ordered regions being the more mobile. These motions take place at rates of 10^9 to 10^{10} sec^{-1}. The properties at the center of the bilayer can be said to approach those of a liquid, whereas nearer the glycerol moiety they are more like semisolids. What then of the glycolipid headgroups? Our measurements indicate that they are well ordered, and moving up to ten times more slowly than are the upper regions of the acyl or alkyl chains in bilayers.

Thus, although the sugar moieties do interconvert rapidly enough to average a variety of interactions and to produce an axially symmetric state, they do so relatively slowly. We should emphasize that each physical technique has its own definition of fast and slow. For longitudinal relaxation measurements with ^{2}H, slow means $<50 \times 10^6$ sec^{-1}. Molecular diffusion in non-viscous solvents takes place at about 10^{12} sec^{-1}. Thus, small molecule perturbants approaching a membrane surface can sample that surface as many as a million times before a surface conformational change occurs. The conformations of the surface glycolipid receptors are thus crucial in determining the success or failure of a collision with a small molecule approaching from extracellular space.

The Influence of Anesthetics on Glycolipid Conformation and Phase

In earlier studies of phospholipid membrane systems, it was found that local anesthetics such as tetracaine can cause large changes in quadrupole splitting of the head group species.[7,8] Glycolipids are major lipid components of both nerve and brain, and, therefore, we might expect significant alterations in their behaviour due to membrane-active anesthetics. Such is indeed the case.

FIGURE 5 shows the ^{2}H NMR spectra of the ^{2}H- labelled β-glucolipid in the absence and presence of neutral tetracaine, pH 9.5. The anesthetic causes a change in phase from lamellar to hexagonal, as seen by the factor of two drop in the quadrupole splitting. At pH 5.5, where tetracaine is charged, a transition to an isotropic phase is seen. Tetracaine has been shown to locate between the acyl chains of lipids in bilayers.[9] The transition from lamellar to nonlamellar phase may be due to the effectively greater interlipid separation, which increases the average head group separation.

Are these interactions relevant to the *in vivo* situation? Recent infrared spectroscopic studies on the interaction of tetracaine with frog sciatic nerve or lobster walking leg nerve

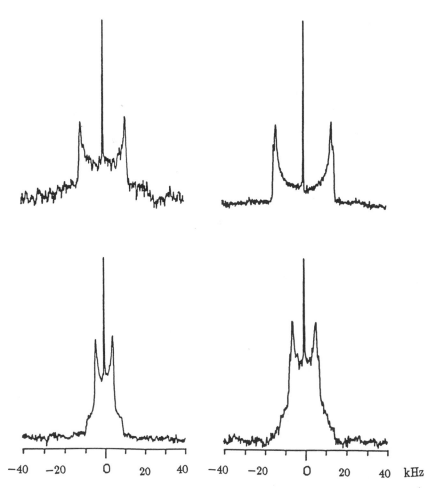

−40   −20   0   20   40   −40   −20   0   20   40   kHz

FIGURE 5. Deuterium NMR spectra of the β-glycolipid labelled with ^{2}H at C1 of the glucosyl moiety *(left)* and at C3 of the glyceryl moiety *(right)* in the absence *(top spectra)* and presence *(bottom spectra)* of tetracaine, pH 9.5, 52°C.

suggest that the depth of penetration of the anesthetic into the bilayer is comparable with that found in model systems, and that the overall behaviour is similar, including the pressure-induced exclusion of the anesthetic.[10] The concentrations used in the NMR experiments are very similar to those thought to arise during local anesthesia.

Are there regions of hexagonal or isotropic phase lipid present in biological membranes? None has been seen to date in the presence or absence of anesthetics, but the hexagonal phase has been observed for erythrocyte ghosts in the presence of the fusogen, oleic acid.[11] If all the nerve membrane lipid were to become hexagonal, the nerve would no longer exist morphologically. Our view is that at any given time the population of such nonbilayer phases in natural membranes is probably very low, and, therefore, is masked by the dominant spectrum of the lamellar phase. We shall attempt to observe these nonlamellar structures using techniques which discriminate against the bilayer spectrum.[12]

Is the In Vivo Situation Amenable to This Approach?

In the anesthetic studies we have observed the deuterated local anesthetics in isolated natural nerve membranes by infrared and NMR spectroscopy.[8,10] By means of ^{19}F NMR, halothane has been observed in the brain of a live rabbit.[13] Using gradient and pulse methods, it is possible to localize a volume within an animal, and to characterize it both anatomically and biochemically. A localization experiment to observe [2]H-labelled anesthetics or other drugs in the nerve of an intact animal is, although technically demanding, now possible.

To observe lipids in intact organs or animals by this technique is more difficult. For isolated membranes, or for the membranes of microorganisms, it is possible to intercalate lipids from vesicles[14] or from films deposited on vessel walls. With simple microorganisms a biosynthetic approach to labelling is not unreasonable.[15]

For mammalian systems we shall have to resort to intercalation from vesicles of labelled lipids borne by the circulation, or exposed by local injection. By tailoring the nature of the lipid to the membrane of interest, we may be able to gain specificity. The problem may well be the amount of labelled compound required to achieve a reasonable level of incorporation. These experiments could be very expensive. Nonetheless, we are considering trials in small animals to test the viability of the approach.

CONCLUSION

The methods described here are very powerful for obtaining the information and dynamics of membrane constituents in isolated membrane systems. As little as one atomic position in these large complex structures can be observed. The information can be obtained from membranes in suspension or as precipitates. For the membranes of microorganisms it is undoubtedly the best method to obtain information at the atomic level.

When we step up to mammalian systems two problems remain to be overcome. The first is delivery of a labelled species to the organ(elle) of interest. With membrane-active drugs this should be possible. For membrane components of low population density this may be very difficult, although a variety of possible approaches are available.

Twenty years ago we were only able to obtain spectra of ^{13}C at natural abundance from neat liquids in 12-mm spheres. Today we do it at concentrations of 10^{-3} M. NMR improves its versatility and detection sensitivity on a steady upward curve.

Therefore, we have confidence that the elegant level of information presently available via ²H NMR of membrane components will soon be derived from the *in vivo* situation with animals.

REFERENCES

1. I. C. P. SMITH. 1985. Structure and dynamics of cell membranes as revealed by NMR techniques. *In* Structure and Properties of Cell Membranes, Vol. III. G. Benga, Ed.: 237–260. CRC Press. Boca Raton, FL.
2. H. C. JARRELL, K. W. BUTLER, R. A. BYRD, R. DESLAURIERS, I. EKIEL & I. C. P. SMITH. 1982. A ²H NMR study of Acholeplasma laidlawii membranes highly enriched in myristic acid. Biochim. Biophys. Acta **688**: 622–636.
3. H. C. JARRELL, J. B. GIZIEWICZ & I. C. P. SMITH. 1986. Structure and dynamics of a glyceroglycolipid: A ²H NMR study of head group orientation, ordering and the effect on lipid aggregate structure. Biochemistry, **25**: 3950–3957.
4. H. C. JARRELL, A. J. WAND, J. B. GIZIEWICZ & I. C. P. SMITH. 1987. The dependence of glyceroglycolipid orientation and dynamics on head group structure. Biochim. Biophys. Acta **897**: 69–82.
5. J.-P. RENOU, J. B. GIZIEWICZ, I. C. P. SMITH & H. C. JARRELL. 1989. Glycolipid membrane surface structure: Orientation, conformation, and motion of a disaccharide headgroup. Biochemistry **28**: 1804–1814.
6. PERLY, B., I. C. P. SMITH & H. C. JARRELL. 1985. Acyl chain dynamics of phosphatidylethanolamines containing oleic acid and dihydrosterculic acid: ²H NMR relaxation studies. Biochemistry **24**: 4659–4665.
7. Y. BOULANGER, S. SCHREIER & I. C. P. SMITH. 1981. Molecular details of anesthetic-lipid interaction as seen by deuterium and phosphorus-31 nuclear magnetic resonance. Biochemistry **20**: 6824–6830.
8. I. C. P. SMITH & K. W. BUTLER. 1986. Molecular details of anesthetic-lipid interaction as seen by nuclear magnetic resonance. In Molecular and Cellular Mechanisms of Anesthetics. S. H. Roth K. W. Miller, Eds.: 309–318. Plenum. New York, NY.
9. M. AUGER, H. C. JARRELL & I. C. P. SMITH. 1988. Interactions of the local anesthetic tetracaine with membranes containing phosphatidylcholine and cholesterol: A ²H NMR study. Biochemistry **27**: 4660–4667.
10. M. AUGER, H. C. JARRELL, I. C. P. SMITH, P. T. T. WONG, D. J. SIMINOVITCH & H. H. MANTSCH. 1987. Pressure-induced exclusion of a local anesthetic from model and nerve membranes. Biochemistry **26**: 8513–8516.
11. P. R. CULLIS & M. J. HOPE. 1978. Effects of fusogenic agent on membrane structure of erythrocyte ghosts and the mechanism of membrane fusion. Nature **271**: 672–674.
12. R. DESLAURIERS, I. EKIEL, R. A. BYRD, H. C. JARRELL & I. C. P. SMITH. 1982. A ³¹P NMR study of structural and functional aspects of phosphonate and phosphate distribution in tetrahymena. Biochim. Biophys. Acta **720**: 329–337.
13. A. M. WYRWICZ, C. B. CONBOY, K. R. RYBACK, B. G. NICHOLS & P. EISELE. 1989. *In vivo* ¹⁹F NMR study of isoflurane elimination from brain. Biochim. Biophys. Acta. In press.
14. Y. SUGIMOTO, H. SAITO, R. TABETA & M. KODAMA. 1986. A ²H NMR study on alteration of the membrane organization of mouse B16 melanoma cells treated with 12-0-tetradecanoylphorbol-13-acetate. J. Biochem. **100**: 867–874.
15. L. C. STEWART, M. KATES, I. EKIEL & I. C. P. SMITH. 1989. NMR studies of the lipids of *Halobacterium cutirubrum*. Submitted.

Effects of Calcium, ATP, and Lipids on Human Erythrocyte Sugar Transport[a]

A. CARRUTHERS,[b] A. L. HELGERSON, D. N. HEBERT,
R. E. TEFFT, JR., S. NADERI, AND D. L. MELCHIOR

Department of Biochemistry
University of Massachusetts Medical Center
55 Lake Avenue North
Worcester, Massachusetts 01605

INTRODUCTION

Human red cell sugar transport shows profound changes upon cellular aging.[1] As the human erythrocyte glucose transport protein shares more than 98% primary, structural homology with the rat and rabbit brain glucose transport proteins[2-5] can be isolated from human red cells in large (mg) quantities and, owing to the relative ease of manipulation of red cell contents by forming resealed ghosts, erythrocyte sugar transport may provide a useful experimental model for the study of the effects of the aging process on a protein-mediated membrane process. In this short paper we examine how the catalytic activity of the human erythrocyte glucose transport protein is modified by intracellular ATP and calcium, by extracellular calcium, and by the physical and compositional properties of its resident membrane bilayer.

BACKGROUND

The facilitated diffusion of sugars across cell membranes assumes a pivotal role in cellular energy metabolism. The complete, cellular, oxidative phosphorylation of a single glucose molecule results in the net synthesis of 36 ATP molecules. In some cells (*e.g.*, muscle), the rate of protein-mediated glucose uptake is regulated by the metabolic state of the cell.[6] In this way, the energy requirements of the cell determine the rate of glucose uptake and thus the net rate of synthesis of high energy metabolic intermediates. Certain tissues (*e.g.*, mammalian erythrocytes and brain) are almost completely dependent upon glucose as their source of carbon for ATP synthesis. In these tissues, glucose deprivation can lead to metabolic depletion.

Protein-mediated, facilitated diffusion of sugars across the cell membrane is mediated by a family of integral membrane proteins called the Glucose Transport Proteins.[7-9] The rat brain glucose transport protein shares extensive homology with the HepG2 and human erythrocyte glucose transport proteins.[2,3,5] Hydropathy analysis suggests that this glucose transport protein contains twelve strongly hydrophobic domains that are sufficiently large to form transmembrane alpha-helices[2] (FIG. 1) although this conclusion is not supported by Chou and Fasman's analysis of the primary structure of the glucose transport protein (Carruthers, unpublished). Circular dichroism and Fourier Transform Infra-Red Spectros-

[a]This work was supported by National Institutes of Health Grant RO1 DK36081 (A.C.) and by National Science Foundation Grant DMB-8717601 (D.L.M.). We gratefully acknowledge this support.
[b]To whom correspondence should be addressed.

52

OUTSIDE

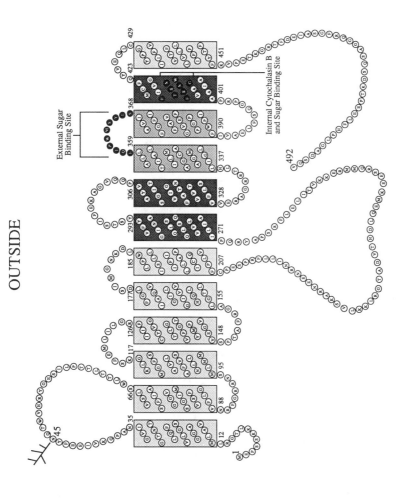

FIGURE 1. The primary structure and suggested secondary structure of the HEP G2 glucose transport protein.[2] Single letter amino acid codes are used. The suggested cytochalasin B and extracellular glucose binding sites[13,16] are indicated.

copy studies indicate that the human erythrocyte glucose transport protein is predominantly alpha-helical in structure with some beta sheet structure and that the majority of the alpha-helical content in oriented normal to bilayer.[10–12]

The catalytic activity of human erythrocyte glucose transport protein (HEGTP) and other proteins in this family is strongly inhibited by cytochalasin B (CCB). CCB binds with high affinity ($K_{d(app)}$ = 140 nM) to a site on the HEGTP that faces the interior of the cell. This site bears all the hallmarks of the glucose efflux site of the HEGTP. [^{3}H]-CCB can form a covalent adjunct to the HEGTP upon photo-irradiation. This has allowed workers to localize the CCB binding site of the glucose transport protein[13] (FIG. 1). Holman and co-workers have synthesized a photoreactive, radiolabelled bis-manose derivative that binds to the extracellular glucose binding site of the HEGTP and, in an analogous fashion to the studies performed with CCB, have localized the extracellular glucose binding site to a region in the C-terminal half of the HEGTP[13–15] (FIG. 1).

The HEGTP can be isolated in a highly purified form from red cells and reconstituted into artificial lipid bilayers.[16] In this way the functional properties of the protein can be studied in isolation from other cellular constituents. The catalytic properties of this protein are the subject of some controversy at this time. The studies of several laboratories[17,18] support the view that the HEGTP is a classic example of the Alternating Conformation model for transport in which sugar influx and efflux sites are mutually exclusive being only alternately accessible to the external and internal milieu.[19] The studies of other laboratories support the view that these sugar binding sites are not mutually exclusive but rather display negative cooperativity[20–24] (FIG. 2). An interesting synthesis of these opposing views is the hypothesis that the HEGTP may be a functional dimer—each monomer being able to bind substrate via an alternating conformation type of mechanism but where substrate interaction with one monomer may influence the ability of the second monomer to react with substrate.

Effects of Lipid Bilayer on Sugar Transporter Function

In order to systematically examine how the composition of the membrane lipid bilayer can influence the activity of the HEGTP, we adopted the following strategy. The purified glucose transport protein was reconstituted into lipid bilayers of predetermined composition. Michaelis ($K_{m(app)}$) and velocity (V_{max}) parameters for glucose transport were obtained over a wide range of temperatures (0 to 64°C) and the number of functionally reconstituted HEGTPs determined by CCB binding measurements (each functional HEGTP contains a single CCB binding site[25]). As the turnover number of the HEGTP (k_{cat}) is given by

$$k_{cat} = V_{max}/[HEGTP]$$

where [HEGTP] = [CCB binding sites] we can calculate k_{cat} for HEGTP-mediated sugar transport in a variety of lipid bilayers. This permits a direct comparison of catalytic efficiency in bilayers of different composition.

We examined how a number of bilayer features can modify k_{cat}. These features were the following: bilayer order/disorder; bilayer acyl chain carbon number and saturation/unsaturation; bilayer lipid headgroup composition; and bilayer cholesterol content.

Hydrocarbon Chains

Lipid bilayers formed from a single lipid species undergo a phase transition from an ordered to disordered state (the bilayers melt) at a characteristic temperature as temper-

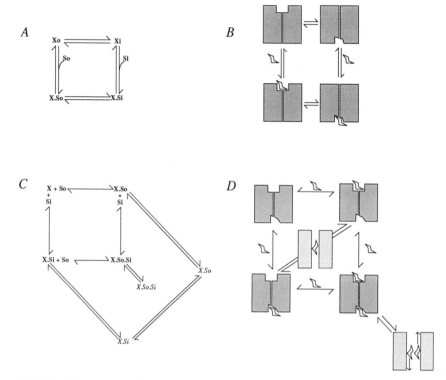

FIGURE 2. Suggested models for sugar transport. The alternating conformation Model (**A** and **B**) is shown in King-Altman (**A**) and diagrammatic form (**B**). The carrier can exist in one of four possible states: Xo (sugar binding site facing the cell's exterior), Xi (sugar binding site facing the cell's interior), X.So (carrier occupied by extracellular sugar) or X.Si (carrier occupied by intracellular sugar). Each of these forms is mutually exclusive. The Fixed or Two-site model is shown in King-Altman (**C**) and diagrammatic (**D**) forms. The carrier X contains two nonexclusive substrate binding sites and can exist in 7 possible forms: X (unoccupied carrier), X.So (carrier complexed with extracellular sugar), X.Si (carrier complexed with intracellular sugar), X.Si.So (carrier complexed with both intra- and extracallular sugar), *X.So* (a transition state in which extracellular sugar has penetrated the pore and is committed to translocation to the opposite side of the membrane), *X.Si* (a transition state in which intracellular sugar has penetrated the pore and is committed to translocation to the opposite side of the membrane), and *X.Si.So* (a transition state in which both intra- and extracellular sugar have penetrated the pore and are simultaneously committed to translocation to the opposite side of the membrane). The diagrammatic form of this model (shown as **D**) is simplified to show only a single transition state when carrier is occupied by a single sugar molecule.

ature is raised. During the phase transition, the packing of acyl chains in the bilayer hydrocarbon core becomes disordered. This is reflected as increased acyl chain mobility and bilayer thinning. In the absence of extrinsic ordering forces (*e.g.*, anchoring to cytoskeletal elements), lipids and proteins in the disordered bilayer have greater motional freedom.

Upon reconstitution into a variety of disaturated phosphatidylcholine (PC) bilayers, the turnover number (k_{cat}) of the glucose transporter increases during the bilayer phase transition.[26-28] Indeed, catalytic activity is not detectable in crystalline (ordered) dimyristoyl PC (DMPC, C_{14}) and dipalmitoyl PC (DPPC) membranes in spite of apparently normal ligand binding. These observations are consistent with the notions that solvent (bilayer) order governs the catalytic activity of the carrier to a greater extent than it does the ligand-binding properties of the transport protein and that substrate binding is not rate limiting for transport. However, a number of findings indicate that bilayer order is not a primary determinant of transporter function. (a) The phase transition in phosphatidylglycerol (PG) bilayers is not associated with altered transporter k_{cat}[29] (see below). (b) k_{cat} for transport in ordered distearoyl PC (DSPC, a C_{18} disaturate) and diarachidonoyl PC (DAPC, a C_{20} disaturate) bilayers is as great or even greater than that observed in disordered DMPC and dioleoyl PC (DOPC, C_{18} a cis-9,10 diunsaturate) bilayers.

Close examination of k_{cat} for transport and the temperature dependence of transport in PC bilayers indicates a linear dependence of k_{cat} and E_a (the activation energy for transport) on PC acyl chain carbon number up to at least C_{18}. Bilayer thickness is, to a large extent, determined by bilayer lipid acyl chain carbon number and saturation/unsaturation.[30] If bilayer thickness were the major determinant of glucose transporter activity in PC bilayers, the expected activities in the above membranes would be DMPC < DPPC $\leq$ DOL < dielaidoyl PC (DEPC, C_{18} a trans-9,10 diunsaturate) < DSPC < DAPC. This is almost exactly the activity relationship observed, with the exception that activity in DSPC bilayers is greater than that observed in DAPC bilayers. It is possible that an activity optimum is achieved in C_{18} disaturates. The increase in ordered bilayer thickness per C2 addition to disaturated PC acyl chains is approximately 4 Å[30]—a value approximately equivalent to 1 turn of a proteinaceous alpha helix. Similarly, upon melting DPPC and DMPC bilayers, bilayer thickness falls by approximately 11 and 7 Å, respectively.[30] This thinning of the bilayer is approximately equivalent to 2 and 1 turns, respectively, in a polypeptide alpha helix. Although these considerations indicate that altered bilayer order and lipid acyl chain carbon number could have drastic consequences upon protein structure within the bilayer, they also show that the relationship between bilayer thickness and carrier activity, if real, is not simple. Activity increases not only with increasing acyl chain carbon number in ordered bilayers (and presumably bilayer thickness) but also during the bilayer crystalline-to-fluid phase transition (and presumably bilayer thinning).

Head Group

The influence of bilayer lipid head group on glucose transporter activity has been studied using a series of lipids of homologous acyl chain composition (myristate, a C_{14} disaturate).[29] These lipids are PC, PG, phosphatidylserine (PS), and phosphatidic acid (PA). The turnover number of the glucose carrier at both 20 and 50°C in proteoliposomes formed from these lipids is PC < PG < PS < PA. Similar but less pronounced findings are obtained using lipids with heterogeneous acyl chain composition (egg PC, egg PG, egg PS, and egg PA). The surface potential of bilayers formed from these lipid species also increases in the order PC < PG, PS < PA,[31] suggesting that lipid head groups could modify the catalytic activity of the carrier protein through bilayer surface charge effects. However, titration of surface charge using protons or Na^+ has quantitatively similar effects on transport in PC and PG bilayers in spite of the clear demonstration of modified phase behavior of DMPC bilayers upon titration of surface potential.[29]

The temperature dependence of transporter activity in PC and PG bilayers has been examined for the series dimyristoyl, dipalmitoyl, and distearoyl PC and PG. Unlike DMPG and DPPG bilayers, DMPC and DPPC bilayers are unable to support detectable transport activity in the crystalline, ordered state. During the pre- and main transitions of DMPC and DPPC bilayers, transport activity becomes measurable and increases thereafter with increasing temperature. DMPG and DPPG bilayers, however, support measurable transport activity in the ordered state. In addition, there is no significant increase in transporter turnover number during the crystalline-to-fluid phase transition of PG bilayers. Rather $K_{m(app)}$ for transport displays a reversible decrease during the phase transition. With both PC and PG bilayers, the carrier turnover number increases with increasing acyl chain carbon number. These rather dramatic differences between carrier activities in PC and PG bilayers are interesting. PC and PG bilayers display very similar phase transition temperatures when formed from lipids with homologous acyl chain composition, which indicates that the forces determining bilayer structure are very similar. While head group compositional differences are presumably sufficient to account for alterations in transporter activity, the influence of acyl chain composition on transporter properties is still discernable.

Backbone

The influence of lipid backbone on sugar transporter activity in proteoliposomes has been studied only in membranes formed from lipids containing glycerol and sphingosine backbones.[27–29] Egg sphingomyelin (containing a sphingosine backbone) bilayers support 10-fold lower activity than do egg PC, PG, and PA (containing a glycerol backbone) bilayers. $K_{m(app)}$ for transport in sphingomyelin bilayers is some 30-fold greater than that observed in PC bilayers, and carrier turnover number increases reversibly during the phase transition of sphingomyelin bilayers.

Cholesterol

Cholesterol is a major lipid component of eukaryotic plasma membranes accounting for as much as 42 mol% bilayer lipid.[32] Unlike the aforementioned lipids, cholesterol belongs to a class of lipids that by themselves are unable to form bilayers upon hydration but can interdigitate into bilayers to modify bilayer structure.[30] Cholesterol modifies the packing properties of bilayer lipids. At 30–50 mol% bilayer content, cholesterol acts as a membrane plasticizer and transforms membranes to a state intermediate between ordered and disordered.[33] Under such conditions, the bilayer phase transition is suppressed.[34,35] Increasing the cholesterol content of DPPC bilayers from 0 to 20 mol% progressively reduces the expansion of bilayer volume occurring during the bilayer phase transition.[36] At approximately 17.5–18.5 mol% cholesterol, an abrupt decrease in ordered DPPC bilayer volume occurs, but is reversed at 20 mol% cholesterol. At 20 mol% cholesterol, a marked change in bilayer lipid packing occurs. The ordered bilayer is expanded and the disordered bilayer condensed. This progresses to approximately 30 mol% cholesterol when an additional rearrangement of packing is observed (expansion and condensation of ordered and disordered bilayer, respectively) that continues monotonically with cholesterol concentrations up to 50 mol%. Above this concentration, cholesterol forms unstable associations with the bilayer.

The effects of bilayer cholesterol on sugar transporter activity in DPPC membranes are complex.[37,38] Cholesterol is without effect on transport at concentrations of up to 10 mol%. Between 10 and 20 mol% cholesterol, a marked and reversible decrease in carrier turnover number is observed for disordered bilayers and a reversible increase in turnover number for ordered bilayers. Cholesterol is without significant effect on transport between 20 and 30 mol%. Thereafter, cholesterol increases k_{cat} in ordered bilayers and reduces k_{cat} in disordered bilayers. These findings are consistent with the view that carrier activity is related to bilayer order and that cholesterol's actions at high sterol levels reflect its ability to modify lipid packing. At low concentrations (typical of those found in intracellular membranes), cholesterol's effect on transport are less consistent with the sterol's action on bilayer lipid packing. If there is a systematic relationship between bilayer order and transporter activity, transporter turnover number at any bilayer cholesterol concentration should be governed largely by bilayer apparent partial specific volume (a parameter directly related to bilayer disorder). This is not the case.[37] In addition, cholesterol at 40 mol% while acting to expand ordered DSPC bilayers in fact reduces the catalytic activity of the transporter under these conditions. These findings argue for a transport-modulating role of bilayer cholesterol that is independent of the sterol's ability to act as a membrane plasticizer.

Bioactive Lipids

The work described up to this point has been concerned with how lipids modulate the activity of the human sugar transporter by altering overall bilayer properties. Certain bilayer lipids appear to be able to act on the transporter at low bilayer concentrations, behaving as lipid soluble "co-factors" or "allosteric regulators."

Lysophospholipids are present in red cell membranes at concentrations less than 2 mol%. Intermediates in phospholipid metabolism, lysolipids can be generated by the actions of phospholipase types A1 and A2, which cleave acyl chains from phospholipids at positions 1 and 2, respectively.[39] At high concentrations, lyso compounds can act as membrane lytic agents. At lower concentrations, the presence of lyso lipids in the membrane is presumably a consequence of lipid metabolism. Monopalmitoyl PC (MPPC) acts as a reversible, partial uncompetitive inhibitor of exchange glucose transport in human red cells ($K_{m(app)}$ and V_{max} for transport are reduced).[40] The consequence of this action is a stimulation of transport at physiological glucose levels and an inhibition at higher sugar concentrations. $K_{i(app)}$ for MPPC action is reduced by increasing saturation of the glucose carrier with sugar, which indicates a specific interaction of lyso PC with the carrier hexose complex. At 400 mM D-glucose (when the carrier is approximately 85–95% saturated with sugar), half-maximal inhibition of transport is reached at a MPPC bilayer lipid molar ratio of 0.07 mol% and a MPPC-carrier molar ratio of approximately 0.5:1. Similar results have been obtained using reconstituted glucose carrier in a variety of PC bilayers. In addition, this effect has been shown to be specific to lyso lipids with specific head groups and chain lengths. (Inactive PCs include monostearoyl PC, monooleoyl PC; active PEs include monomyristoyl PE; inactive PEs include monopalmitoyl PE, monostearoyl PE; monooleoyl PE produces a small increase in carrier T_n and sphingosine is inactive.) These findings argue strongly against an indirect metabolic or bilayer modulating action of MPPC on transport, but rather indicate a direct mode of action of MPPC on the transport system.

Effects of ATP on Sugar Transport

The ATP content of red cells falls upon prolonged cold storage.[41] As a number of sugar transport parameters are modified in cold-stored blood, it seemed worthwhile to examine whether cellular ATP can interact with the sugar transport system of human red cells. This possibility was strengthened by observations from a number of laboratories indicating fundamental differences between sugar transport kinetics in intact cells versus red cell ghosts.[42-45] Carruthers and Melchior[45] observed that $K_{m(app)}$ and V_{max} for sugar uptake were increased and $K_{m(app)}$ for exit decreased in red cell ghosts versus intact cells. Moreover, red cell cytosol obtained upon hypotonic lysis of intact cells restores normal (intact cell) transport properties to ghosts when incorporated into ghosts during the re-sealing process.

Hebert and Carruthers[41] discovered that neutralized, acid extracts of red cell cytosol obtained upon hypotonic lysis of intact cells could mimic the ability of whole cytosol to restore normal transport properties to red cell ghosts. The net effect of this extract was to reduce $K_{m(app)}$ and V_{max} for sugar uptake by ghosts and to increase $K_{m(app)}$ for sugar exit. The ATP content of the extract appeared to be the active principle in mediating this effect. Apyrase treatment of extracted cytosol (AEL) reduced the ATP content of AEL and concommittantly the ability of AEL to modify transport in ghosts. Myokinase (adenylate kinase) could partly restore the ATP content of apyrase-treated AEL and its ability to modify transport in ghosts. In addition, ATP but not AMP, ADP, GTP, UTP or CTP could mimic the action of AEL on transport in inside-out red cell membrane vesicles (IOVs). ATP was half-maximal in its ability to modify transport at ≈ 50 μM.

Subsequently it was found[22,23] that ATP could interact with the glucose transport protein to reduce the affinity of the sugar efflux site and to increase the affinity of the uptake site for D-glucose. This action was half-maximal at 50–60 μM ATP. Some insight into the mechanism of ATP on sugar transport was obtained when three key studies were performed. First it was found that relative to the NaK,ATPase of dog kidney, the purified glucose transport protein from human red cells contained no detectable ATPase activity. Second we find that the glucose transport protein binds Mg.ATP with $K_{d(app)} = 60$ μM at an apparent stoichiometry of 1 site per transport protein monomer. Third, azidoATP can be photoincorporated into purified carrier in a stereospecific fashion—ATP inhibits photoincorporation of the label with $K_{i(app)} = 45$ μM. These observations confirm that the glucose transport protein binds ATP (probably at a single site) and that once bound, the nucleotide is not hydrolyzed. In addition, they suggest that ATP modulation of transport involves an allosteric regulatory mechanism. The net effect of ATP depletion would be increased V_{max} for sugar uptake resulting from ATP-dissociation from the carrier. Our preliminary experiments suggest that the point(s) of photoincorporation of azidoATP into the carrier lies between residues 331 and 456.

While these observations suggest a simple negative feedback mechanism for regula-tion of sugar transport protein activity by ATP, it seems unlikely that such a mechanism (as described) could result in physiologic control of sugar transport. Intracellular ATP levels lie in the range 1 to 4 mM and rarely fall to μM levels. However, we have also found that ADP and, to a lesser extent, AMP can competitively inhibit azidoATP binding to the carrier with $K_{i(app)}$ of 132 and 79 μM respectively. As neither AMP nor ADP interacts with the carrier protein to modify transport but do interfere with the ability of azidoATP to bind to the carrier, this suggests that a decrease in ATP coupled to a concommittant increase in AMP and ADP could amplify the sensitivity of the glucose carrier to a decline in cellular ATP. A model summarizing these effects is illustrated in FIG. 3.

Effects of Calcium on Sugar Transport

TABLE 1 summarizes the results of experiments in which the effects in intra- and extracellular Ca^{2+} on sugar transport in red cells and red cell ghosts was examined in the presence and nominal absence of Mg.ATP$_i$. Ghosting cells increases $K_{m(app)}$ and V_{max} for 3-O-methyl-α-D-glucopyranoside (3OMG, a nonmetabolized but transported sugar) up-take. This effect is reversed by addition of ATP to the resealing medium. Inclusion of $CaCl_2$ in the ATP resealing medium further reduces V_{max} for 3OMG uptake. External Ca^{2+} reduces V_{max} for uptake by intact cells and ghosts containing ATP. 3-O-Methyl-

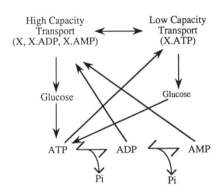

High Capacity
Transport
(X, X.ADP, X.AMP)

Low Capacity
Transport
(X.ATP)

FIGURE 3. A model for ATP regulation of sugar transport in red blood cells. The carrier can exist in one of two possible states: a high transport capacity state (ATP is not bound to the carrier) and a low transport capacity state (ATP is bound to the carrier). Glucose entry results in ATP synthesis which shifts the equilibrium to the low capacity state through ATP binding to the carrier. ATP degradation to ADP and AMP favors a shift to the high capacity state through ATP dissociation from the carrier and competition between ATP, ADP, and AMP for binding to the carrier. Although ADP and AMP bind to the carrier, they lack the ability of ATP to modify the intrinsic activity of the carrier.

glucose uptake by red cell ghosts lacking or containing 2 mM ATP is half-maximally inhibited at ≈ 2.4 μM Ca_i^{2+} and at 2.2 μM Ca_o^{2+}.

Inhibition of 3OMG uptake by $Ca^{2+}{}_o$ does not result from increased $[Ca^{2+}]_i$ and seems not to require ATP_i. TABLE 2 shows that Ca_o inhibits 3OMG uptake in red cell ghosts containing 2 mM EGTA, 0 $CaCl_2$ ± 4 mM ATP although uptake inhibition is somewhat blunted in the presence of ATP_i. Inhibition by $Ca^{2+}{}_i$ plus $Ca^{2+}{}_o$ is additive (TABLE 2). Inhibition by $Ca^{2+}{}_i$ does not require the simultaneous presence of exogenous

TABLE 1. Effects of ATP_i and Ca^{2+} on Zero-Trans 3-O-Methylglucose Uptake in Red Cells and Red Cell Ghosts[a]

$[ATP]_i$ mM	$[Ca^{2+}]_o$ μM[b]	$[Ca^{2+}]_i$ μM[c]	$K_{m(app)}$ mM	V_{max} μmol/L/min	n[d]
Red cells					
1 ± 0.1	0	?	0.4 ± 0.1	114 ± 20	6
			$(0.12–0.9)^e$	(74–190)	
0.8 ± 0.1	101	?	0.36 ± 0.03	23 ± 1*	4
			(0.25–0.41)	(19–27)	
Red cell ghosts					
4	0	0	0.42 ± 0.11	110 ± 15	5
			(0.19–0.8)	(70–151)	
4	101	0	0.28 ± 0.04	43 ± 5*	4
			(0.17–0.36)	(38–49)	
4	0	15	0.32 ± 0.05	68 ± 6*	4
			(0.21–0.45)	(46–77)	
0	0	0	7.1 ± 1.8*	841 ± 191*	5
			(2.3–13)	(204–1107)	
0	101	0	2.9 ± 0.4*	36 ± 7*	4
			(1.7–3.8)	(26–48)	
0	0	11	2.7 ± 0.7*	52 ± 5*	4
			(1.9–4.3)	(42–61)	

[a]Results are shown mean ± SE.
[b]Ca:EGTA ratio = 2.1 : 2 at 2 mM $MgCl_2$, pH 7.4.
[c]Ca:EGTA ratio = 2.2 : 2 at 2 mM $MgCl_2$, 4 mM ATP, pH 7.4, and 2 : 2 at 2 mM $MgCl_2$, 0 ATP, pH 7.4.
[d]n is the number of separate experiments.
[e]The range of measured values is shown in parenthesis.
*Indicates that the results are significantly different from control $K_{m(app)}$ and V_{max} values (red blood cell, 0 Ca^{2+}) at the $p < 0.05$ level (two-tailed t test).

ATP_i (TABLE 2). Inhibition by Ca^{2+}_o in Ca-free, ATP-free ghosts is unaffected by the presence of 2 mM adenylyl-imidodiphosphate (AMP-PNP, a nonmetabolizable analogue of ATP, TABLE 2).

It is important to remember in examining TABLE 2 that in the absence of Ca^{2+} and ATP, uptake by ghosts at 0.2 mM 3OMG is reduced and uptake at 5 mM 3OMG is increased. ATP acts to reduce $K_{m(app)}$ and V_{max} for uptake in ghosts and thus stimulates uptake at subsaturating (≤ 1 mM) 3OMG but inhibits uptake at saturating [3OMG].

TABLE 2. Effects of Ca^{2+} on 3OMG Uptake in Red Cell Ghosts

	3OMG Uptake μmol/L/min		
	[3OMG] of Uptake Determination		
Condition	0.2 mM	5 mM	n^a
Cells	24.0 ± 1.2	112 ± 5	4,4
Cells + 100 μM $Ca^{2+}_o{}^b$	8.1 ± 1.3*	26 ± 3*	4,4
Ghosts 0 ATP_i	8.1 ± 0.5*	236 ± 13**	4,4
+ 11 μM $Ca^{2+}_i{}^c$	3.2 ± 0.1*	57.5 ± 6.0*	4,4
+ 100 μM $Ca^{2+}_o{}^b$	3.3 ± 0.9*	43 ± 2.8*	4,4
+ $Ca^{2+}_i + Ca^{2+}_o$	3.4 ± 0.2*	26.5 ± 1.8***	4,4
Ghosts + 4 mM $ATP_i{}^a$	24.5 ± 0.4	146 ± 6**	4,4
+ 15 μM $Ca^{2+}_i{}^c$	15.6 ± 0.4*	87 ± 1.4*	4,4
+ 100 μM $Ca^{2+}_o{}^b$	14.8 ± 0.2*	88.5 ± 1.8*	4,4
+ $Ca^{2+}_i + Ca^{2+}_o$	13.2 ± 0.6*	68.1 ± 0.7***	4,4
Ghosts + 2 mM AMPPNP	23.1 ± 0.7		
+ 100 μM Ca^{2+}_o	9.6 ± 0.4*		

^{a}n is the number of separate duplicate determinations shown first for uptake at 0.2 mM 30 MG then for uptake at 5 mM 3OMG.
^{b}Ca:EGTA ratio = 2.1 : 2 at 2 mM $MgCl_2$, pH 7.4.
^{c}Ca:EGTA ratio = 2.2 : 2 at 2 mM $MgCl_2$, 4 mM ATP, pH 7.4 and 2 : 2 at 2 mM $MgCl_2$, 0 ATP, pH 7.4.
*Significantly less than control values at the $p < 0.01$ level (1-tailed t test).
**Significantly greater than control values at the $p < 0.01$ level (1-tailed t test).
***The inhibition by Ca^{2+}_i plus Ca^{2+}_o is significantly greater than that by Ca^{2+}_i or Ca^{2+}_o alone.

Reversibility

TABLE 3 summarizes experiments that examine the reversibility of Ca^{2+}-effects on transport. Ca-inhibition of RBC 3OMG uptake is largely reversed by incubation of Ca-treated cells in Ca-free medium for 30 min. If Ca-treated and untreated cells are lysed then resealed in ATP-free medium, ghosts from Ca-treated cells display lower uptake rates than do ghosts from untreated cells. If membranes from these cells are resealed in the presence of ATP, uptake in ghosts from Ca-treated and untreated cells is indistinguishable. These data suggest that reversal of Ca-effects on transport requires intracellular ATP. AMP-PNP cannot substitute for ATP in this reversal of Ca-inhibition of transport (TABLE 3). Modulation of uptake in the absence of Ca^{2+} by intracellular ATP seems not to require ATP-hydrolysis as the nonhydrolyzable analogue of ATP (AMP-PNP) can substitute for ATP in this action (TABLE 3).

TABLE 3. Reversibility of Ca^{2+} Action on 3OMG Uptake

Origin[a]	$[Ca^{2+}]_o$ μM	$[ATP]_i$[b] mM	$[AMPPNP]_i$ mM	3OMG Uptake at 1 mM $\mu mol/L/min$	n[c]
Cells					
A	0	1.0 ± 0.1	0	53.0 ± 6.2	4
B	100[d]	0.8 ± 0.1	0	24.6 ± 4.3*	4
B	100≥0	?	0	47.9 ± 4.0	4
Ghosts					
A	0	0	0	29.8 ± 1.0*	4
B	0	0	0	13.7 ± 0.1*,**	4
A	0	4	0	61.5 ± 4.6	4
B	0	4	0	51.7 ± 4.1	4
A	0	0	4	55.5 ± 3.2	4
B	0	0	4	18.1 ± 1.0*,**	4

[a]Origin refers to original treatment of cells. A cells were originally Ca-naive. B cells were initially exposed to 100 μM Ca^{2+} for 30 min at 24°C. $[Ca^{2+}]_o$ refers to the concentration of Ca^{2+} present during uptake measurements.

[b]For intact cells $[ATP]_i$ refers to the measured ATP content (mmol/kg cell water) of cells and for ghosts refers to the initial [ATP] of the resealing medium.

[c]n refers to the number of separate measurements made. Results are shown as mean ± SE.

[d]Ca:EGTA ratio = 2.1 : 2 at 2 mM $MgCl_2$, pH 7.4.

*Indicates that uptake is significantly lower than uptake measured in Ca^{2+}-naive intact red cells (p <0.01, 1-tailed t test).

**Indicates that uptake by B cell ghosts is significantly less (p <0.01, 1-tailed t test) than uptake by A cell ghosts resealed under identical conditions.

Red Cell Lipid Content

TABLE 4 summarizes lipid analyses of RBCs and ghosts exposed to varying $Ca^{2+}_{i/o}$ and ATP_i. Other than Ca_i-induced increases in the phosphatidic acid content of ATP-containing ghosts, no systematic effects of Ca and ATP treatment on red cell membrane lipid composition were observed.

Conclusions

These findings support the view that human red cell sugar transport is tonically inhibited by physiological levels of Ca^{2+} ($\approx$ 1 mM). The effect is mediated by a Ca-induced change in membrane structure/composit on which is reversed only upon ATP-hydrolysis. While the molecular basis of these phenomena remains to be established a number of possible mechanisms may be excluded by the experimental data. These include: 1. Ca-activation of phospholipases A_2 and C; 2. Ca-activation of diacylglycerol dependent protein kinase C; 3. Ca-activation of proteases; 4. Ca-dependent changes in the phospholipid composition of the red cell membrane.

A general effect on bilayer lipid cannot be excluded. Ghosting cells in the presence of Ca^{2+} reduces bilayer lipid asymmetry.[46] Monkey red blood cell bilayer lipid asymmetry is reduced following incubation in glucose-free saline.[46] This effect can be prevented by addition of EGTA or glucose to saline.[47]

GENERAL CONCLUSIONS

Protein mediated sugar transport is sensitive to membrane lipid composition, to intracellular ATP content, and, in human red cells, to prevailing levels of intracellular and extracellular Ca^{2+}. FIGURE 4 summarizes these findings.

Sugar transport in synaptosomes and peripheral nerve is also sensitive to $[ATP]_i$[48–52] which acts to increase the affinity of the transport system for sugar. Studies with nerve, however, indicated that transport is insensitive to $[Ca^{2+}]_i$ in the range 50 nM–5 μM.[51] Sugar transport in both nerve and human red cells is asymmetric ($K_{m(app)}$ and V_{max} for sugar uptake are lower than the corresponding parameters for sugar efflux[21,51,52]), although asymmetry in red cells increases, while asymmetry in nerve decreases with falling temperature. While these findings suggest subtle differences in the behavior of neuronal and erythrocyte sugar transport systems, the similarities are striking and suggest that human erythrocyte sugar transport and its modulation by bilayer lipid and cellular ATP and Ca^{2+} content may serve as a useful model for the study of sugar transport regulation.

TABLE 4. Analysis of the Lipid Content of Red Cell Membranes

	RBC(A)	RBC(B)	[a]B→0 Ca_o	[b]A ghosts			[c]B ghosts		
ATP_i[d]	+	+	+	−	+	+	−	+	+
Ca_o	−	+	±	−	−	−	−	−	−
Ca_i	−	−	−	−	−	+	−	−	+
Lipid			% of Total Measured Lipid[h]						
Total[e]									
Phospholipid	47.0	47.7	47.8	52.2	50.3	49.1	50.0	46.6	47.5
Cholesterol	45.8	44.4	45.6	40.8	43.3	44.9	42.2	44.5	43.9
DAG	7.2	7.9	6.6	7.0	6.4	6.0	7.8	8.9	8.6
Phospholipid									
species[f]									
PC	14.1	14.0	13.8	14.5	15.1	15.2	15.0	15.9	14.2
PE	13.1	13.9	13.0	15.1	12.8	12.3	12.2	11.7	12.3
Sp	12.2	12.1	13.2	14.2	13.4	13.0	13.8	11.1	12.5
PA	1.2	1.2	1.1	1.3	1.2	1.9	1.1	1.3	2.6
PS/PI[g]	5.0	4.9	5.3	5.3	5.5	4.7	5.4	5.0	4.3
LPC	1.4	1.6	1.4	1.8	2.3	2.0	2.5	1.6	1.6

[a]B cells (previously incubated in Ca-medium) were incubated in Ca-free medium for 30 min.
[b]Ghosts were made from A cells (Ca-naive).
[c]Ghosts were made from B cells (Ca-exposed).
[d]The ATP content of the ghost resealing medium was 2 mM. The ATP content of red cells was not determined.
[e]Total lipid content includes phospholipids, cholesterol, and diacylglycerol (DAG).
[f]The phospholipid species are: PC, phosphatidylcholine; PE, phosphatidylethanolamine; Sp, spingomyelin; PA, phosphatidic acid; PS, phosphatidylserine; PI, phosphatidylinositol; LPC, lysophosphatidylcholines.
[g]PS and PI co-migrate as a single spot using this chromatography system.
[h]Lipid species (acyl chain composition) were not identified. The results are shown as mean of two separate determinations.

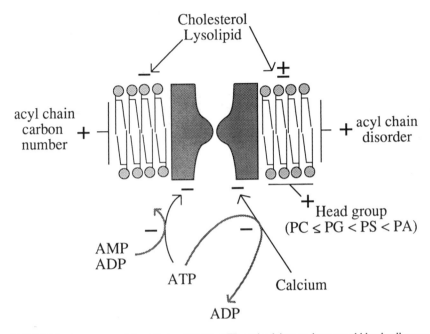

FIGURE 4. A summary of the effects of lipids, ATP, and calcium on human red blood cell sugar transport. The glucose carrier is shown embedded in the membrane lipid bilayer in diagrammatic form. The lipid component of the membrane can affect the activity of the carrier in a variety of forms. Increased lipid acyl chain disorder and carbon number stimulate (+) sugar transport. The bilayer lipid head group can also stimulate (+) transport in the order PC ≤ PG < PS < PA. In addition bilayer modifying lipids (cholesterol and lysolipids) can exert stimulatory (+) and inhibitory (−) actions on transport. ATP binds to the carrier to inhibit (−) transport—an effect competitively inhibited by ADP and AMP (shown by the *gray line*). Ca²⁺ inhibits transport—an effect reversed by ATP hydrolysis (also shown by a *gray line*).

REFERENCES

1. WEISER, M. B., M. RAZIN & W. D. STEIN. 1983. Kinetic tests of models for sugar transport in human erythrocytes and a comparison of fresh and cold-stored cells. Biochim. Biophys. Acta **727:** 379–388.
2. MUECKLER, M., C. CARUSO, S. A. BALDWIN, M. PANICO, I. BLENCH, H. R. MORRIS, W. J. ALLARD, G. E. LIENHARD & H. F. LODISH. 1985. Sequence and structure of a human glucose transporter. Science **229:** 941–945.
3. BIRNBAUM, M. J., H. C. HASPEL & O. M. ROSEN. 1986. Cloning and characterization of a cDNA encoding the rat brain glucose transporter protein. Proc. Natl. Acad. Sci. USA **83:** 5784–5788.
4. ASANO, T., Y. SHIBASAKI, M. KASUGA, Y. KANAZAWA, F. TAKAKU, Y. AKANUMA, Y. OKA. 1988. Cloning of a rabbit brain glucose transporter cDNA and alteration of glucose transporter mRNA during tissue development. Biochem. Biophys. Res. Commun. **154:** 1204–1211.

5. CAIRNS, M. T., J. ALVAREZ, M. PANICO, A. F. GIBBS, H. R. MORRIS, D. CHAPMAN & S. A. BALDWIN. 1987. Investigation of the structure and function of the human erythrocyteglucose transporter by proteolytic dissection. Biochim. Biophys. Acta **905:** 295–310.

6. CARRUTHERS, A. 1984. Sugar transport in animal cells: The passive hexose transfer system. Prog. Biophys. Molec. Biol. **43:** 33–69.

7. THORENS, B., H. K. SARKAR, H. R. KABACK & H. F. LODISH. 1988. Cloning and functional expression in bacteria of a novel glucose transporter present in liver, intestine, kidney, and beta-pancreaticislet cells. Cell **55:** 281–290.

8. FUKUMOTO, H., S. SEINO, H. IMURA, Y. SEINO, R. L. EDDY, Y. FUKUSHIMA, M. G. BYERS, T. B. SHOWS & G. I. BELL. 1988. Sequence, tissue distribution, and chromosomal localization of mRNA encoding a human glucose transporter-like protein. Proc. Natl. Acad. Sci. USA **85:** 5434–5438.

9. KAYANO, T., H. FUKUMOTO, R. L. EDDY, Y. S. FAN, M. G. BYERS, T. B. SHOWS & G. I. BELL. 1988. Evidence for a family of human glucose transporter-like proteins. J. Biol. Chem. **263:** 15245–15248.

10. CHIN, J. J., E. K. JUNG & C. Y. JUNG. 1986. Structural basis of human erythrocyte glucose transporter function in reconstituted vesicles. J. Biol. Chem. **261:** 7101–7104.

11. CHIN, J. J., E. K. JUNG, V. CHEN & C. Y. JUNG. 1987. Structural basis of human erythrocyte glucose transporter function in proteoliposome vesicles: Circular dichroism measurements. Proc. Natl. Acad. Sci. USA **84:** 4113–4116.

12. ALVAREZ, J., D. C. LEE, S. A. BALDWIN & D. CHAPMAN. 1987. Fourier transform infrared spectroscopic study of the structure and conformational changes of the human erythrocyte glucose transporter. J. Biol. Chem. **262:** 3502–3509.

13. KARIM, A. R., W. D. REES & G. D. HOLMAN. 1987. Binding of cytochalasin B to trypsin and thermolysin fragments of the human erythrocyte hexose transporter. Biochim. Biophys. Acta **902:** 402–405.

14. HOLMAN, G. D., B. A. PARKAR & P. J. MIDGLEY. 1986. Exofacial photoaffinity labelling of the human erythrocyte sugar transporter. Biochim. Biophys. Acta **855:** 115–126.

15. HOLMAN, G. D. & W. D. REES. 1987. Photolabelling of the hexose transporter at external and internal sites: Fragmentation patterns and evidence for a conformational change. Biochim. Biophys. Acta **897:** 395–405.

16. KASAHARA, M. & P.C. HINKLE. 1977. Reconstitution and purification of the D-glucose transporter from human erythrocytes. J. Biol. Chem. **253:** 7384–7390.

17. GORGA, F. R. & G. E. LIENHARD. 1981. Equilibria and kinetics of ligand binding to the glucose carrier. Evidence for an alternating conformation model for transport. Biochemistry **20:** 5108–5113.

18. LOWE, A.G. & A. R. WALMSLEY. 1986. The kinetics of glucose transport in human red blood cells. Biochim. Biophys. Acta **857:** 146–154.

19. WIDDAS, W. F. 1952. Inability of diffusion to account for placental glucose transfer in the sheep and consideration of the kinetics of a possible carrier transfer. J. Physiol. (London) **118:** 23–39.

20. BAKER, G.F. & W. F. WIDDAS. 1973. The asymmetry of the facilitated transfer system for hexoses in human red cells and the simple kinetics of a two component model. J. Physiol. (London) **231:** 143–165.

21. NAFTALIN, R.J. & G. D. HOLMAN. 1977. Transport of sugar in human red cells. *In* Membrane Transport in Red Cells. J. C. Ellory & V. L. Lew, Eds: 257–300. Academic Press. New York, NY.

22. CARRUTHERS, A. 1986. Anomalous asymmetric kinetics of human red cell hexose transfer: Role of cytosolic adenosine 5'-triphosphate. Biochemistry **25:** 3592–3602.

23. CARRUTHERS, A. 1986. ATP regulation of the human red cell sugar transporter. J. Biol. Chem. **261:** 11028–11037.

24. HELGERSON, A. L. & A. CARRUTHERS. 1987. Equilibrium ligand binding to the human erythrocyte sugar transporter. Evidence for two sugar binding sites per carrier. J. Biol. Chem. **262:** 5464–5475.

25. BALDWIN, J. M., J. C. GORGA & G. E. LIENHARD. 1981. The monosaccharide transporter of the human erythrocyte. Transport activity upon reconstitution. J. Biol. Chem. **256:** 3685–3689.

26. CARRUTHERS, A. & D. L. MELCHIOR. 1984. Human erythrocyte hexose transporter activity is
     governed by bilayer lipid composition in reconstituted vesicles. Biochemistry 23: 6901–
     6911.
27. CARRUTHERS, A. & D. L. MELCHIOR. 1986. How bilayer lipids affect membrane protein
     activity. TIBS 11: 331–335.
28. CARRUTHERS, A. & D. L. MELCHIOR. 1987. Effect of lipid environment on membrane trans-
     port: The human erythrocyte sugar transport protein/lipid bilayer system. Annu. Rev. Physiol.
     50: 527–571.
29. TEFFT, R. E., JR., A. CARRUTHERS & D. L. MELCHIOR. 1986. Reconstituted human eryth-
     rocyte sugar transporter activity is determined by bilayer lipid head groups. Biochemistry
     25: 3709–3718.
30. SMALL, D. M. 1986. Handbook of Lipid Research: The Physical Chemistry of Lipids. Plenum.
     New York, NY.
31. HAUSER, H. & M. C. PHILLIPS. 1979. Interaction of the polar groups of phospholipid bilayer
     membranes. Prog. Surf. Membr. Sci. 13: 297–413.
32. CARRUTHERS, A. & D. L. MELCHIOR. 1983. Studies of the relationship between bilayer water
     permeability and bilayer physical state. Biochemistry 22: 5797–5807.
33. DEMEL, R. A. & B. DEKRUYFF. 1976. The function of sterols in membranes. Biochim.
     Biophys. Acta 406: 97–107.
34. CHAPMAN, D., R. M. WILLIAMS & B. D. LADBROOKE. 1967. Physical studies of phospholip-
     ids. Chem. Phys. Lipids 1: 445–476.
35. MABREY, S., P. L. MATEO & J. M. STURTEVANT. 1978. High sensitivity scanning calorimetric
     study of mixtures of cholesterol with dimyristoyl and dipalmitoylphosphatidylcholines. Bio-
     chemistry 17: 2464–2468.
36. MELCHIOR, D. L., F. J. SCAVITTO & J. M. STEIM. 1980. Dilatometry of dipalmitoylphos-
     phatidylcholine bilayers. Biochemistry 19: 4828–4834.
37. CONNOLLY, T. J., A. CARRUTHERS & D. L. MELCHIOR. 1985. Effects of bilayer cholesterol
     content on reconstituted human erythrocyte sugar transporter activity. J. Biol. Chem. 260:
     2617–2620.
38. CONNOLLY, T. J., A. CARRUTHERS & D. L. MELCHIOR. 1985. Effects of bilayer cholesterol
     on human erythrocyte hexose transport protein activity in synthetic lecithin bilayers. Bio-
     chemistry 24: 2865–2873.
39. WELTZIEN, H. U. 1979. Cytolytic and membrane-perturbing properties of lysophosphatidy-
     choline. Biochim. Biophys. Acta 559: 259–287.
40. NADERI, S., A. CARRUTHERS & D. L. MELCHIOR. 1989. Inhibition of red blood cell glucose
     transport by lysolipid. Biochim. Biophys. Acta. In press.
41. HEBERT, D. N. & A. CARRUTHERS. 1986. Direct evidence for ATP modulation of sugar
     transport in human erythrocyte ghosts. J. Biol. Chem. 261: 10093–10099.
42. JUNG, C. Y. 1971. Evidence of high stability of the glucose carrier function in human red cell
     ghosts extensively washed in various media. Arch. Biochim. Biophys. 146: 215–226.
43. TAVERNA, R. D. & R. G. LANGDON. 1973. A new method for measuring glucose translocation
     through biological membranes and its application to human erythrocyte ghosts. Biochim.
     Biophys. Acta 298: 412–421.
44. TAVERNA, R. D. & R. G. LANGDON. 1973. Glucose transport in white erythrocyte ghosts and
     membrane derived vesicles. Biochim. Biophys. Acta 298: 422–428.
45. CARRUTHERS, A. & D. L. MELCHIOR. 1983. Asymmetric or symmetric? Cytosolic modulation
     of human erythrocyte hexose transfer. Biochim. Biophys. Acta 728: 254–266.
46. SCHLEGEL, R. A., L. MCEVOY & P. WILLIAMSON. 1985. Membrane phospholipid asymmetry
     and the adherence of loaded red blood cells. Bibl. Haematol. 51: 150–156.
47. ALAM, A. 1985. Red cell membrane cholesterol levels and their effects on transbilayer phos-
     pholipid asymmetry. Indian J. Biochem. Biophys. 22: 38–42.
48. DIAMOND, I. & R. A. FISHMAN 1973. High affinity transport and phosphorylation of 2-
     deoxy-D-glucose in synaptosomes. J. Neurochem. 20: 1533–1542.
49. DIAMOND, I. & R. A. FISHMAN. 1973. Development of sodium stimulated glucose oxidation
     in synaptosomes. J. Neurochem. 21: 1043–1050.
50. BAKER, P. F. & A. CARRUTHERS. 1981. Sugar transport in giant axons of Loligo. J. Physiol.
     316: 481–502.

51. BAKER, P. F. & A. CARRUTHERS. 1981. 3-O-methylglucose transport in internally dialysed giant axons of Loligo. J. Physiol. **316:** 503–525.
52. BAKER, P. F. & A. CARRUTHERS. 1984. Transport of sugars and amino acids. Curr. Top. Membr. Transp. **22:** 91–130.

Calcium Homeostasis and Aging: Role in Altered Signal Transduction

GEORGE S. ROTH

Molecular Physiology and Genetics Section
Laboratory of Cellular and Molecular Biology
Gerontology Research Center
National Institute on Aging
Francis Scott Key Medical Center
4940 Eastern Avenue
Baltimore, Maryland 21224

The Role of Calcium in Physiological Processes, Mechanisms of Action, and Age Changes

Calcium is almost ubiquitous in biological systems. Proper movement on this ion is a prerequisite for numerous physiological processes, ranging from cell division to muscle contraction to secretion to neurotransmission.[1,2] Most of these processes also exhibit functional decline during aging.[3] We first became aware of the possibility that impaired calcium mobilization might account for some types of decreased hormonal responsiveness in 1980 when we observed that impaired beta adrenergic stimulation of aged myocardial contraction could be reversed simply by increasing media calcium concentration.[4] No age changes were observed in beta adrenergic receptor levels, cyclic AMP generation, or protein kinase activation—all precursor events to calcium flux.

Since this initial observation, a steady stream of similar findings in many types of calcium-dependent systems has appeared in the literature (for reviews REFS. 3, 5). Two rather remarkable phenomena have been associated with most of these reports. First, age changes in responsiveness can be partially, if not fully, reversed by appropriate manipulation of cellular calcium fluxes and/or levels. Second, the mechanisms responsible for normal calcium mobilization in these systems have been quite varied. Thus, it is conceivable that some generalized aging mechanism impacts upon many specialized calcium transport systems.

For example, Meldolesi and Pozzan have recently categorized at least nine separate systems by which calcium is translocated by cells.[2] These include influx from the extracellular fluid by voltage-operated channels, receptor-operated channels, and second-messenger-operated channels, while efflux is affected by a calcium ATPase and sodium-calcium exchange. Calcium is stored intracellularly in the endoplasmic reticulum and mitochondria, entering the former through a pump and a calcium ATPase and the latter through a calcium channel. Calcium is in turn released from the endoplasmic reticulum by the action of inositol trisphosphate on its receptor and from the mitochondria by sodium-calcium exchange. Since different cell types utilize these transport processes in different ways, yet exhibit similar net impairments in calcium transfer during aging, either multiple defects occur or some common link between these systems is affected by age.

TABLE 1 lists some of the calcium-dependent systems which have been reported to exhibit altered responsiveness during aging. A wide variety of stimuli, types of responses, and species are represented. In addition, studies have examined cells and/or tissues, both *in vivo* and *in vitro* as well as complex responses, such as psychomotor function, in whole animals. Again, quite varied calcium sources and mechanisms of mobilization are utilized in the systems listed in TABLE 1, yet in most cases age differences can be at least partially reversed if sufficient calcium can be introduced to the site of its action. Thus, old cells and tissues often maintain an innate capacity to respond, yet apparently lose selective mechanisms by which to regulate calcium movement.

TABLE 1. Impaired Stimulation of Calcium Mobilization; Generalized Manifestation of Aging

Stimulus	Species	Tissue	Response	References
Alpha adrenergic	rat	parotid	electrolyte secretion	6, 7
Alpha adrenergic	rat	parotid	glucose oxidation	8, 9
Alpha adrenergic	rat	aorta	contraction	10
Beta adrenergic	rat	heart	contraction	4
Cholinergic	rat	brain (striatum)	dopamine release	11
Depolarization	rat	heart	contraction	12
Depolarization	rat	brain (forebrain and cortex)	acetylcholine release	13, 14
Depolarization	mouse	brain (forebrain)	acetylcholine release	15
Depolarization	mouse	whole animal	motor function	16
Depolarization	rat	whole animal	maze learning	17
Serotonin	rat	whole animal	contraction	10
Gonadotropin releasing hormone	rat	pituitary	gonadotropin secretion	18
Lectin	rat	lymphocyte	mitogenesis	19, 20
Lectin	mouse	lymphocyte	mitogenesis	21–23
Lectin	human	lymphocyte	mitogenesis	24
Compound 48-80	rat	mast cell	histamine release	25
Formyl-methionyl leucyl-phenylalinine	human	neutrophil	superoxide generation	26, 27
Thyroid hormones	human	erythrocyte	activation of calcium ATPase	28
Low density lipoprotein	human	polymorphonuclear leukocytes	release of β-glucuronidase	29
Cytochalasin B	human	polymorphonuclear leukocytes	release of β-glucuronidase	29
Immune complexes	human	polymorphonuclear leukocytes	release of β-glucuronidase	29
Phosphatidylserine	rat	brain	protein kinase activation	30

The Alpha$_1$-Adrenergic Stimulated Parotid Gland as a Model for One Class of Age-Impaired Calcium Mobilizaton Systems

In our laboratory, the most intensely studied of the age-impaired calcium mobilization systems has been the alpha$_1$-adrenergic stimulated rat parotid cell aggregate. Both stimulated glucose oxidation[8,9] and potassium release[6,7] are reduced with increasing age, although the temporal pattern of change differs somewhat. Nevertheless, both processes are dependent on stimulated mobilization of calcium from intracellular and extracellular sources and age-related decrements are completely reversible in the presence of the ionophore, A23187, and extracellular calcium.[6,9] Both responses are also dependent on the binding of adrenergic agents to the alpha$_1$-adrenergic receptor subtype, and this interacton is unaltered during aging.[6]

Subsequent to receptor binding, the enzyme, phospholipase C, is activated to cleave phosphatidylinositol-4,5-bisphosphate to inositol trisphosphate and diacylglycerol. The former is considered to be a "second messenger" in the mobilization of calcium from intracellular stores.[2] We have not detected any significant changes in inositol trisphosphate production during aging.[31] In contrast, the ability of inositol trisphosphate to stimulate release of calcium from intracellular sites is reduced by approximately 50% over the adult rat lifespan.[31] This reduction is consistent with overall age-related reductions in alpha$_1$-adrenergic-stimulated calcium efflux,[6,7,9,31] glucose oxidation,[8,9] and potassium release.[6,7] In addition, a post receptor defect in the ability to mobilize calcium is consistent with the ability of A23187 to reverse age changes in glucose oxidation and potassium release by allowing sufficient calcium flux to appropriate cellular sites.[6,7]

Thus, future studies on this system must examine the functionality of the microsomal inositol-trisphosphate receptor which regulates intracellular calcium release, as well as the content of mobilizable calcium during the aging process. It is quite possible that age-associated alterations occurring at either or both of these levels may be responsible for overall dysfunctions in control of secretion and energy metabolism.

Muscarinic Stimulation of Striatal Dopamine Release as a Model of Impaired Calcium Mobilization in the Aging Brain

Another system currently under investigation in our laboratory is the muscarinic-stimulated rat striatal slice preparation. Dopamine release therein is controlled by inhibitory dopamine autoreceptors which are themselves inhibited by nearby muscarinic cholinergic heteroreceptors.[32] Muscarinic activation by carbachol or oxotremorine enhances potassium-evoked release of dopamine from striatal slices from mature (6 months old) but not senescent (24 months old) rats.[11] However, when receptors are bypassed by administration of the ionophore A23187, which allows calcium to flood the intracellular space, striatal slices from senescent rats can release dopamine as well as those from younger counterparts. Obliteration of age differences by this manipulation is also dependent on the presence of extracellular calcium and potassium.[11] Thus, these results are probably not due to direct toxic effects of the ionophore allowing dopamine to leak out of the cells.

Interestingly, when inositol trisphosphate, which is believed to mediate calcium mobilization in this system,[33] is administered directly to striatal slices only a slight (statistically nonsignificant) age-related reduction in dopamine release occurs.[11] These results are somewhat different from those for alpha$_1$-adrenergic stimulation of parotid cell aggregates in which age differences are completely retained upon direct inositol trisphosphate stimulation.[31]

It is thus probable that the bulk of the age associated defect in the striatal slice system lies proximal to inositol-phosphate-stimulated calcium mobilization. Possible candidate events include impaired inositol trisphosphate generation or receptor/effector-binding/coupling as has been previously reported by others.[34,35] Examination of these possibilities is currently underway.

CONCLUSIONS AND FUTURE DIRECTIONS

Altered calcium mobilization has recently emerged as a fairly generalized manifestation of aging with serious physiological consequences. Despite the multiplicity of processes which regulate cellular calcium flow, age-related impairments in these systems appear to be quite widespread and particularly important in many responses to stimuli such as hormones, neurotransmitters, etc.

Most remarkable is the fact that in almost every documented case of decreased calcium-dependent responsiveness during aging, impairments can be partially or fully reversed if sufficient calcium can be moved to the site of its action. Thus, it may be possible to devise novel therapeutic strategies for the amelioration of such dysfunctions based on appropriate manipulation of selective calcium movements.

Future work in this area must continue to elucidate 1) the basic mechanisms by which calcium is transported, independent of the aging process; 2) the mechanisms by which aging affects these basic transport mechanisms; and 3) methods to halt, prevent, or reverse such age effects by molecular pharmacological intervention.

REFERENCES

1. CARAFOLI, E. & J. T. PENNISTON. 1987. The calcium signal. Sci. Am. **257:** 70–78.
2. MELDOLESI, J. & T. POZZAN. 1987. Pathways of Ca^{++} influx at the plasma membrane: Voltage-, receptor-, and second messenger-operated channels. Exp. Cell Res. **171:** 271–283.
3. ROTH, G. S. 1988. Mechanisms of altered hormone and neurotransmitter action during aging: The role of impaired calcium mobilization. Ann. N.Y. Acad. Sci. **521:** 170–176.
4. GUARNIERI, T., C. R. FILBURN, G. ZITNIK, G. S. ROTH & E. G. LAKATTA. 1980. Mechanisms of altered cardiac inotropic responsiveness during aging in the rat. Am. J. Physiol **239:** H501–H508.
5. GIBSON, G. E. & C. PETERSON. 1987. Calcium and the aging nervous system. Neurobiol. Aging **8:** 329–343.
6. ITO, H., B. J. BAUM, T. UCHIDA, M. T. HOOPES, L. BODNER & G. S. ROTH. 1982. Modulation of rat parotid cell α-adrenergic responsiveness at a step subsequent to receptor activation. J. Biol. Chem. **257:** 9532–9538.
7. BODNER, L., M. T. HOOPES, M. GEE, H. ITO, G. S. ROTH & B. J. BAUM. 1983. Multiple transduction mechanisms are likely involved in calcium mediated exocrine secretory events in rat parotid cells. J. Biol. Chem. **258:** 2774–2777.
8. ITO, H., M. T. HOOPES, G. S. ROTH & B. J. BAUM. 1981. Adrenergic and cholinergic mediated glucose oxidation by rat parotid gland acinar cells during aging. Biochem. Biophys. Res. Commun. **398:** 275–282.
9. GEE, M. V., Y. ISHIKAWA, B. J. BAUM & G. S. ROTH. 1986. Impaired adrenergic stimulation of rat parotid cell glucose oxidation during aging: The role of calcium. J. Gerontol. **41:** 331–335.
10. COHEN, M. L. & B. A. BERKOWITZ. 1976. Vascular contraction: Effect of age and extracellular calcium. Blood Vessels **67:** 139–149.
11. JOSEPH, J. A., T. K. DALTON, G. S. ROTH & W. A. HUNT. 1988. Alterations in muscarinic control of striatal dopamine autoreceptors in senescence: A deficit at the ligand-muscarinic receptor interface? Brain Res. **454:** 149–155.
12. ELFELLAH, M. S., A. JOHNS & A. M. M. SHEPHERD. 1986. Effect of age on responsiveness of isolated rat atria to carbachol and on binding characteristics of atria muscarinic receptors. J. Cardiovasc. Pharmacol. **8:** 873–877.
13. MEYER, E. M., F. T. CREW, D. H. OTERO & K. LARSON. 1986. Aging decreases the sensitivity of rat cortical synaptosomes to calcium ionophore-induced acetylcholine release. J. Neurochem. **47:** 1244–1246.
14. PETERSON, C. & G. E. GIBSON. 1983. Aging and 3, 4-diaminopyridine alter synaptosomal calcium uptake. J. Biol Chem. **258:** 11482–11486.
15. PETERSON, C., D. G. NICHOLS & G. E. GIBSON. 1985. Subsynaptosomal distribution of calcium during aging and 3, 4 diaminopyridine treatment. Neurobiol. Aging **6:** 297–304.
16. PETERSON, C. & G. E. GIBSON. 1983. Amelioration of age-related neurochemical and behavioral deficits by 3, 4-diaminopyridine. Neurobiol. Aging **4:** 25–30.
17. DAVIS, H. P., A. IDOWU & G. E. GIBSON. 1983. Improvement of 8-arm maze performance in aged Fischer 344 rats with 3, 4-diaminopyridine. Exp. Aging Res. **9:** 211–214.

18. CHUKNYISKA, R. S., M. R. BLACKMAN & G. S. ROTH. 1987. Ionophore A23187 partially
     reverses LH secretory defect of pituitary cells from old rats. Am. J. Physiol. **258:** E233–
     E237.
19. WU, W., M. PAHLAVANI, A. RICHARDSON & H. T. CHEUNG. 1985. Effect of maturation and
     age on lymphocyte proliferation induced by A23187 through an interleukin independent
     pathway. J. Leukocyte Biol. **38:** 531–540.
20. SEGAL, J. 1986. Studies on the age-related decline in the response of lymphoid cells to
     mltogens: Measurements of concanavalin A binding and stimulation of calcium and sugar
     uptake in thymocytes from rats of varying ages. Mech. Ageing Dev. **33:** 295–303.
21. MILLER, R. A., B. JACOBSON, G. WEIL & E. R. SIMONS. 1987. Diminished calcium influx in
     lectin-stimulated T cells from old mice. J. Cell. Physiol. **132:** 337–342.
22. MILLER, R. A. 1986. Immunodeficiency on aging: Restorative effects of phorbol ester com-
     bined with calcium ionophore. J. Immunol. **137:** 805–808.
23. PROUST, J. J., C. R. FILBURN, S. A. HARRISON, M. A. BUCHHOLZ & A. A. NORDIN. 1987.
     Age-related defect in signal transduction during lectin activation of murine T lymphocytes. J.
     Immunol. **139:** 1472–1478.
24. CHOPRA, R., J. NAGEL & W. ALDER. 1987. Decreased response of T cells from elderly
     individuals to phytohemogglutinin (PHA) stimulation can be augmented by phorbo myristate
     acetate (PMA) in conjunction with Ca-ionophore A23187. Gerontologist **27:** 204A.
25. ORIDA, N. & J. D. FELDMAN. 1982. Age related deficiency in calcium uptake by mast cells.
     Fed. Proc. **41:** 822.
26. LIPSCHITZ, D. A., K. B. UDUPA & L. A. BOXER. 1988. The role of calcium in the age related
     decline of neutrophil function. Blood In press.
27. LIPSCHITZ, D. A., K. B. UDUPA & L. A. BOXER. 1987. Evidence that microenvironmental
     factors account for the age-related decline in neutrophil function. Blood **70:** 1131–1135.
28. DAVIS, P. J., F. B. DAVIS & S. D. BLAS. 1987. Donor age-dependent decline in response of
     human red cell Ca^{++}-ATP'ase activity to thyroid hormone *in vitro*. J. Clin. Endocrinol.
     Metab. **64:** 921–925.
29. FULOP, T., G. FARIS, I. WORCUM, G. PARAGH & A. LEOVEY. 1985. Age related variations of
     some polymorphonuclear leukocyte functions. Mech. Ageing Dev. **29:** 1–8.
30. CALDERINI, G., F. BELLINI, A. C. BONETTI, E. GALBAITI, S. TEOLOTO & G. T OFFANO. 1986.
     Effect of aging on phospholipid sensitive Ca^{++} dependent protein kinase in the rat brain.
     Abstr. Soc. Neurosci. **12:** 275.
31. ISHIKAWA, Y., M. V. GEE, I. S. AMBUDKAR, B. J. BAUM & G. S. ROTH. 1988. Age-related
     impairment in rat parotid cell α-1-adrenergic action at the level of inosItol trisphosphate
     responsiveness. Biochim. Biophys. Acta **968:** 203–210.
32. LEHMAN, J. & S. J. LANGER. 1982. Muscarinic receptors on dopamine terminals in the cat
     caudate nucleus: Neuromodulation of ^3H-dopamine release *in vitro* by endogenous acetyl-
     choline. Brain Res. **248:** 61–69.
33. FISHER, S. K. & B. W. AGRANOFF. 1987. Receptor activation and inosItol lipid hydrolysis in
     neurol tissues. J. Neurochem. **48:** 999–1018.
34. LIPPA, A. S., C. C. LOULLIS, J. ROTROSEN, D. M. CORDASCO, D. J. CRITCHETL & J. A.
     JOSEPH. 1985. Conformational changes in muscarinic receptors may produce diminished
     cholinergic neurotransmission and memory deficits in aged rats. Neurobiol. Aging **6:** 317–
     325.
35. STRONG, R., J. C. WAYMIRE, T. SAMORAJSKI & Z. GOTTESFELD. 1984. Regional analysis of
     neostriatal cholinergic and dopaminergic receptor binding and tyrosine hydroxylase activity as
     a function of aging. Neurochem. Res. **9:** 1641–1653.

Cycling of Ca^{2+} across the Plasma Membrane as a Mechanism for Generating a Ca^{2+} Signal for Cell Activation[a]

HOWARD RASMUSSEN, PAULA BARRETT,
WALTER ZAWALICH, CARLOS ISALES, PETER STEIN,
JOAN SMALLWOOD, RICHARD McCARTHY, AND
WENDY BOLLAG

Departments of Internal Medicine and Physiology
Yale University School of Medicine
333 Cedar Street
New Haven, Connecticut 06510
and
Yale University School of Nursing

When studies of Ca^{2+} messenger function were initially carried out in nerve and muscle, it seemed that Ca^{2+} played its messenger role in a very simple manner: increases and decreases in intracellular Ca^{2+} concentration led directly to increases and decreases in cellular activity. That is, there was a direct correspondence between changes in Ca^{2+} concentration and changes in cellular response. However, as studies of Ca^{2+} messenger function expanded to include a variety of other tissues, it became evident that the Ca^{2+} messenger system is considerably more complex than originally envisioned.[1] In the present discussion, we will present a brief account of the classic view of Ca^{2+} messenger function, and then consider the intimate relationship between the Ca^{2+} and cAMP messenger systems.[2] This will be followed by a consideration of Ca^{2+} as a cellular toxin and the important constraint that the threat of Ca^{2+} intoxication imposes on the use of Ca^{2+} as intracellular messenger. Finally, the complexity of Ca^{2+} messenger function will be emphasized by discussing two systems: aldosterone secretion from adrenal glomerulosa cells and insulin secretion from beta cells.[3] In the course of this discussion, a particular thesis will be developed: during sustained cellular responses, it is a cycling of Ca^{2+} across the plasma membrane that is the mechanism by which a specific Ca^{2+} messenger, an increase in the Ca^{2+} concentration in a subdomain of the plasma membrane, $[Ca^{2+}]_{sm}$, is generated. The way in which this Ca^{2+} messenger leads to a specific cellular response is determined by the nature and number of the Ca^{2+}-sensitive transducers associated with the plasma membrane. In many cases, the cAMP messenger system must be activated appropriately in order for the Ca^{2+} messages to be effective.

Ca^{2+} as Messenger: The Classic View

The classic systems in which a messenger function for Ca^{2+} was first described were neurosecretion, skeletal muscle contraction, and cardiac muscle contraction.[4] In each cell

[a]The work described in this report was supported by a grant (DK 19813) from the National Institutes of Diabetes and Digestive and Kidney Disease.

73

type, activation by an appropriate agonist leads to an increase in intracellular Ca^{2+} concentration. This rise occurs either as a result of an increase in rate of Ca^{2+} entry into the cell across the plasma membrane or a release of Ca^{2+} from an intracellular pool. The magnitude and duration of the response correlates with the magnitude and duration of the Ca^{2+} signal. In these early studies, the only recognized Ca^{2+} receptor protein was troponin C (in skeletal and cardiac muscle). Even at this early stage of knowledge, evidence for spatial domains of Ca^{2+} distribution and action was obtained: neurotransmission ceased immediately if extracellular Ca^{2+} was removed and the site of Ca^{2+} action was at the endoplasmic surface of the plasma membrane. Skeletal muscle contraction and relaxation could be induced repetitively in the absence of extracellular Ca^{2+}, because the Ca^{2+} released from the sarcoplasmic reticulum activated contractile systems in close proximity to the site of release and was then taken back up into the sarcoplasmic reticulum.

Ca^{2+} and cAMP as Synarchic Messengers

At this time (1960s), the second messenger function of cAMP was also discovered.[5] For a brief time, it seemed that the messenger function of Ca^{2+} might be confined to excitable and that of cAMP to nonexcitable tissues. However, our studies of the action of parathyroid hormone in renal tubules, and of serotonin in the blowfly salivary gland showed that Ca^{2+} and cAMP served as interrelated second messengers in the actions of these hormones.[6] At the same time, it was found that Ca^{2+} regulated the activity of a crucial enzyme in cAMP metabolism, phosphodiesterase, and shortly thereafter that Ca^{2+} could also regulate the activity of adenylate cyclase. These discoveries led to the identification of a Ca^{2+}-dependent regulator protein, or Ca^{2+} binding subunit of the phosphodiesterase. When isolated this protein was found to be widely distributed in animal and plant cells and to be a structural homolog of troponin C. A nearly universal intracellular Ca^{2+} receptor protein had been identified. It was eventually named calmodulin.[7]

These discoveries led to the demonstration that Ca^{2+} and cAMP interact in a number of ways:[2] 1) they each can control the other's concentrations by regulating either its generation or disposal; 2) they can regulate the activity of a particular protein via its phosphorylation by separate cAMP- and Ca^{2+}-dependent protein kinases; 3) they can regulate ion fluxes via different channels and/or pumps in a system carrying out transcellular transport; 4) they can act at sequential steps in a protein kinase cascade, as exemplified in the control of glycogenolysis; or 5) they can arise simultaneously as a consequence of the activation of a single receptor.

The relationship between the Ca^{2+} and cAMP messenger systems has become even more intimate by the recent demonstration that a single receptor type is linked both to adenylate cyclase and to either Ca^{2+} channels or the phosphoinositide system, which is an important component of the Ca^{2+} messenger system.[8,9] In fact, there is a nearly universal synarchic relationship between the Ca^{2+} and cAMP messenger systems.[2] The behavior and function of the Ca^{2+} messenger system can only be understood by appreciating its relationship to the cAMP messenger system in the particular cellular response under study. These two messenger systems can interact either in a cooperative, hierarchical, antagonistic, sequential, or redundant fashion in controlling cellular responses. A particularly common relationship is one in which cAMP determines the sensitivity of the cell to the actions of a given Ca^{2+} messenger as will be described below in our discussions of aldosterone and insulin secretion.

Ca^{2+} Cycling: Mechanism

Another kind of complexity arises when cells employ Ca^{2+} as an intracellular messenger to regulate a sustained response. In classic tissues (nerve and muscle), both responses are brief and changes in Ca^{2+} concentration are brief. However, many cellular responses are sustained in the sustained presence of agonist. Thus, for example, aldosterone secretion, insulin secretion, or smooth muscle contraction may last hours or even days in the presence of the appropriate stimulus. In each of these tissues Ca^{2+} plays a critically important messenger function. There is a sustained increase in Ca^{2+} influx rate despite the fact that excessive Ca^{2+} is a cellular toxin. The potential for a progressive accumulation of Ca^{2+} and Ca^{2+} toxicity must exist in any cell in which a sustained increase in Ca^{2+} influx rate occurs. Yet, in such cells, evidence of toxicity is rare, because these cells possess autoregulatory mechanisms by which any change in Ca^{2+} influx rate is balanced by a compensatory increase in efflux rate: using Ca^{2+} cycling these cells can generate a specific Ca^{2+} message, a change in $[Ca^{2+}]_{sm}$, and still maintain cellular Ca^{2+} homeostasis.[10,11]

A simple example of this behavior is illustrated in the case of insulinlike growth factor ((IGF-II) acting on competent-primed Balb 3T3 cells.[12] Addition of IGF-II to these cells induces a proliferative response which is preceded by a transient increase in intracellular free Ca^{2+} (lasting 30–50 sec) but a sustained increase in Ca^{2+} influx rate. The sequence appears to occur as follows: 1) IGF-II stimulates Ca^{2+} influx; and 2) this, in turn, causes an immediate increase in intracellular Ca^{2+} concentration which acts as a signal to increase Ca^{2+} efflux rate so that despite a continued high rate of influx, the global intracellular Ca^{2+} concentration falls back to its original value as a result of enhanced efflux. Thus, during the sustained phase of this response, there is an increase in the rate of Ca^{2+} cycling across the plasma membrane without any net accumulation of Ca^{2+} by the cell, and without a significant increase in the global intracellular Ca^{2+} concentration. We propose that this Ca^{2+} cycling does lead, however, to an increase in the Ca^{2+} concentration in a subdomain of the plasma membrane, $[Ca^{2+}]_{sm}$, and that it is this change in $[Ca^{2+}]_{sm}$ which acts as the messenger to regulate the activities of Ca^{2+}-sensitive, plasma membrane-associated transducers.

The molecular basis for this autoregulation of cellular Ca^{2+} metabolism is found in the properties of the $Ca^{2+}/2 H^{+}$-ATPase, or Ca^{2+} pump, in the plasma membrane. This pump is regulated by calmodulin (CaM).[13,14] When the $[Ca^{2+}]$ rises, Ca^{2+} associates with CaM to form Ca_4CaM which interacts with the Ca^{2+} pump to lower the K_m for Ca^{2+} and raise the V_{max} of the pump. By this mechanism, there is a nearly instantaneous and compensatory increase in Ca^{2+} efflux rate to balance any increase in influx rate. Besides this basic feedback control of pump activity, additional controls operate. A particularly important one is the Ca^{2+}-dependent activation of protein kinase C.[14] This enzyme catalyzes the phosphorylation of the pump, resulting in a 4-fold increase in its V_{max}. Additionally, in specific cell types, such as the cardiac myocyte, cAMP-dependent protein kinase can phosphorylate the pump and so activate it.[15] Also, in certain cells cGMP may serve to stimulate pump activity. By these means different cells can maintain Ca^{2+} homeostasis under conditions of changing activity and changing rates of plasma membrane Ca^{2+} influx.

Ca^{2+} Cycling: The Ca^{2+}-Sensitive Plasma Membrane Transducers

In our view, there is a fundamental difference between brief responses or the initial phases of sustained responses in which a rise in global cytosolic Ca^{2+} concentration, $[Ca^{2+}]_c$, is transient, and the plateau phases of such responses in which there is no significant change in $[Ca^{2+}]_c$ but an increase in Ca^{2+} influx rate and $[Ca^{2+}]_{sm}$. During the sustained response, a change in Ca^{2+} concentration throughout the cytosolic domain is no longer the mechanism by which information is carried from cell surface to cell interior. Rather, the change in $[Ca^{2+}]_{sm}$ activates Ca^{2+}-sensitive plasma membrane (PM)

transducers. The output from these transducers generates the signals that convey information from cell surface to cell interior. A list of some of the common Ca^{2+}-sensitive,
PM-associated transducers is given in TABLE 1. In addition, our proposed requirement for
the presence of one or more Ca^{2+}-sensitive transducers on the plasma membrane suggest
a corollary: a rise in $[Ca^{2+}]_{sm}$ is an ineffective messenger if such transducers are absent
or inhibited, or their output is rapidly inactivated, e.g., rapid dephosphorylation of proteins.

Ca^{2+} Cycling: Its Role in Sustained Cellular Responses

The regulation of aldosterone secretion can serve as a prototype for a cellular response
system in which Ca^{2+} plays a critically important role during a sustained response.
Adrenal glomerulosa cells respond to one of several agonists: angiotensin II (AII), adreno-
corticotrophin (ACTH) and/or extracellular K^+ with a sustained increase in Ca^{2+} influx
rate (or Ca^{2+} cycling) across the plasma membrane and a sustained increase in aldosterone secretory rate.[3,10,16] However, there is no correlation between the magnitude of the
change in Ca^{2+} cycling rate and the magnitude of the change in aldosterone secretory rate
induced by these different agonists. Furthermore, these agonists do not have a common

TABLE 1. Ca^{2+}-Sensitive, Plasma Membrane-Associated Transducers

Element	New Signal
Adenylate cyclase	↑ cAMP
Phosphodiesterase	↓ cAMP
? Guanylate cyclase	↑ cGMP
Phospholipase A_2	↑ eicosanoids
CaM-dependent protein kinase	↑ phosphoproteins
PI-PLC	↑ IP_3, DAG
Protein kinase C	↑ phosphoproteins

mode of action in terms of other messengers generated.[17] Additionally, simply increasing
Ca^{2+} cycling rate is not a sufficient stimulus to induce a sustained increase in aldosterone
secretory response. Nonetheless, inhibition of any of these agonist-induced increases in
Ca^{2+} influx rate leads to an inhibition of aldosterone secretion. These results indicate that
a sustained change in Ca^{2+} cycling rate is a necessary but not sufficient condition for the
stimulation of aldosterone production. A second condition is necessary: the activation of
one or more Ca^{2+}-sensitive plasma membrane transducers on the plasma membrane.
In the case of both ACTH and K^+, this transducer is adenylate cyclase. The increase
in Ca^{2+} cycling rate leads to a CaM-dependent activation of adenylate cyclase. (ACTH,
but not K^+ also activates the cyclase via a receptor-G protein pathway). It is likely, but
not yet proved, that the ACTH- and/or K^+-induced Ca^{2+} cycling act on additional
PM-associated Ca^{2+} transducers.
In the case of AII, an important PM transducer is protein kinase C (PKC). In contrast
to adenylate cyclase which is continually plasma membrane-associated, PKC is not membrane-associated in the unstimulated cell but becomes associated during the initial phase
of AII action. This association is facilitated by Ca^{2+}. During AII action there are two
distinct phases during which Ca^{2+} serves a messenger function but it does so in different
cellular domains acting on different molecular targets. During the initial phase, AII
activates PI-specific phospholipase C (PI-PLC) so that phosphatidylinositol 4,5-bisphos-

phate is hydrolyzed to inositol (1,4,5)-trisphosphate (Ins(1,4,5)P$_3$) and diacylglycerol (DAG).[18] The Ins(1,4,5)P$_3$ induces the release of intracellular Ca^{2+} and hence a transient increase in cytosolic Ca^{2+} concentration. This Ca^{2+} transient activates CaM-dependent enzymes including the Ca^{2+} pump (see above) and CaM-dependent protein kinases. As a consequence, a specific subset of intracellular proteins become phosphorylated and their phosphorylation is linked to the initial phase of the aldosterone secretory response. Also during this phase, the Ca^{2+} transient, along with an increase in plasma membrane DAG content act in a concerted fashion to cause the shift of the cytosolic, Ca^{2+}-insensitive C-kinase to a plasma membrane-associated, Ca^{2+}-sensitive form.

During the sustained response, there is a continued elevation of DAG and hence of PKC associated with the membrane, and a sustained increase in Ca^{2+} influx (cycling) across the plasma membrane. This Ca^{2+} cycling, by increasing $[Ca^{2+}]_{sm}$, stimulates the activity of the PM-associated PKC. As a consequence, a different subset of cellular proteins become phosphorylated and mediate the sustained secretory response.

An additional property of this system is that it displays a type of short-term memory.[19] If cells are treated with AII for 20 min and the AII then removed, the aldosterone secretory response decreases to baseline within 10–15 min. Readdition of the same concentration of AII to these cells leads to a greater aldosterone secretory response than that seen in cells continuously exposed to AII. The basis of this memory and this enhanced response appears to be the sustained association of PKC with the membrane. Upon removal of AII, the aldosterone secretory response declines because of a prompt decrease in Ca^{2+} cycling across the plasma membrane, and thereby a decrease in PKC activity. However, the PKC still remains in its Ca^{2+}-sensitive, membrane-associated form. Upon readdition of AII and a restimulation of Ca^{2+} influx rate, this PKC is immediately stimulated along with the usual CaM-dependent pathway, so that a greater increase in aldosterone secretory rate is observed.[3]

Ca^{2+} Cycling: Plasticity of Messenger Generation and Action

Even though there are distinct spatial and temporal domains of Ca^{2+} messenger function in the case of AII action in the adrenal glomerulosa cell, the pattern of response is quite stereotyped: all the relevant changes in PI, DAG, and Ca^{2+} metabolism occur synchronously in response to hormone-receptor interaction. However, in other cell systems this is not true. A striking example of the plastic nature of basically similar control mechanisms is provided by the interactions of neurotransmitters, enteric hormones, and nutrients in the regulation of insulin secretion from the beta cell of the islets of Langerhans.[20] All the messengers that operate in the glomerulosa cell, Ca^{2+}, Ins(1,4,5)P$_3$, cAMP, and DAG, operate in a highly plastic and integrated manner to regulate insulin secretion.

An understanding of how these different messengers are generated and how they interact can be gained by appreciating the fact that a marked increase in glucose concentration per se acts to generate all these messengers and stimulate insulin secretion. Nevertheless, the usual physiologically observed increase in glucose concentration (4.5 to 6.5 mM) has only small effects on many of these messengers, and only a small effect on insulin secretion unless other agents either act prior to or simultaneously with this small change in glucose concentration. These other agents fall into two classes: those such as gastric inhibitory peptide (GIP) or glucagonlike peptide-1 (GLP-1) that activate adenylate cyclase, and those such as cholecystokinin (CCK-8S) or acetylcholine (Ach) that activate PI-PLC.

In combination, these two classes of agonists mimic many (but not all) of the effects of glucose. Thus, for example, at a low glucose concentration (2.75 mM) CCK-8S will stimulate PIP_2 hydrolysis and GIP will stimulate cAMP production, but neither alone or in combination will stimulate insulin secretion. Hence, glucose has unique effects on beta cell function. Furthermore, glucose exerts these effects through nonreceptor-mediated pathways.

An increase in glucose concentration alone, if great enough (*e.g.*, from 5.0 to 10.0 mM), leads to a biphasic increase in the rate of insulin secretion: an initial rapid burst reaching a peak within 3–5 min, followed by a fall and then a more slowly developing increase in insulin secretory rate to a sustained plateau. This response to glucose is mediated by two intermediates or products of glucose metabolism: DAG and ATP. An increase in ATP (or the ATP/ADP) acts to inhibit K^+ flux through a specific K^+ channel in the plasma membrane. As a consequence, the membrane is depolarized, thereby activating voltage-dependent Ca^{2+} channels. The resulting influx of Ca^{2+} leads to the activation of another type of K^+ channel so that K^+ efflux increases and the membrane repolarizes. This sequence occurs repetitively so that glucose induces repetitive bursts of Ca^{2+} currents lasting 10–15 s that are followed by periods of no currents lasting 10–15 s. During periods of activity there is a net uptake of Ca^{2+} by the cell and during the quiescent periods a loss of Ca^{2+} such that Ca^{2+} homeostasis is maintained: there is a periodic cycling of Ca^{2+} across the plasma membrane. This Ca^{2+} cycling acts as a messenger in activating at least three plasma membrane transducers: adenylate cyclase, PI-PLC, and protein kinase C. Activation of the cyclase via Ca^{2+}-CaM leads to an increase in cAMP. Activation of PI-PLC leads to the production of $Ins(1,4,5)P_3$ and DAG. The increase in $Ins(1,4,5)P_3$ leads to the release of Ca^{2+} from an intracellular pool resulting in a transient rise in cytosolic Ca^{2+} concentration. This Ca^{2+} transient by activating appropriate protein kinases initiates the first phase of insulin secretion. At the same time, there is an increase in DAG content of the plasma membrane caused both by the *de novo* synthesis of DAG from glucose and by the Ca^{2+}-dependent activation of PI-PLC. This rise in DAG along with the Ca^{2+} transient promotes the translocation of PKC to the plasma membrane where it is now regulated by Ca^{2+} cycling, the $[Ca^{2+}]_{sm}$. As a result, a group of proteins becomes phosphorylated, and these phosphoproteins regulate the sustained or second phase of insulin secretion.

The increase in cAMP plays at least two roles: it acts as a positive feedback regulator of Ca^{2+} influx; and it determines the sensitivity of the islets to the messengers generated by PI-PLC action. This latter effect can be best illustrated in islets incubated in 7.0 mM glucose and then exposed to 5 nM CCK-8S with or without GIP. Administration of CCK-8S alone stimulates PI hydrolysis and Ca^{2+} mobilization but has no sustained effect on insulin secretion, *i.e.*, the messengers have been generated but the message is not read. If GIP is given along with CCK-8S, then a biphasic pattern of insulin secretion is seen even though the effect of CCK-8S on PI turnover is less in the presence of GIP than with CCK-8S alone, *i.e.*, cAMP inhibits messenger generation yet enhances the readout of the messages.

Just as adrenal glomerulosa cells display short term memory to a prior exposure to AII, islets display memory to a prior exposure to glucose. If one group of islets is exposed periodically to 10 mM glucose for 30-min periods followed by 20 min at low glucose, and then exposed to 10 mM glucose for a third time, both the first and second phases of insulin secretion are considerably greater than those seen during an initial exposure to 10 mM glucose. As in the glomerulosa cells, so the islets this type of memory also seems to be explained by a continued association of PKC with the plasma membrane after the initial stimulus has ended. Any agent which enhances Ca^{2+} influx will restimulate this membrane-associated PKC and reinitiate insulin secretion.

The normal control of insulin secretion involves an additional type or regulation: the proegeiretic sensitization of the beta cell to the effects of small changes in glucose concentration by acetylcholine and/or CCK-8S. If islets, incubated in media containing

5.5 mM glucose, are exposed to either acetylcholine or CCK-8S, there is no significant increase in insulin secretion even though each agonist stimulates PI-PLC activity and causes the mobilization of Ca^{2+} and the production of DAG (but does not cause an increase in Ca^{2+} influx). However, if after a 5–20-min exposure, the agonist is removed and 15–30 min later, the glucose concentration is raised from 5.5 to 7.5 mM, the resulting change in insulin secretory rate is greater in the pretreated (with acetylcholine or CCK-8S) than in control unprimed islets. The prior exposure to acetylcholine (or CCK-8S) has sensitized the islets to a subsequent challenge with glucose. Again, our present evidence indicates that it is the translocation of PKC to the membrane which accounts for the enhanced responsiveness to glucose. In other words, acetylcholine, presumably by enhancing DAG production and inducing a Ca^{2+} transient causes the association of PKC with the membrane, but because Ca^{2+} influx rate is not enhanced, this PKC is not active, and little or no secretion of insulin occurs. Nonetheless, the PKC remains associated with the membrane for a period of time after the acetylcholine stimulus is terminated so that a subsequent increase in glucose concentration by increasing Ca^{2+} influx rate activates PKC and insulin secretion.

CONCLUSION

Studies in these endocrine systems illustrate several points. First, Ca^{2+} cycling across the plasma membrane is an important and often the critical mechanism for generating a specific type of Ca^{2+} signal, an increase in $[Ca^{2+}]_{sm}$. Second, this type of Ca^{2+} signal is often of primary importance in regulating long-term cellular responses. Third, in order for this type of Ca^{2+} signal to be effective, the cell must be in a state in which an appropriate Ca^{2+}-sensitive transducer(s) is associated with the plasma membrane. Fourth, in many systems in which the hydrolysis of polyphosphoinositides is linked to changes in Ca^{2+} metabolism in the regulation of the cellular response, a significant aspect of this response is the association of a new Ca^{2+}-sensitive enzyme, PKC, with the plasma membrane such that a two-component plasma membrane transducer is generated consisting of a membrane-associated PKC and increased Ca^{2+} cycling across the plasma membrane. Both components are necessary during a sustained cellular response. An enhancement of Ca^{2+} influx rate alone is ineffective in inducing the response, and an association of PKC with the membrane without the increase in Ca^{2+} influx rate is also ineffective. Fifth, the persistent association of PKC with the membrane, after termination of a particular stimulus, can serve as the pathway by which a type of short-term memory can be achieved. Sixth, the proegeiretic association of PKC with the membrane lasting beyond an initial stimulus, which by itself does not evoke a response, is a mechanism by which an enhanced cellular responsiveness to a subsequent heterologous stimulus can be achieved.

REFERENCES

1. RASMUSSEN, H. 1986. The calcium messenger system. N. Engl. J. Med. **314:** 1094–1101; 1164–1170.
2. RASMUSSEN, H. 1981. Calcium and cAMP as synarchic messengers. John Wiley. New York, NY.
3. ALKON, D. L. & H. RASMUSSEN, 1988. A spatial-temporal model of cell activation. Science **239:** 998–1005.
4. CAMPBELL, A. K. 1983. Intracellular calcium: Its universal role as regulator. John Wiley. New York, NY.

5. SUTHERLAND, E. W., Jr. 1961. The biological role of adenosine-3′,5′-phosphate. Harvey Lect. **57:** 12–33.
6. RASMUSSEN, H., D. B. P. GOODMAN & A. TENENHOUSE. 1972. The role of cyclic AMP and calcium in cell activation. CRC Crit. Rev. Biochem. **2:** 95–148.
7. CHEUNG, W.Y. 1980. Calmodulin plays a pivotal role in cellular regulation. Science **207:** 19–27.
8. LIMBIRD, L. L. 1988. Receptors linked to inhibition of adenylate cyclase: Additional signalling mechanisms. FASEB J. **2:** 2686–2695.
9. YATANI, A., Y. IMOTO, J. CODINA, S. L. HAMILTON, A. M. BROWN & L. BIRNBAUMER. 1988. The stimulatory G protein of adenylyl cyclase, G_s, also stimulates dihydropyridine-sensitive Ca^{2+} channels. J. Biol. Chem. **263:** 9887–9895.
10. RASMUSSEN, H., I. KOJIMA, K. KOJIMA, W. ZAWALICH & W. APFELDORF. 1984. Calcium as intracellular messenger: Sensitivity modulation, C-kinase pathway, and sustained cellular response. Adv. Cyclic Nucleotide Res. Prot. Phos. Res. **18:** 159–193.
11. NICHOLLS, D. G. 1986. Intracellular calcium homeostasis. Br. Med. Bull. **42:** 353–358.
12. NISHIMOTO, I., Y. OHKUNI, E. OGATA & I. KOJIMA. 1987. Insulin-like growth factor II increases cytoplasmic free calcium in competent Balb/c 3T3 cells treated with epidermal growth factor. Biochem. Biophys. Res. Commun. **142:** 275–286.
13. VINCENZI, F. F., T. R. HINDS & B. U. RAESS. 1980. Calmodulin and the plasma membrane calcium pump. N.Y. Acad. Sci. **356:** 232–244.
14. SMALLWOOD, J. I., B. GÜGI & H. RASMUSSEN. 1988. Regulation of erythrocyte Ca^{2+} pump activity by protein kinase C. J. Biol. Chem. **263:** 2195–2202.
15. CAROFOLI, E. 1987. Intracellular calcium homeostasis. Annu. Rev. Biochem. **56:** 395–433.
16. RASMUSSEN, H. & P. Q. BARRETT. 1984. Calcium messenger system: An integrated view. Physiol. Rev. **64:** 938–984.
17. RASMUSSEN, H., P. BARRETT, W. APFELDORF, Y. TAKUWA, N. TAKUWA, J. SMALLWOOD, C. ISALES, W. BOLLAG & P. STEIN. 1987. Multiplicity of Ca^{2+} signal transduction pathways. In Calcium-Dependent Processes in the Liver. E. C. Heilmann, Ed.: 9–29. MTP Press. Lancester.
18. CATT, K. J., T. BALLA, A. J. BAUKAL, W. P. HAUSDORFF & G. AQUILERA. 1988. Control of glomerulosa cell function by angiotensin II: Transduction by G-proteins and inositol polyphosphates. Clin. Exp. Pharmacol. Physiol. **15:** 501–515.
19. BARRETT, P. Q., I. KOJIMA, K. KOJIMA, K. ZAWALICK, C. M. ISALES & H. RASMUSSEN. 1986. Short term memory in the calcium messenger system. Biochem. J. **238:** 905–912.
20. RASMUSSEN, H. & W. ZAWALICH. 1988. Control of insulin secretion: Its elegance and complexity. Submitted.

Regulation of Calcium in Isolated Nerve Terminals (Synaptosomes): Relationship to Neurotransmitter Release[a]

DAVID G. NICHOLLS

Department of Biochemistry
University of Dundee
Dundee DD1 4HN, Scotland, U.K.

The synaptosome is the simplest system in which the events from plasma membrane depolarization to transmitter release can be observed, and at the same time the most complex neuronal preparation which is sufficiently homogenious for detailed biochemical analysis of these events, since it largely avoids contributions from the cell body or from glial cells inherent in cell culture or slice preparations respectively. This paper will discuss the regulation of Ca^{2+} in the nerve terminal, the Ca^{2+}-dependent release of the excitatory neurotransmitter glutamate, and possible loci for the modulation of the coupling between depolarization and elevation in cytosolic free Ca^{2+} concentration ($[Ca^{2+}]_c$), and between increased $[Ca^{2+}]_c$ and glutamate release. Results from our own laboratory refer to synaptosomes prepared from the cerebral cortices of Dunkin-Hartley strain guinea pigs incubated at either 30°C or 37°C.[1]

The Maintenance of Resting $[Ca^{2+}]_c$ in Synaptosomes

Synaptosomes loaded with either quin-2 or fura-2 as indicators of Ca^{2+} maintain a resting $[Ca^{2+}]_c$ of 0.1–0.3 μM when suspended in media containing millimolar Ca^{2+}.[2–6] There is a continuous inward leakage of Ca^{2+} into the terminals of 0.3 nmol/min^{-1}/mg^{-1} at 30°C,[7] which is counteracted by an equivalent efflux. This continual cycling of Ca^{2+} across the plasma membrane is an important feature for the regulation of $[Ca^{2+}]_c$. In the absence of an inward leak, the efflux mechanism would lower $[Ca^{2+}]_c$ until thermodynamic equilibrium would be attained between the energy input for the efflux process and the Ca^{2+}-electrochemical gradient across the membrane. Such a single pathway is unsuitable for two reasons: firstly approach to equilibrium would be asymptotically slow, and secondly any alteration in the energy available for the efflux pathway would cause wild fluctuations in the Ca^{2+} gradient. In contrast, a dynamic balance between a pump and a leak allows for the precise regulation of $[Ca^{2+}]_c$ at the level where the pump can exactly balance the inward leak.

Two mechanisms co-exist in the plasma membrane which are in theory capable of expelling Ca^{2+}, a Ca^{2+}-ATPase[8,9] and 3 Na^+/Ca^{2+} exchange.[10–12] We favour a major role for the former for the following reasons:

1. ATP depletion initiates an immediate linear increase in $[Ca^{2+}]_c$.[13]
2. Synaptosomes can still maintain submicromolar $[Ca^{2+}]_c$ in low Na^+-media[4] or in

[a]Work in the author's laboratory is supported by Merck, Sharp and Dohme Research Laboratories, Harlow, U.K., by the Wellcome Trust and by the British Medical Research Council.

81

the presence of concentrations of the Na^+-channel activator veratridine sufficient to collapse the Na^+-electrochemical gradient across the plasma membrane.[6]

3. A 3 Na^+/Ca^{2+} exchanger has barely enough thermodynamic potential to maintain the resting Ca^{2+}-electrochemical gradient of some 36 kJ/mole Ca^{2+},[14] whereas a 1 Ca^{2+} per ATP plasma membrane pump has at its disposal an adenine nucleotide phosphorylation potential which, by analogy to other cells, should be in the range 50–60 kJ/mole. This means that the Ca^{2+}-ATPase is the more powerful mechanism, and raises the possibility that the 3 Na^+/Ca^{2+} exchanger may be more involved with the inward flux of Ca^{2+} under certain circumstances, as will be discussed below.

4. The activity of the Ca^{2+}-ATPase is highly dependent on $[Ca^{2+}]_c$, both because of the normal Michaelis-Menton saturation of the Ca^{2+} binding site on the cytoplasmic face, and because of the Ca^{2+}-dependent interaction of calmodulin with the

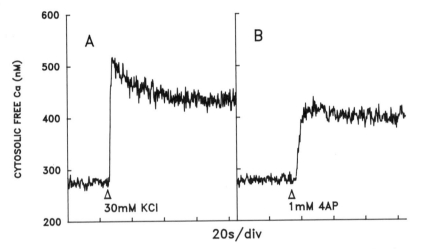

FIGURE 1. Cytosolic free Ca^{2+} within guinea pig cerebrocortical synaptosomes following plasma membrane depolarization by either 30 mM KCl (**A**) or 1 mM 4-aminopyridine (**B**). $[Ca^{2+}]_c$ was determined by fura-2 with sampling at 200 ms. Traces are the means of three experiments. For details see Kauppinen *et al.*, 1988.[13]

enzyme.[15] Any fluctuation in $[Ca^{2+}]_c$ will thus cause a steep change in the activity of the enzyme, aiding the restoration of the "set-point" where uptake and efflux of Ca^{2+} are equal and opposite.

It should be noted that, due to their finite capacity to store Ca^{2+}, internal organelles such as mitochondria cannot influence the *steady-state* Ca^{2+} in the resting terminal.

$[Ca^{2+}]_c$ Following Plasma Membrane Depolarization

When synaptosomes are depolarized by elevated KCl, there is an extremely rapid initial uptake of Ca^{2+} into the terminal within the first second, which then declines to a slower rate still substantially above the basal leak.[16,17] The initial transient produces a spike in $[Ca^{2+}]_c$ which then declines to a stable elevated plateau (FIG. 1A).

The decline from the peak to the plateau is not due to the net extrusion of Ca^{2+} from the terminal, but rather to the sequestration of Ca^{2+} within intrasynaptosomal organelles. Thus when mitochondria are rapidly isolated from KCl-depolarized synaptosomes, their content of Ca^{2+} is found to have increased 300% above controls.[18] The role of intraterminal mitochondria in the regulation of $[Ca^{2+}]_c$ therefore appears to be as a temporary sink for the Ca^{2+} taken up during depolarization. As in the case of the plasma membrane, mitochondria cycle Ca^{2+} continuously between uptake and efflux pathways (reviewed in REF. 19). The uptake pathway is highly dependent on $[Ca^{2+}]_c$, and the mitochondria will rapidly accumulate Ca^{2+} from the cytosol when $[Ca^{2+}]_c$ rises above the mitochondrial set-point where uptake and efflux balance. Under physiological conditions the set-point is in the range $0.5-1$ μM, *i.e.*, at the upper end of the physiological range. Mitochondria in nerve terminals (and indeed other cells) may play the major role in limiting the rise in $[Ca^{2+}]_c$ to nonpathological levels, and in providing a temporary sink within the terminal for the Ca^{2+} which enters to trigger exocytosis. Since the set-point for the plasma membrane is lower than for the mitochondrion, during periods of electrical inactivity the plasma membrane will lower $[Ca^{2+}]_c$ sufficiently to deplete the mitochondrial Ca^{2+}.

The Nature of the Depolarization-Dependent Ca^{2+} Entry

Ca^{2+} channels in neuronal cell bodies have been classified as L ("long-lasting"), T ("transient") or N ("neuronal")-type on the basis of their electrophysiology and pharmacology.[20] The extent to which this classification can be applied to the nerve terminal is unclear. Depolarization-dependent Ca^{2+} entry into synaptosomes is extremely insensitive to verapamil and dihydropyridines,[21] diagnostic inhibitors of L-type channels. The snail toxin Ω-conotoxin, which potently inhibits N-type channels in neuronal cell bodies has little effect on synaptosomes; in our hands no inhibition of the $[Ca^{2+}]_c$ increase can be seen even at 1 μM toxin (unpublished).

There appears to be an intimate relationship between the Ca^{2+}-channels which trigger exocytosis and the exocytotic sites themselves, such that Ca^{2+} binding site for the exocytotic apparatus may be located close to the mouth of the Ca^{2+}-channel (reviewed in REF. 22). This has two substantial kinetic advantages. Firstly, it means that the Ca^{2+} binding site can have a relatively low affinity for Ca^{2+}, since it will see the high concentration of Ca^{2+} emerging from the channel; and a low affinity implies a high off-rate when $[Ca^{2+}]_c$ falls. Secondly, as soon as the Ca^{2+} channel closes, $[Ca^{2+}]_c$ in the region of the binding site will fall as the spatial gradient collapses and the Ca^{2+} diffuses out into the cytosol. Thus the next exocytotic event is not dependent on the kinetics of sequestration of cytosolic Ca^{2+} by, for example, the mitochondria. In view of this special relationship it may not be surprising that synaptosomal Ca^{2+} channels do not readily fall into the categories which can be observed in cell bodies.

Despite the uncertain pharmacology, the initial transient Ca^{2+} entry on KCl depolarization has the appearance of resulting from a rapidly inactivating Ca^{2+} channel. Inactivation is a function of the time of depolarization rather than the entry of Ca^{2+} since the rapid phase of influx and the spike in $[Ca^{2+}]_c$ (but not the plateau elevation) disappear if the synaptosomes are pre-depolarized for about 10 s prior to addition of Ca^{2+}. The nature of the plateau in $[Ca^{2+}]_c$ is controversial. A stable increase in $[Ca^{2+}]_c$, without an initial spike, may be induced by low concentrations of the Na^+-channel activator veratridine,[3,6] by the K^+-channel blocker 4-aminopyridine (FIG. 1B), and by the Ca^{2+}-ionophore ionomycin. Veratridine inhibits the closure of voltage-dependent Na^+-channels; since Na^+-channels will only occasionally open in polarized synaptosomes the onset of depolarization should be much slower than for high KCl. Consistent with this, $[Ca^{2+}]_c$ takes >10 s to increase to a plateau after veratridine, as opposed to KCl, when $[Ca^{2+}]_c$ spikes within 1 s (FIG. 1A). The absence of a transient spike in the case of veratridine can thus be ascribed to inactivation of the transient channel during the slow depolarization.

A similar kinetic is observed with the K^+-channel inhibitor 4-aminopyridine (FIG. 1B). A surprising feature of the 4-aminopyridine increase in $[Ca^{2+}]_c$ is that it is almost completely sensitive to the Na^+-channel inhibitor tetrodotoxin (unpublished observation). This indicates that 4-aminopyridine is inducing the synaptosome to fire repetitive action potentials, and we suggest that this in turn is due to an increased statistical fluctuation in the membrane potential of these tiny (1 μm diameter) organelles when the voltage-clamping effect of the dominant K^+-conductance is attenuated. In individual synaptosomes, when the membrane potential "noise" encompasses both the threshold for the firing of the Na^+-channel and a sufficiently high potential for reactivation of the channel, spontaneous repetitive firing will result.

Both in the case of veratridine and 4-aminopyridine it is possible that a reversed 3 Na^+/Ca^{2+} exchange (*i.e.,* in the direction of Ca^{2+} entry) contributes to the elevation in $[Ca^{2+}]_c$, as has been suggested for the plateau following KCl depolarization.[23,24] However for 4-aminopyridine this is difficult to test, since Na^+ would be required both for Na^+/Ca^{2+} exchange and for the Na^+-channel effect discussed above.

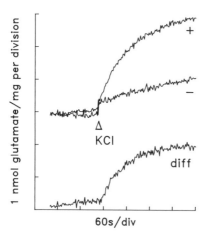

FIGURE 2. Release of glutamate from guinea pig cerebrocortical synaptosomes following plasma membrane depolarization by 30 mM KCl in the presence ($+$) or absence ($-$) of 1.3 mM $CaCl_2$. Traces were obtained by continuous fluorometric assay of released glutamate.[27] The bottom trace is the computed difference between the releases in the presence and absence of Ca^{2+} and represents the exocytotic release of the transmitter.

The Coupling of $[Ca^{2+}]_c$ to Glutamate Exocytosis

The release mechanism for amino acid neurotransmitters is contentious and little molecular information is known (reviewed in REF. 25). In 1986 we described a technique for the continuous assay of glutamate released from isolated nerve terminals in suspension by coupling release to NADPH fluorescence.[26] This has subsequently enabled us to investigate the kinetics, mechanism and regulation of release in considerable detail.[13,27–33] The mixed population of cerebrocortical synaptosomes release 8–10% of the total glutamate (which includes metabolic glutamate in both glutamatergic and non-glutamatergic terminals) in a Ca^{2+}- and energy-dependent manner.[13,27,34] Ca^{2+}-dependent release occurs from a slowly exchanging noncytosolic pool, and has therefore the characteristics of an exocytotic process.[35] FIGURE 2 shows that the phasic, Ca^{2+}-dependent release of glutamate may be readily distinguished from the Ca^{2+}-independent release of glutamate from the cytosol, which occurs by reversal of the electrogenic acidic amino acid carrier on depolarization of the plasma membrane. Glutamate exocytosis may be inhibited by botulinum neurotoxin type A.[29]

By monitoring fura-2 and NADPH fluorescence in parallel experiments it is possible to investigate the relationship between $[Ca^{2+}]_c$ and glutamate release. With the depolarization induced by 30 mM KCl, both the initial spike in $[Ca^{2+}]_c$ (occurring in the first second after depolarization) and the subsequent plateau (FIG. 1A) are coupled to glutamate exocytosis (note the extended time-course in FIG. 2). Thus the extent of Ca^{2+}-dependent release is much greater when Ca^{2+} is present throughout the KCl depolarization, compared with the release when KCl is followed after 2–5 s by EGTA to chelate external Ca^{2+} and inhibit further Ca^{2+} entry (unpublished). Similarly, glutamate is released under conditions where the Ca^{2+} spike is absent (e.g., Ca^{2+} addition to pre-depolarized synaptosomes, veratridine addition or 4-aminopyridine addition (FIG. 1b)). This argues that the Ca^{2+} entering during the peak and the plateau has equal access to the exocytotic trigger. It has been suggested that the plateau phase may be due to Na^+/Ca^{2+} exchange;[23,24] however, if this were so, it would imply that these carriers were also selectively located in extremely close proximity to the exocytotic sites. It can be shown with our preparation that when $[Ca^{2+}]_c$ is increased by mechanisms which are manifestly unrelated to voltage-dependent Ca^{2+}-channels (i.e., release of mitochondrial Ca^{2+}[28,34] and addition of Ca^{2+}-ionophore (unpublished) the increase in $[Ca^{2+}]_c$ is coupled to glutamate release with much diminished efficiency.

When a partial depolarization is performed (e.g., by 10 mM KCl), the most noticeable effect on glutamate release is that it is the final extent, rather than the rate, of the Ca^{2+}-dependent release which is diminished.[33] It appears therefore that a given elevation in $[Ca^{2+}]_c$ predestines a set proportion of synaptic vesicles for release during an indefinitely extended depolarization.

The Modulation of Glutamate Exocytosis

Glutamatergic synaptosomes are unaffected by N-methyl-D-aspartate, suggesting the absence of functional presynaptic autoreceptors of the NMDA-type.[33] Kainate, in the concentration range 0.1–1 mM, is known to inhibit the Na^+-coupled acidic amino acid carrier in the plasma membrane responsible for maintaining the >10,000-fold concentration gradient of glutamate across the plasma membrane.[36] As a result, glutamate slowly leaks out of the synaptosomes by a Ca^{2+}-independent mechanism.[30] However this does not influence the coupling of depolarization to the Ca^{2+}-dependent release of transmitter glutamate.[30]

Quisqualate and a range of quisqualate agonists added at 100 μM induce the slow release of glutamate from synaptosomes.[33] This release is independent of external Ca^{2+} and is not associated with plasma membrane depolarization. Unlike kainate, the net release of glutamate caused by quisqualate is not due to inhibition of reuptake by the acidic amino acid carrier, and the exact mechanism remains to be established.

Glutamate, and other substrates for the plasma membrane acidic amino acid carrier, cause a novel form of facilitation of glutamate release from guinea pig synaptosomes. The acidic amino acid carrier is so active in glutamatergic terminals that the electrogenic reuptake of glutamate together with Na^+ can cause a depolarization of the plasma membrane which can be detected even in the mixed cerebral cortical preparation.[33] This depolarization can be sufficient to activate voltage-dependent Ca^{2+}-channels and induce the exocytotic release of transmitter from the synaptosomes.[33] The implications are that during periods of synaptic activity, when glutamate would accumulate in the synaptic cleft, the reuptake of the transmitter into the terminal would cause a partial depolarization of the terminal, causing a short-term facilitation of further transmitter release.

Glutamate Release during Energy Deprivation

When energy production is impaired in the brain, as a result of anoxia, hypoglycaemia or ischaemia, there is a massive release of glutamate into the extracellular milieu. The energetic consequences of these conditions can be mimicked in isolated nerve terminals by inhibiting respectively respiration, glycolysis or both means of ATP production. Three major requirements for ATP in the nerve terminal are to maintain the Na^+-electrochemical gradient through the operation of the $Na^+ + K^+$-ATPase, to maintain a low $[Ca^{2+}]_c$ through the Ca^{2+}-ATPase and for the exocytotic mechanism itself. Depending on the temporal sequence of these events, the observed release of glutamate could be due to an elevation of $[Ca^{2+}]_c$ triggering exocytotic release, or a blockade of exocytosis due to the ATP drop, followed by a Ca^{2+}-independent release of glutamate from the cytosol due to the decay in the Na^+-electrochemical potential allowing the acidic amino acid carrier to reverse. We find that glutamate exocytosis has a very high requirement for ATP; even slight falls in ATP/ADP ratio drastically decrease the extent of release.[13] In contrast, $[Ca^{2+}]_c$ rises only relatively slowly on energy deprivation. Thus exocytosis of glutamate is inhibited by energy lack before $[Ca^{2+}]_c$ rises to levels which would induce exocytosis. Instead, there is a steady increase in the rate of release from the cytosol. The conclusion is therefore that the glutamate released during energy deprivation originates from the cytosol rather than from transmitter stores.[13]

CONCLUSIONS

It is only when the full complexity of the coupling between the action potential and transmitter release is understood that meaningful hypotheses can be formulated concerning the way in which this relationship might be altered in disease states and in aging. As the main excitatory transmitter in the CNS, glutamate is playing an increasing role in such hypotheses. However, while molecular genetic and patch-clamping techniques have allowed postsynaptic events to be defined in exquisite detail, our knowledge of presynaptic mechanisms is, by comparison, primitive. The central unknown is the mechanism of exocytosis itself, but also the nature of the Ca^{2+}-channels in the terminal, which are not readily accessible to electrophysiological techniques, and the mechanism by which vesicles are selected for release requires much further work.

REFERENCES

1. NICHOLLS, D. G. 1978. Calcium transport and proton electrochemical potential gradient in mitochondria from guinea-pig cerebral cortex and rat heart. Biochem. J. **170:** 511–522.
2. RICHARDS, C. D., J. METCALFE & T. R. HESKITH. 1984. Changes in free Ca^{2+} levels and pH in synaptosomes during transmitter release. Biochim. Biophys. Acta **803:** 215–220.
3. HANSFORD, R. G. & F. CASTRO. 1985. Role of Ca^{2+} in pyruvate dehydrogenase interconversion in brain mitochondria and synaptosomes. Biochem. J. **227:** 129–136.
4. NACHSHEN, D. A. 1985. Regulation of cytosolic calcium concentrations in presynaptic nerve endings isolated from rat brain. J. Physiol. (London) **363:** 87–101.
5. ASHLEY, R. H. 1986. External calcium, intrasynaptosomal free calcium and neurotransmitter release. Biochim. Biophys. Acta **854:** 207–212.
6. ADAM-VISI, V. & R. H. ASHLEY. 1987. Relation of acetylcholine release to Ca^{2+} uptake and intra-terminal Ca^{2+} concentration in guinea-pig cortex synaptosomes. J. Neurochem. **49:** 1013–1021.
7. SNELLING, R. & D. G. NICHOLLS. 1985. Calcium efflux and cycling across the synaptosomal plasma membrane. Biochem. J. **226:** 225–231.

8. GILL, D. L., E. F. GROLLMAN & L. D. KOHN. 1981. Calcium transport mechanism in membrane vesicles from guinea-pig synaptosomes. J. Biol. Chem. **256:** 184–192.

9. MICHAELIS, E. K., M. L. MICHAELIS, H. H. CHANG & T. E. KITOS. 1983. High-affinity Ca^{2+}-stimulated Mg^{2+}-ATPase in rat brain synaptosomes, synaptic membranes and microsomes. J. Biol. Chem. **258:** 6106–6114.

10. SWANSON, P. D., K. ANDERSON & W. L. STAHL. 1974. Uptake of Ca^{2+} ions by synaptosomes from rat brain. Biochim. Biophys. Acta **356:** 174–183.

11. BLAUSTEIN, M. P. & C. J. OBORN. 1975. The influence of sodium on Ca^{2+} fluxes in pinched-off nerve terminals in vitro. J. Physiol. (London) **247:** 657–686.

12. BLAUSTEIN, M. P. & A. C. ECTOR. 1976. Carrier-mediated Na^+-dependent and Ca^{2+}-dependent Ca^{2+} efflux from pinched-off presynaptic nerve terminals (synaptosomes) in vitro. Biochim. Biophys. Acta **419:** 295–308.

13. KAUPPINEN, R. A., H. MCMAHON & D. G. NICHOLLS. 1988. Ca^{2+} dependent and Ca^{2+}-independent glutamate release, energy status and cytosolic free Ca^{2+} concentration in isolated nerve terminals following in vitro hypoglycaemia and anoxia. Neuroscience **27:**175–182.

14. ÅKERMAN, K. E. O. & D. G. NICHOLLS. 1981. Ca transport by intact synaptosomes: Influence of ionophore A23187 on plasma membrane potential, plasma membrane Ca transport, mitochondrial membrane potential, cytosolic free Ca concentration and noradrenaline release. Eur. J. Biochem. **115:** 67–73.

15. CARAFOLI, E. 1987. Intracellular calcium homeostasis. Annu. Rev. Biochem. **56:** 395–433.

16. DRAPEAU, P. & M. P. BLAUSTEIN. 1983. Initial release of ^3H-dopamine from rat striatal synaptosomes: Correlation with calcium entry. J. Neurosci. **3:** 703–713.

17. NACHSHEN, D. A. 1985. The early time-course of potassium stimulated calcium uptake in presynaptic nerve terminals isolated from rat brain. J. Physiol. (London) **361:** 251–268.

18. ÅKERMAN, K. E. O. & D. G. NICHOLLS. 1981. Intra-synaptosomal compartmentation of Ca during depolarization-induced Ca uptake across the plasma membrane. Biochim. Biophys. Acta **645:** 41–48.

19. NICHOLLS, D. G. & K. E. O. ÅKERMAN. 1982. Mitochondrial Ca transport. Biochim. Biophys. Acta **683:** 57–88.

20. FOX, A. P., M. C. NOWYCKY & R. W. TSIEN. 1987. Kinetic and pharmacological properties distinguishing three types of calcium currents in chick sensory neurones. J. Physiol. (London) **394:** 149–172.

21. SUSZKIW, J. B., M. E. O'LEARY, M. M. MURAWSKY & T. WANG. 1986. Presynaptic Ca^{2+} channels in rat cortical synaptosomes: Fast-kinetics of phasic Ca^{2+} influx, channel inactivation and relationships to nitrendipine receptors. J. Neurosci. **6:** 1349–1357.

22. SMITH, S. J. & G. J. AUGUSTINE. 1988. Calcium ions, active zones as synaptic transmitter release. Trends Neurosci. **11:** 458–464.

23. SUSZKIW, J. B. 1988. Properties of presynaptic voltage-sensitive calcium channels in rat synaptosomes. *In* Cellular and Molecular Basis of Transmission. H. Zimmermann, Ed. Vol. 21: 285–291. NATO ASI Series. Springer, Berlin.

24. CARVALHO, A. P., M. S. SANTOS, A. O. HENRIQUES, P. TAVARES & C. M. CARVALHO. 1988. Calcium channels and Na^+/Ca^{2+} exchange in synaptosomes. *In* Cellular and Molecular Basis of Transmission. H. Zimmermann, Ed. Vol. 21: 263–284. NATO ASI Series. Springer, Berlin.

25. NICHOLLS, D.G. 1989. The release of glutamate, aspartate and GABA from isolated nerve terminals. J. Neurochem. **52:** 331–341.

26. NICHOLLS, D. G. & T. S. SIHRA. 1986. Synaptosomes possess an exocytotic pool of glutamate. Nature (London) **321:** 772–773.

27. NICHOLLS, D. G., T. S. SIHRA & J. SANCHEZ-PRIETO. 1987. Calcium dependent and independent release of glutamate from synaptosomes monitored by continuous fluorometry. J. Neurochem. **49:** 50–57.

28. SANCHEZ-PRIETO, J., T. S. SIHRA & D. G. NICHOLLS. 1987. Characterization of the exocytotic release of glutamate from guinea pig cerebral cortical synaptosomes. J. Neurochem. **49:** 58–64.

29. SANCHEZ-PRIETO, J., T. S. SIHRA, D. EVANS, A. ASHTON, J. O. DOLLY & D. G. NICHOLLS. 1987. Botulinum toxin A blocks glutamate exocytosis from guinea pig cerebral cortical synaptosomes. Eur. J. Biochem. **165:** 675–681.

30. POCOCK, J. M., H. MURPHIE & D. G. NICHOLLS. 1988. Kainic acid inhibits the synaptosomal plasma membrane glutamate carrier and allows glutamate leakage from the cytoplasm, but does not affect glutamate exocytosis. J. Neurochem. **50:** 745–751.
31. DIAZ-GUERRA, M. J. M., J. SANCHEZ-PRIETO, L. BOSCA, J. POCOCK, A. BARRIE & D. G. NICHOLLS. 1988. Phorbol ester translocation of protein kinase C in guinea-pig synaptosomes and the potentiation of calcium-dependent glutamate release. Biochim. Biophys. Acta **970:** 157–165.
32. TIBBS, G., J. O. DOLLY & D. G. NICHOLLS. 1989. Dendrotoxin, 4-aminopyridine and β-bungarotoxin act at common loci but by two distinct mechanisms to induce calcium-dependent release of glutamate from guinea-pig cerebral cortical synaptosomes. J. Neurochem. **52:** 201–206.
33. MCMAHON, H.T., A. P. BARRIE, M. LOWE & D. G. NICHOLLS. 1989. Glutamate release from guinea-pig synaptosomes: Stimulation by reuptake-induced depolarization: J. Neurochem. **53:** 71–79.
34. NICHOLLS, D. G., T. S. SIHRA & J. SANCHEZ-PRIETO. 1987. The role of plasma membrane and intra-cellular organelles in synaptosomal calcium regulation. In Cell Calcium and the Control of Membrane Transport. L. J. Mandel & D. C. Eaton, Eds.: 31–44. Soc. Gen. Physiol. Series, Vol. 42. Rockefeller University Press, New York, NY.
35. WILKINSON, R. & D. G. NICHOLLS. 1989. Compartmentation of glutamate and aspartate within cerebral cortical synaptosomes: Evidence for a non-cytoplasmic origin for the Ca^{2+}-releasable pool of glutamate. Neurochem. Int. In press.
36. JOHNSTON, G. A. R., S. M. E. KENNEDY & B. TWITCHEN. 1979. Action of the neurotoxin kainic acid on high affinity uptake of L-glutamic acid in rat brain slices. J. Neurochem. **32:** 121–127.

Ca^{2+} Handling Systems and Neuronal Aging[a]

MARY L. MICHAELIS

Department of Pharmacology/Toxicology
University of Kansas
Lawrence, Kansas 66045

The idea that some alterations in the cellular disposition of Ca^{2+} occur as a function of aging has been around for some time.[1,2] Certainly the early observations that inadequate control of free intracellular Ca^{2+} levels leads to cell injury and cell death has received substantial experimental support in studies involving metabolic stress with liver, heart, and nerve cells, to name a few well-known examples.[3-6] However, in trying to determine whether the aging process leads to compromised Ca^{2+} regulation capable of producing cellular dysfunction and cell death, we are not likely to see toxicity of the magnitude observed with ischemia or anoxia. Instead, we are likely to be looking for very subtle changes in the ability of aging cells to respond to normal stimuli. In fact the changes may be so subtle that they manifest themselves clearly only when cells are severely stressed by events such as anoxia or excessive stimulation. Presumably the consequences of such subtle aging-induced changes in the brain for the organism as a whole are simply marginal decreases in the efficiency of neuronal processing and the subsequent behavioral output. Consideration of the multiplicity of roles that Ca^{2+} plays in neurotransmission and intracellular signalling suggests that it is quite conceivable that minor changes in Ca^{2+}-regulating processes in neurons could contribute significantly to the cognitive and behavioral changes frequently associated with the aging process.

As Gibson and Peterson[2] document so well in their extensive review, there is a very significant body of experimental literature suggesting that the aging process affects neuronal Ca^{2+} regulation, but the mechanisms underlying the reported observations are for the most part unknown. The incredible complexity of the overall cellular regulation of Ca^{2+} disposition presents a formidable task for those interested in identifying the precise nature of alterations in Ca^{2+} handling systems that may underlie some changes in neuronal and, ultimately, in cognitive and behavioral performance. A definitive characterization of age-dependent changes in neuronal Ca^{2+} regulation will first require a detailed knowledge of the properties of the systems that participate in the influx, binding, sequestration, intracellular release, and extrusion of Ca^{2+}.

FIGURE 1 presents a schematic representation of what are believed to be the major systems responsible for maintaining Ca^{2+} homeostasis in nerve terminals, but the same processes are likely to exist throughout the entire neuron and in other cells as well. The resting free intracellular Ca^{2+} [Ca^{2+}]$_i$ appears to be controlled primarily by sequestration into organelles such as the smooth endoplasmic reticulum (SER) or perhaps into the recently described "calciosomes."[7] Small elevations in [Ca^{2+}]$_i$ are likely to be reversed by Ca^{2+} extrusion via the (Ca^{2+} + Mg^{2+})-ATPase. The large increases in [Ca^{2+}]$_i$ which occur upon stimulation of the cells require several mechanisms to restore prestimulation levels or to signal the establishment of a new steady state for Ca^{2+}.[8] Under

[a]The author's work cited herein was supported by Grant AG 04762 from the National Institutes of Health.

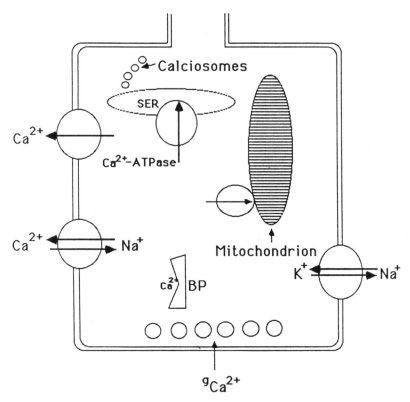

FIGURE 1. Ca^{2+} regulating systems in nerve terminals. This diagram depicts our current state of knowledge about the systems that participate in controlling free intraneuronal Ca^{2+} concentrations in the 10^{-7} M range against an extracellular concentration of 10^{-3} M. When Ca^{2+} channel conductance is increased by depolarization, Ca^{2+} rapidly enters the nerve terminal due to its large electrochemical gradient. The Ca^{2+} signal decays rapidly due to the binding of the ion to high-affinity binding proteins and presumably to rapid sequestration within intracellular organelles such as the SER and possibly the "calciosomes." The Ca^{2+} extrusion systems in the plasma membrane ultimately remove the excess Ca^{2+} from the intracellular compartment. The $(Ca^{2+} + Mg^{2+})$-ATPase hydrolyzes ATP to provide the energy needed for Ca^{2+} extrusion. The Na^+ -Ca^{2+} antiporter utilizes the Na^+ gradient provided by the activity of the Na^+ -K^+ ATPase as a driving force for countertransport of Ca^{2+}. The mitochondria are believed to participate in significant uptake of Ca^{2+} primarily under pathological conditions.

conditions of stimulation both the Na^+ -Ca^{2+} antiporter and the calmodulin-activated $(Ca^{2+} + Mg^{2+})$-ATPase in the plasma membrane are likely to increase their activity. Events involved in the release and re-sequestration of cytosolic Ca^{2+} either in the SER or calciosomes are not yet fully understood, but certainly represent a crucial aspect of both Ca^{2+} regulation and intracellular signalling.[8] Changes occurring in neurons as a function of aging could affect the efficiency, the density, or the molecular nature of any or all of these regulatory systems, leading to a cascade of subtle, and sometimes not so subtle, alterations in intercellular communication within the nervous system.

The major focus of our work is the characterization of Ca^{2+} transporting proteins present in neuronal plasma membranes. Since synaptic terminals are a major site for Ca^{2+}

entry with depolarization and preparations highly enriched in nerve terminals (synaptosomes) can be easily obtained from brain, we have used the plasma membranes isolated from such synaptosome preparations to study the properties of the two major proteins involved in Ca^{2+} extrusion, the (Ca^{2+} + Mg^{2+})-ATPase and the Na$^+$ -Ca^{2+} exchanger. Even though these systems are crucial for the ultimate step in Ca^{2+} regulation, *i.e.*, its extrusion from the cell, very little information is available about the molecular characteristics of these entities and the exact contribution they make to overall Ca^{2+} homeostasis in neurons. Earlier reports from our laboratory have described some of the major properties of the synaptic plasma membrane Na$^+$ -Ca^{2+} antiporter[9,10] and the (Ca^{2+} + Mg^{2+})-ATPase and ATP-dependent Ca^{2+} transport.[11,12] We have used this information to undertake studies in which the kinetic characteristics of these two systems were compared in adult and aged animals.

A summary of some of the comparisons of Ca^{2+} transport activities in membranes from aged and adult animals is shown in TABLE 1. Our initial comparison of the synaptic plasma membrane (SPM) Na$^+$-Ca^{2+} exchange activity in Fisher 344 rats aged 5–7 mo and aged 23–25 mo revealed an increase in the Ca^{2+} concentration required for half maximal activation of Ca^{2+} transport activity with a relatively small decrease in the maximal transport activity (V$_{max}$).[13] In order to see if these changes were even greater in older animals, we conducted a similar series of experiments using membranes from 10–11-mo compared to 27–28-mo animals. Once again there was a consistently lower activity in membranes from the aged versus adults rats, but the nature of the alterations in the kinetic characteristics was essentially the same as we observed in the earlier comparisons, *i.e.*, approximately a 20% increase in the K$_{act}$ for Ca^{2+} in the aged group with a smaller change in the V$_{max}$ (TABLE 1). The ANOVA indicated, however, that the actual curves for the two populations differed significantly from each other. Thus it appears that some change in the efficiency of this transport system had occurred in the brain membranes from aged animals. It is possible that the membrane environment surrounding the protein has been altered leading to a lower affinity for Ca^{2+}. Or it is equally possible that some of the protein molecules themselves are altered. Until adequate molecular probes are developed for studying this transport protein, the molecular mechanism(s) underlying the observed changes cannot be elucidated.

The "Ca^{2+} pump" activity of the plasma membrane can be assessed in two different ways, by examining the enzymatic hydrolysis of ATP or the ATP-dependent transport of Ca^{2+} across the membrane. Our characterization of these processes in highly purified synaptic plasma membranes has revealed the following: both the ATP hydrolysis and Ca^{2+} transport are Mg^{2+}-dependent, both have a very high affinity for Ca^{2+} (K$_{act}$ = 0.2–0.6 μM), and both are very sensitive to inhibition by vanadate and stimulation by calmodulin. In addition, we were also able to identify some properties of this system that distinguished it from the microsomal Ca^{2+} sequestering system within neurons.[12]

TABLE 1. Summary of Age-Related Changes in Synaptic Membrane Ca^{2+} Transport Activities[a]

Transport System	K$_{0.5}$ Aged/K$_{0.5}$ Control	V$_{max}$ Aged/V$_{max}$ Control
Na$^+$ -Ca^{2+} antiporter		
23–25 mo	1.20	0.90
27–28 mo	1.18	0.90
(Ca^{2+} + Mg^{2+})ATPase		
23–25 mo	1.0	0.80
ATP-dependent Ca^{2+} transport		
23–24 mo	1.09	0.86
27–28 mo	1.10	0.77

[a]n = 7 pairs of animals in each set of experiments.

In our initial comparative studies with brain membranes from aged animals, we studied only the Ca^{2+} activated hydrolysis of ATP by the synaptic membranes from adult (5–7 mo) and aged animals (23–25 mo),[13] and found maximal activation of the enzyme by Ca^{2+} was about 20% lower in the aged animals. We then examined the ATP dependent Ca^{2+} transport activity across Ca^{2+} concentrations. Highly purified SPM's were obtained from F-344 rats of 6–8 mo and 23–24 mo with identical protein recoveries at each step of the membrane isolation. Calcium transport activity was measured in the presence of 50 μM Mg^{2+} and 100 μM ATP, for 60 sec at 35°C. The aged animals exhibited a lower transport activity at every Ca^{2+} concentration, and the V_{max} values obtained from Eadie-Hofstee transformation revealed an approximately 14% decrease in maximal transport capacity for the aged animals. In an effort to see whether animals older than 23–24 mo would have greater decreases in Ca^{2+} transport, we conducted similar experiments in the membranes from 11–12-mo and 27–28-mo animals (TABLE 1). Data from 7 pairs of animals showed a consistently lower Ca^{2+} transport activity in the aged animals, again with the major change being a decrease in V_{max} in membranes from the older animals.

At present, our assessment of the kinetic characteristics of these two plasma membrane Ca^{2+} transporting systems is that the antiporter may be operating less efficiently in the aged membranes and that the maximal transport capacity of the $(Ca^{2+} + Mg^{2+})$-ATPase is reduced in membranes from aging brain. In these assays, however, we supply the ionic gradients or the ATP required to drive the transport systems. This may mask the actual situation in the intact cells in aging brain. The fact that we were able to see differences under these conditions suggested that intact nerve terminals from brains of aged animals may be even less effective in maintaining and restoring normal $[Ca^{2+}]_i$ following a massive influx of Ca^{2+} such as might be triggered by depolarization.

In order to see if the decreased Ca^{2+} translocating activity in membranes had any detectable dynamic consequences for intact nerve terminals, we used the fluorescent Ca^{2+}-sensitive dye fura-2 to monitor intrasynaptosomal $[Ca^{2+}]_i$ under various conditions. Intact synaptosomes were loaded internally with fura-2, the resting $[Ca^{2+}]_i$ levels determined, and the change in $[Ca^{2+}]_i$ monitored following various depolarizing stimuli. The resting $[Ca^{2+}]_i$ and the responses to increasing concentrations of three different depolarizing stimuli were monitored as the fluorescence ratios of the dye in the Ca^{2+}-bound and the free state. The average resting $[Ca^{2+}]_i$ in synaptosomes from 23–24-mo animals was slightly higher (approximately 10%) than that in 7–9-mo animals: $[Ca^{2+}]_i = 214 \pm 3$ nM and 194 ± 4 nM for aged and adult animals, respectively, n = 60 determinations from 7 pairs of animals. In addition, the *net* change in $[Ca^{2+}]_i$ following depolarization by various concentrations of KCl, veratridine, and ibotenic acid was consistently greater (magnitude ranging from 30% to >70%) in synaptosomes from aged animals (unpublished data). Given the very complex dynamics of the Ca^{2+} buffering and extrusion mechanisms present in nerve terminals, the reason for the higher Ca^{2+} levels is presently unknown. It is quite conceivable, however, that the changes we have observed in the two plasma membrane Ca^{2+} extrusion systems are contributing to the differences in Ca^{2+} handling observed here with the more intact synaptosomes.

A substantial number of observations have been reported by other investigators which are consistent with our tentative conclusion that Ca^{2+} handling by brain nerve terminals from aged animals is slower and less effective than that in terminals from younger animals. These investigators have been led to a fairly similar conclusion, though each has approached the issue of age-dependent changes in the nervous system from a very different perspective. For example, Landfield and colleagues[14,15] have examined specific electrophysiological parameters in hippocampal brain slice preparations from aged and adult rats. They have observed that the after-hyperpolarization was more prolonged in aged animals, with studies of Ca^{2+}-dependent K^+ conductance suggesting that depolarization-induced elevations in $[Ca^{2+}]_i$ may remain high for up to 50% longer in neurons from aged animals. From a morphological perspective, Fifkova and Cullen-Dockstader[16,17] have observed that the dendritic spine apparatus in sections from the hippocampal region of the brain showed a significantly greater amount of Ca^{2+} sequestered within this organelle in tissue from very old animals. These observations may be reflecting the consequences of elevated free $[Ca^{2+}]_i$, some of which is sequestered in the spine apparatus in a state of equilibrium with the cytosolic Ca^{2+} levels.

Recent studies done in various neuromuscular preparations from adult and aged animals also are strongly suggestive that presynaptic Ca^{2+} signals are prolonged or that Ca^{2+} clearance is slowed in the aged animals.[18] The types of measurements and the nature of the preparations studied are quite different from any of those already mentioned. Nevertheless, the tentative conclusion suggested by Smith's observations[18] is quite consistent with that discussed above, namely, that the kinetics of the buffering or the clearing of Ca^{2+} following stimulation are altered in aged animals such that Ca^{2+} remains elevated for a longer period of time.

It is not possible to say at this point that a more prolonged elevation of $[Ca^{2+}]_i$ is necessarily deleterious for neuronal function. We know only that some aspects of central nervous system activity decline as most organisms age. Certainly in view of the rather protracted Ca^{2+} signalling events stimulated by phosphoinositide hydrolysis, or the hypothesis about NMDA receptor activation of Ca^{2+} influx and long-term potentiation, it is quite possible to argue that prolonged elevations in $[Ca^{2+}]_i$ may enhance rather than disrupt signal transduction from the membrane. Clearly, there are complexities and subtleties about neuronal Ca^{2+} regulation which totally escape our present understanding. These are the events we must seek to uncover so that we can put together an accurate picture of the dynamics of Ca^{2+} regulation both locally and throughout the entire cell. Once that complex scenario can be drawn with reasonable precision, we will be in a position to determine if and why the altered Ca^{2+} disposition in aged neurons has untoward consequences for the organism.

REFERENCES

1. KHATCHATURIAN, Z. 1984. Towards theories of brain aging. In Handbook of Studies in Psychiatry and Old Age. E.W. Kay & G.D. Burrows, Eds.: 7–30. Elsevier. New York, NY.
2. GIBSON, E.G. & C. PETERSON. 1987. Calcium and the aging nervous system. Neurobiol. Aging 8: 329–343.
3. SCHANNE, F.A.X., A.B. KANE, E.E. YOUNG & J.L. FARBER. 1979. Calcium dependence of toxic cell death: A common pathway. Science 206: 700–702.
4. SIESJO, B.K. 1981. Cell damage in the brain: A speculative synthesis. J. Cereb. Blood Flow Metab. 1: 155–185.
5. CHEUNG, J.Y., J.V. BONVENTURE, C.D. MALIS & A. LEAF. 1986. Mechanisms of disease. N. Engl. J. Med. 314: 1670–1676.
6. DESHPANDE, J.K., B.K. SIESJO & T. WIELOCH. 1987. Calcium accumulation and neuronal damage in the rat hippocampus following cerebral ischemia. J. Cereb. Blood Flow Metab. 7: 89–95.
7. VOLPE, P., K.-H. KRAUSE, S. HASHIMOTO, F. ZORZATO, T. POZZAN, J. MELDOLESI & D.P. LEW. 1988. "Calciosome," a cytoplasmic organelle: The inositol 1,4,5-trisphosphate-sensitive Ca^{2+} store of nonmuscle cells? Proc. Natl. Acad. Sci. USA 85: 1091–1095.
8. TAYLOR, C.W. & J.W. PUTNEY, JR. 1987. Phosphoinositides and calcium signaling. In Calcium and Cell Function. Vol. 3. W.Y. Cheung, Ed.: 2–28. Academic Press. Orlando, FL.
9. MICHAELIS, M.L. & E.K. MICHAELIS. 1981. Ca^{2+} fluxes in resealed synaptic plasma membrane vesicles. Life Sci. 28: 37–45.
10. MICHAELIS, M.L. & E.K. MICHAELIS. 1983. Alcohol and local anesthetic effects on Na^{+}-dependent Ca^{2+} fluxes in brain synaptic membrane vesicles. Biochem. Pharmacol. 32: 963–969.

11. MICHAELIS, E.K., M.L. MICHAELIS, H.H. CHANG & T.E. KITOS. 1983. High affinity Ca^{2+}-stimulated Mg^{2+}-dependent ATPase in rat brain synaptic membranes, synaptosomes, and microsomes. J. Biol. Chem. **258:** 6101–6108.
12. MICHAELIS, M.L., T.E. KITOS, E.W. NUNLEY & E.K. MICHAELIS. 1987. Characteristics of Mg^{2+}-dependent, ATP-activated Ca^{2+}-transport in synaptic and microsomal membranes and in permeabilized synaptosomes. J. Biol. Chem. **262:** 4182–4189.
13. MICHAELIS, M.L., K. JOHE & T.E. KITOS. 1984. Age-dependent alterations in synaptic membrane systems for Ca^{2+} regulation. Mech. Ageing Dev. **25:** 215–225.
14. LANDFIELD, P.W. & T.A. PITLER. 1984. Prolonged Ca^{2+}-dependent afterhyperpolarizations in hippocampal neurons of aged rats. Science **226:** 1089–1091.
15. LANDFIELD, P.W. & G.A. MORGAN. 1984. Chronically elevating plasma Mg^{2+} improves hippocampal frequency potentiation and reversal learning in aged and young rats. Brain Res. **322:** 167–171.
16. FIFKOVA, E. & K. CULLEN-DOCKSTADER. 1986. Age-related changes in the distribution of calcium-containing synaptic vesicles in the terminals of the perforant pathway. Soc. Neurosci. Abstr. **12:** 271.
17. FIFKOVA, E. & K. CULLEN-DOCKSTADER. 1986. Calcium distribution in dendritic spines of the dentate fascia varies with age. Brain Res. **376:** 357–362.
18. SMITH, D.O. 1987. Non-uniform changes in nerve-terminal calcium homeostasis during aging. Neurobiol. Aging **8:** 366–368.

Aging-Related Increases in Voltage-Sensitive, Inactivating Calcium Currents in Rat Hippocampus

Implications for Mechanisms of Brain Aging and Alzheimer's Disease

PHILIP W. LANDFIELD, LEE W. CAMPBELL, SU-YANG HAO,
AND D. STEVEN KERR

Department of Physiology and Pharmacology
Bowman Gray School of Medicine
Wake Forest University
300 South Hawthorne Road
Winston-Salem, North Carolina 27103

Age-Related Deficit in Hippocampal Synaptic Plasticity

In studies begun during the mid-to-late 1970s, we found consistent evidence that several aspects of synaptic plasticity were impaired in the hippocampus of aging rats. These effects of aging were observed in both the *in vitro* slice preparation,[1] and the intact, anesthetized rat preparation.[2] Moreover, these experiments showed that the impairment of synaptic potentiation was closely correlated with impaired learning/memory processes, both across groups[1,2] and within aged animals.[3] A subsequent series of studies aimed at analyzing the cellular basis of this impairment led us to focus on altered Ca homeostasis as a likely candidate mechanism for the impaired plasticity, and therefore, perhaps for impaired behavioral functions as well. This paper briefly reviews those studies[4] as well as some more recent findings on the specific nature and membrane sites of altered Ca influx. In addition, the possible implications of these recent findings for age-related structural as well as physiological brain deterioration are considered.

The neurophysiological deficits in synaptic plasticity were most apparent under conditions of repetitive synaptic stimulation, during which the degree of frequency potentiation (FP: the increase of synaptic responses during a train of 3–20 Hz repetitive activation)[5] of hippocampal responses was markedly impaired in aged rats[1,2,6] (cf. FIG. 1). Somewhat analogous age-related impairments of synaptic responses also have been seen during frequency facilitation paradigms at the neuromuscular junction of rats.[7,8] Moreover, in some[2] but not all[3] experimental protocols, it was possible to detect age-related deficits in the rate of development of hippocampal long-term potentiation (LTP: the sustained increase of synaptic responses after cessation of a train of repetitive activation).[9] Other investigators have also found age differences in hippocampal LTP[10,6] although in some cases the effect was observed only in the duration rather than the magnitude of LTP.[10]

Increased Calcium Availability in the Synaptic Deficit

Studies following those outlined led to evidence that the locus of the age-related deficit in FP was very likely in the presynaptic terminal.[11,12,4,13] In addition, a separate

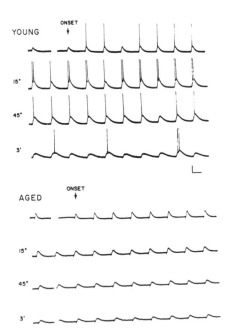

FIGURE 1. Intracellular recordings from CA1 cells in hippocampal slices of young and aged rats during 4 min of continuous 10-Hz synaptic stimulation. 15″: EPSPs during the fifteenth second of stimulation. 45″: EPSPs during the forty-fifth second of 10-Hz stimulation. 3′: EPSPs at the third minute of stimulation. During repetitive stimulation, the EPSP was usually potentiated substantially more in young rat cells. In the examples shown, the EPSP did not reach threshold for triggering an action potential in the aged rat cell, although control EPSPs were set at 75% of spike threshold in both instances. Calibration: 20 mV, 50 ms.[2]

set of studies was directed at clarifying the sites and nature of the molecular mechanisms underlying the deficits in physiological plasticity and function. Based on analogies with peripheral frequency facilitation, we attempted to restore FP in aged rat synapses by increasing the extracellular magnesium/calcium ratio in the bathing medium of hippocampal slices. In the periphery, high Mg improves percentage facilitation by reducing transmitter release on each pulse and retarding depletion.[14]

As illustrated in FIGURE 2, we found that a high Mg/Ca ratio could do the same in hippocampal slices, particularly in those from aged rats,[15,16] and also had a similar effect in intact animals.[17] Moreover, in the intact animals, a high Mg diet counteracted a deficit in maze learning in the aged rats.[17]

The findings that elevated Mg could counteract the age-related deficit in FP, whereas elevated Ca could impair FP, and that the effect of Mg was greater in aged rats, suggested that aged rat hippocampal neurons might be characterized by an excess of voltage-dependent Ca influx.[12,4,16] (The possible mechanisms through which elevated Ca might impair FP include accelerated transmitter depletion, activation of Ca-dependent hyperpolarization in the terminals, or Ca-dependent inactivation of additional Ca influx.[16])

Other Evidence of Altered Calcium Homeostasis in Brain Aging and Alzheimer's Disease

The possibility that altered Ca homeostasis was a key factor in brain aging was emerging from a number of other lines of evidence as well.[18] During the past years, evidence has been obtained of reduced Ca extrusion or buffering in neurons from aged rat brain,[19] reduced clearance of Ca from rat neuromuscular terminals,[20] and elevated Ca in neurons bearing neurofibrillary tangles from brains of demented humans.[21] Moreover, there is growing evidence of the cytotoxic effects of elevated intracellular Ca on a wide range of excitable tissues,[22-26] apparently mediated by Ca-dependent protease activation.[27,9] These and other findings, led to the hypothesis that altered Ca homeostasis might play a key role in brain aging and Alzheimer's disease.[18] Conversely, a number of studies have found evidence that Ca availability in excitable cells might be reduced rather than elevated with aging.[28,29] These latter findings indicate that the nature of altered Ca homeostasis in aging is complex and that its analysis is likely to be difficult.

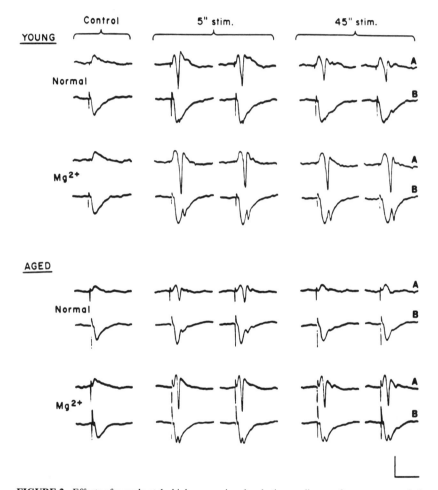

FIGURE 2. Effects of a moderately-high-magnesium incubating medium on frequency potentiation of the extracellular population spike *("A" traces)* and field EPSP *("B" traces)* in hippocampal slices from young and aged rats. Normal: Slices incubated in medium with equal Mg and Ca concentrations. Mg: Slices incubated in medium with a 2:1 ratio of Mg to Ca. Control: Responses obtained at 0.2-Hz synaptic stimulation, before the onset of a continuous 7-Hz stimulation train. 5″ stimulation: Responses obtained in the fifth second of 7-Hz stimulation. 45″ stimulation: Responses obtained in the forty-fifth second of stimulation. Calibration: 2 mV, 10 ms. During 7-Hz stimulation, slices in high Mg exhibited greater potentiation and less depression than slices in normal medium. The effect was greater in aged rat slices.[16]

In order to make a clearer assessment of the nature of age-related alterations in Ca utilization, we tested more directly the hypothesis that the availability of voltage-dependent Ca was increased. That is, synaptic function is dependent upon many factors other than Ca influx, and a response more uniquely dependent upon intracellular Ca concentrations would provide a more specific test.[30,31] A Ca-dependent, K-mediated afterhyperpolarization (AHP) has been described in hippocampus,[32-35] and we utilized this AHP

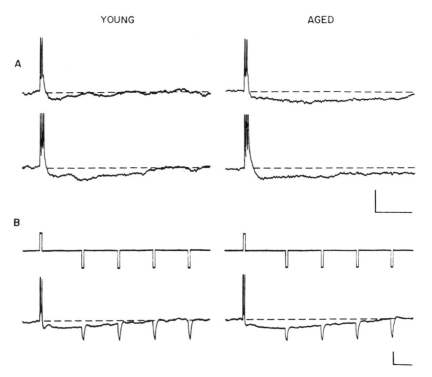

FIGURE 3. Intracellular current-induced bursts of action potentials and subsequent afterhyperpolarizations (AHPs) in CA1 neurons of hippocampal slices from young and aged rats. (**A**) AHPs following a current-induced burst of two spikes *(upper traces)* or three spikes *(lower traces)*, in slices from young or aged rats. (**B**) AHPs and concomitant measures of conductance increases following a 0.4-nA current-induced burst of three spikes. *Dashed lines* show resting potentials before the burst. In the upper trace of (B) are shown the initial intracellular depolarizing current pulse used to induce a spike burst and the subsequent 2-Hz train of 0.4-nA hyperpolarizing pulses used to assess input conductance during the AHP, for cells shown in the *lower trace* of (B).[30]

to test the possibility that aged rats exhibited an increased response of Ca-dependent processes following depolarization. As shown in FIGURE 3, the AHPs were significantly prolonged in aged rat neurons, and could be eliminated by the same levels of Mg/Ca elevation that were able to strengthen FP.[30]

However, this study did not demonstrate definitively that the AHP prolongation was due to an increase of voltage-dependent Ca current (as opposed to reduced Ca buffering or clearance[18-20,28]). Consequently, we examined Ca potentials (Ca spikes)[36,37] and Ca currents in cesium-loaded hippocampal neurons treated with tetrodotoxin (TTX) and tetraethylammonium (TEA). Under these conditions, voltage-dependent Na, and most K currents are blocked, and the remaining currents are carried largely by Ca ions.[38,39]

In hippocampal slice neurons treated with these agents, we found that the duration of the Ca spike was prolonged in aged rat neurons,[40,41] showing increased Ca influx. Therefore, while changes in Ca buffering and clearance may also contribute significantly to alterations in Ca homeostasis, the age-related increase in the AHP appears to be accounted for parsimoniously by changes in voltage-sensitive Ca currents.

Multiple Calcium Currents in Hippocampus: Effects of Aging

It is now well recognized that there are several types of voltage-dependent Ca channels in neurons. These are generally classified as low-voltage activated (LVA) or high voltage activated (HVA), in terms of the amount of depolarization needed to reach threshold for activation.[42,39,36,43] Recently, three classes of voltage-sensitive Ca channels have been described in peripheral neurons (*e.g.*, chick dorsal root ganglion (DRG) cells).[44] These classes and their general properties (which may differ considerably in other cell types; *e.g.*, sympathetic neurons) are outlined below.[43]

1. T Channels: Low-voltage activated (positive to -70 mV), and rapidly inactivating (time constant of inactivation under 50 ms); inactivation range, -100 to -60 mV; insensitive to dihydropyridine (DHP) blocking agents.
2. N Channels: High-voltage activated (positive to -20 mV); moderate rate of inactivation (time constant under 100 ms); inactivation range, -100 to -40 mV; insensitive to DHPs.
3. L Channels: High-voltage activated (generally positive to -10 mV); slow rate of inactivation (time constant greater than 500 ms); partial inactivation range, -60 to -10 mV; blocked by DHPs.

Multiple Ca currents have been described in mammalian brain neurons,[36,39] and recently, three currents similar to those in DRG neurons were described in dissociated hippocampal neurons;[45,46] nevertheless, it should be emphasized that these three currents are not the only kinds of voltage-activated Ca currents present in excitable cells. Several neuron or muscle cell types exhibit slowly inactivating or non-inactivating forms of Ca currents with gating characteristics different from those of the T, N, and L Channels.

It should also be emphasized that essentially all of the recent work on Ca channels has been conducted under highly nonphysiological conditions (*e.g.*, dissociated or disrupted cells, 10 mM Ca or barium, exogenous Ca buffers, etc.) and it is therefore not clear what the properties of these currents may be under more physiological conditions. In recent studies using single-electrode voltage-clamp techniques with Cs-loaded and TTX- and TEA-treated neurons in hippocampal brain slices (which retain much of their structural integrity and normal physiological properties) we have also seen multiple Ca currents, as defined by activation and inactivation properties. These currents share some properties in common with the T-, N- and L-like currents seen in dissociated cells, but also exhibit some notable differences.

In these brain slice preparations, we have found that age differences can be detected quantitatively and statistically, if care is taken to reduce variability, define baseline values, establish clear criteria for cellular "health," and maintain stable conditions from preparation to preparation. At present, our data indicate that the main age differences may lie in L-like currents and that these currents are larger in neurons from aged rat hippocampal slices. However, the data also indicate that other Ca currents with faster kinetics may be affected by aging (Hao et al.; Campbell et al., in preparation). These voltage-clamp results are highly compatible with our prior findings on the effects of age on the duration of the Ca-dependent AHP[30] and of the Ca spike.[40,41]

Calcium-Dependent Inactivation of Calcium Currents: A Possible Role in Aging Changes

Although a form of Ca-dependent inactivation of Ca currents has been described for invertebrate neurons,[47] early studies in mammalian central neurons found Ca currents to be relatively non-inactivating.[48] However, because of certain properties of synaptic depression in hippocampal slices,[12] we tested the possibility of Ca-dependent inactivation directly in hippocampal neurons, using Ca-dependent AHPs,[49] Ca-spikes,[50] and isolated Ca currents.[51,52] In each case, the data were fully consistent with the operation of a negative feedback form of Ca-dependent inactivation of Ca currents in hippocampal neurons. That is, the inactivation was more rapid in high Ca, was not present in barium, was not due to voltage-dependent inactivation (since it occurred during trains of depolarizing pulses separated by repolarization intervals, and did not occur in barium), was not due to altered equilibrium potentials (since it did not occur in barium), and was not due to slow Ca-dependent outward K currents (since these are reduced by Cs-loading, and no outward currents could be detected following inactivating pulses, and the Ca-dependent AHP also exhibited inactivation).[49,50] In addition, recent studies show that a fast K conductance (I_c) which is both Ca- and voltage-sensitive does not influence the Ca current inactivation process, since inactivation is unchanged by TEA,[52] and TEA blocks this Ic.[53] Thus, there is clear evidence of Ca-dependent inactivation of some Ca currents in the mammalian brain.[49-52]

Since Ca currents appear prolonged or larger in aged rat neurons, and are subject to strong inactivation control by the Ca-dependent mechanism, the hypothesis that aging may increase Ca currents by reducing inactivation processes seems to be a clear possibility.[31] An alternative hypothesis, however, is that inactivation mechanisms are normal, and that Ca influx increases with age due to changes in the activation properties of the Ca channels. These possibilities are currently under investigation.

Implications of Alterations in Voltage-Sensitive, Inactivating Calcium Channels for Age-Related Brain Pathology

The potential relevance of the cytotoxic effects of elevated intracellular Ca concentrations ($[Ca]_i$) to brain aging and Alzheimer's disease has been considered in some depth[18] and the results reviewed in the present paper, as well as data from other laboratories, indicate that this Ca hypothesis is still very much a possibility. However, it seems clear that excessive Ca influx can be terminated quickly by both Ca- and voltage-dependent inactivation processes. Therefore, it might be argued that these inactivation mechanisms protect neurons against increased influx through voltage-sensitive channels, by negative feedback regulation of further influx.

Nonetheless, FIGURE 4 illustrates a simplified schema which shows that neurons can restore the voltage sensitivity of Ca channels rapidly following either Ca-dependent or voltage-dependent inactivation. That is, the same intracellular Ca elevation against which the cell protects itself with inactivation of Ca channels, also leads to the removal of these inactivation processes by several mechanisms. In the case of channels sensitive to Ca inactivation (FIG. 4A), Ca clearance and buffering increase following Ca influx, leading

A. Ca-Inactivated Ca Channels

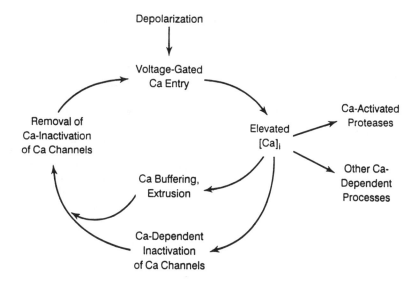

B. Voltage-Inactivated Ca Channels

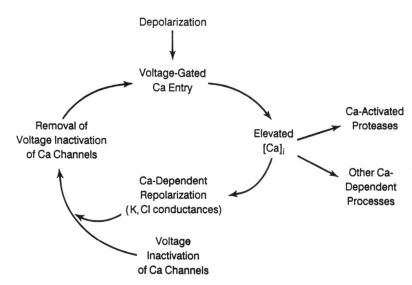

FIGURE 4. Illustration of processes involved in the rapid restoration of voltage-sensitivity to inactivating Ca channels. **(A)** Ca-Inactivated Channels. Following depolarization-dependent elevation of intracellular Ca, Ca buffering and clearance mechanisms reduce $[Ca]_i$ below the level necessary for sustained Ca-dependent inactivation of Ca channels, allowing renewed Ca influx during subsequent depolarization. **(B)** Voltage-Inactivated Channels. Following depolarization-dependent elevation of $[Ca]_i$, Ca-activated K and Cl conductances repolarize the membrane, removing voltage-inactivation and allowing renewed Ca influx during depolarization. This repriming of Ca channels can occur even in the presence of continuously elevated intracellular Ca.

to restoration of sensitivity to Ca-inactivated Ca channels. After $[Ca]_i$ is reduced below the elevated level necessary to maintain channel inactivation, the Ca channels would again be sensitive to activation by normal activity. Since the level of $[Ca]_i$ needed to maintain inactivation is itself an elevated one, $[Ca]_i$ could remain moderately elevated for very prolonged periods. Of course, if the Ca inactivation mechanism were impaired with aging,[31] then the setpoint at which inactivation is achieved could involve a still higher level of $[Ca]_i$.

In the case of Ca channels that are voltage-inactivated (FIG. 4B) (some of which are also inactivated by Ca), elevated $[Ca]_i$ acts strongly to remove inactivation by activating a series of Ca-dependent outward currents (K- and Cl-mediated) which repolarize the membrane and restore voltage sensitivity well before the internal Ca has been cleared. Thus, voltage-inactivation can limit the duration of Ca influx through voltage-sensitive Ca channels for each single depolarization, but, unless there is continuous depolarization, the sensitivity of these channels is restored rapidly by the elevated $[Ca]_i$. Therefore, this mechanism is unlikely to be effective in controlling a prolonged rise in intracellular Ca that develops either from greater Ca influx for a given depolarization, or from reduced Ca clearance/buffering.

In summary, voltage-sensitive Ca channels are subject to powerful forms of Ca- and voltage-dependent inactivation, which can protect against massive Ca influx during any single depolarization. If these inactivation mechanisms were impaired with aging, Ca influx could increase. However, even if inactivation processes function normally, they seem unable to protect against a prolonged, moderate rise in intracellular Ca resulting from an increased influx through voltage-sensitive Ca channels.

Some brain neurons normally fire as rapidly as 70/sec,[54] and hippocampal neurons fire consistently in the 2–20 Na-spikes/sec range.[55,56] Since the Ca-dependent AHP (and presumably, Ca elevation) can persist for 0.3–1 second following a Na spike in rat hippocampal neurons, it seems clear that the age-related increase of 25–50% that occurs in hippocampal Ca influx[30,40] could result in a nearly continuous elevation of intracellular Ca above normal values during 2–20 Hz activity.

The critical importance of Ca regulatory functions suggests that even moderate elevations of $[Ca]_i$ could activate Ca-dependent cellular processes inappropriately, including those involved in cytotoxic actions (e.g., proteases; FIG. 4). Thus, it seems highly possible that repeatedly prolonging the elevation of intracellular Ca following each action potential, could result in the slow deterioration of brain cellular structural integrity. Increased Ca current through voltage-sensitive channels might develop gradually with aging. However, in a minority of aging individuals, a greater failure of Ca inactivation or clearance processes might occur (perhaps as a result of genetic, physiological, viral, or toxic factors) leading to the inability to control Ca elevation even within the moderately-elevated range, and, in turn, to full blown Alzheimer's disease.

REFERENCES

1. LANDFIELD, P. W. & G. LYNCH. 1977. Impaired monosynaptic potentiation in in vitro hippocampal slices from aged, memory-deficient rats. J. Gerontol. 32: 523–533.
2. LANDFIELD, P. W., J. L. MCGAUGH & G. LYNCH. 1978. Impaired synaptic potentiation processes in the hippocampus of aged, memory-deficient rats. Brain Res. 150: 85–101.
3. LANDFIELD, P. W. 1980. Correlative studies of brain neurophysiology and behavior during aging. In The Psychobiology of Aging. Stein, D.G., Ed.: 227–252. Amsterdam. Elsevier.

4. LANDFIELD, P. W., T. A. PITLER & M. D. APPLEGATE. 1986. The aged hippocampus: A model system for studies on mechanisms of behavioral plasticity and brain aging. *In* The Hippocampus, Vol. 3. R.L. Isaacson & K.H. Pribram, Eds.: 323–367. Plenum. New York, NY.

5. ANDERSEN, P. & T. LØMO. 1967. Control of hippocampal output by afferent volley frequency. *In* Progress in Brain Research: Structure and Function of the Limbic System. W.R. Adey & T. Tokizane, Eds. Vol 27: 400–412. Elsevier. Amsterdam.

6. TIELEN, A.M., W. J. MOLLEVANGER, F. H. LOPES DA SILVA & C. F. HOLLANDER. 1983. Neuronal plasticity in hippocampal slices of extremely old rats. *In* Aging of the Brain. W. H. Gispen & J. Traber, Eds.: 73–84. Elsevier. Amsterdam.

7. SMITH, D. O. 1984. Acetylcholine storage, release and leakage at the neuromuscular junction of mature adult and aged rats. J. Physiol. (London) **347:** 161–176.

8. SMITH, D.O. & J. L. ROSENHEIMER. 1984. Aging at the neuromuscular junction. *In* Aging and Cell Structure. J.E. Johnson, Ed.: 113–138. Plenum. New York, NY.

9. LYNCH, G. & M. BAUDRY. 1984. The biochemistry of memory: A new and specific hypothesis. Science **224:** 1057–1063.

10. BARNES, C. A. 1979. Memory deficits associated with senescence: A behavioral and neurophysiological study in the rat. J. Comp. Physiol. Psychol. **93:** 74–104.

11. APPLEGATE, M. D. & P. W. LANDFIELD. 1988. Synaptic vesicle redistribution during hippocampal frequency potentiation and depression in young and aged rats. J. Neurosci. **8:** 1096–1111.

12. LANDFIELD, P. W., T. A. PITLER, M. D. APPLEGATE & J. H. ROBINSON. 1983. Intracellular studies of the aged-related deficit in hippocampal frequency potentiation: Apparent calcium saturation in synapses of aged rats. Soc. Neurosci. Abstr. **9:** 232.

13. PITLER, T. A. & P. W. LANDFIELD. 1987. Postsynaptic membrane shifts during frequency potentiation of the hippocampal EPSP. J. Neurophysiol. **58:** 866–882.

14. MARTIN, A. R. 1977. Presynaptic mechanisms. *In* Handbook of Physiology I: The Nervous System. J.M. Brookhart, V.B. Mountcastle, Eds: 329–355. American Physiological Society. Bethesda, MD.

15. LANDFIELD, P. W. 1981. Age-related impairment of hippocampal frequency potentiation: Evidence of an underlying deficit in transmitter release from studies of Mg^{+2}-bathed hippocampal slices. Soc. Neurosci. Abstr. **7:** 371.

16. LANDFIELD, P. W., T. A. PITLER & M. D. APPLEGATE. 1986. The effects of high Mg^{2+} to Ca^{2+} ratios on frequency potentiation in hippocampal slices of young and aged rats. J. Neurophysiol. **56:** 797–811.

17. LANDFIELD, P. W. & G. MORGAN. 1984. Chronically elevating plasma Mg^{2+} improves hippocampal frequency potentiation and reversal learning in aged and young rats. Brain Res. **322:** 167–171.

18. KHACHATURIAN, Z. S. 1984. Towards theories of brain aging. *In* Handbook of studies on psychiatry and old age. D. Kay, G. D. Burrows, Eds.: 7–30. Elsevier. Amsterdam.

19. MICHAELIS, M. L., K. JOHE & T. E. KITOS. 1984. Age-dependent alterations in synaptic membrane systems for Ca^{2+} regulation. Mech. Ageing Dev. **25:** 215–225.

20. SMITH, D. O. 1988. Muscle-specific decrease in presynaptic calcium dependence and clearance during neuromuscular transmission in aged rats. J. Neurophysiol. **59:** 1069–1082.

21. PERL, D. P., D. C. GAJDUSEK, R. M. GARRUTO, R. T. YANAGIHARA & C. J. GIBBS. 1982. Aluminum accumulation in amyotrophic lateral sclerosis and Parkinsonism-dementia of Guam. Science **217:** 1053–1055.

22. CHOI, W. 1987. Ionic dependence of glutamate neurotoxicity. J. Neurosci. **7:** 369–379.

23. GRIFFITH, T., M. C. EVANS & B. S. MELDRUM. 1983. Intracellular calcium accumulation in rat hippocampus during seizures induced by bicuculline or L-allylglycine. Neuroscience **10:** 383–395.

24. SIESJO, B. K. 1981. Cell damage in the brain: A speculative synthesis. J. Cereb. Blood Flow Metab. **1:** 155–185.

25. NAYLER, W. G., P. A. POOLE-WILSON & A. WILLIAMS. 1979. Hypoxia and calcium. J. Molec. Cell Cardiol. **11:** 683–706.

26. ROTHMAN, S. & J. W. OLNEY. 1986. Glutamate and the pathophysiology of hypoxic-ischemic brain damage. Ann. Neurol. 19: 105–111.
27. SCHLAEPFER, W. W. & M. B. HASLER. 1979. Characterization of the calcium-induced disruption of neurofilaments in rat peripheral nerve. Brain Res. 168: 299–309.
28. GIBSON, G. E. & C. PETERSON. 1987. Calcium and the aging nervous system. Neurobiol. Aging 8: 329–344.
29. ROTH, G. S. 1988. Mechanisms of altered hormone and neurotransmitter action during aging: The role of impaired calcium mobilization. Ann. N.Y. Acad. Sci. 521: 170–176.
30. LANDFIELD, P. W. & T. A. PITLER. 1986. Prolonged Ca^{2+}-dependent afterhyperpolarizations in hippocampal neurons of aged rats. Science 226: 1089–1092.
31. LANDFIELD, P. W. 1987. "Increased calcium current" hypothesis of brain aging. Neurobiol. Aging 8: 346–347.
32. ALGER, B. E. & R. A. NICOLL. 1980. Epileptiform burst afterhyperpolarization: Calcium-dependent potassium potential in hippocampal CA1 pyramidal cells. Science 210: 1122–1124.
33. HOTSON, J. R. & D. A. PRINCE. 1980. A calcium-activated hyperpolarization follows repetitive firing in hippocampal neurons. J. Neurophysiol. 43: 409–419.
34. LANCASTER, B. & P. R. ADAMS. 1986. Calcium-dependent current generating the afterhyperpolarization of hippocampal neurons. J. Neurophysiol. 55: 1268–1282.
35. SCHWARTZKROIN, P. A. & C. A. STAFSTRÖM. 1980. Effects of EGTA on the calcium-activated afterhyperpolarization in hippocampal CA3 pyramidal cells. Science 210: 1125–1126.
36. LLINAS, R. & Y. YAROM. 1981. Electrophysiology of mammalian inferior olivary neurones in vitro. Different types of voltage-dependent ionic conductances. J. Physiol. (London) 315: 549–567.
37. SCHWARTZKROIN, D. A. & M. A. SLAWSKY. 1977. Probable calcium spikes in hippocampal neurones. Brain Res. 135: 157–161.
38. JOHNSTON, D., J. J. HABLITZ & W. A. WILSON. 1980. Voltage-clamp discloses slow inward current in hippocampal burst-firing neurones. Nature 286: 391–393.
39. HALLIWELL, J. W. 1983. Caesium loading reveals two distinct Ca-currents in voltage-clamped guinea-pig hippocampal neurones in vitro. J. Physiol. (London) 341: 10–11.
40. LANDFIELD, P. W. & T. A. PITLER. 1987. Calcium spike duration: Prolongation in hippocampal neurons of aged rats. Soc. Neurosci. Abstr. 13: 718.
41. PITLER, T. P. & P. W. LANDFIELD. Age related prolongation of calcium spikes in rat hippocampal slice neurons. Brain Res. In press.
42. CARBONE, E. & H. D. LUX. 1984. A low voltage-activated fully inactivating Ca channel in vertebrate sensory neurons. Nature 310: 501–502.
43. MILLER, R. J. 1987. Multiple calcium channels and neuronal function. Science 235: 46–52.
44. NOWYCKY, M. C., A. P. FOX & TSIEN, R. W. 1985. Three types of neuronal calcium channel with different agonist sensitivity. Nature 316: 440–443.
45. GRAY, R. & D. JOHNSTON. 1986. Multiple types of calcium channels in acutely-exposed neurons from adult hippocampus. Biophys. J. 49: 432a.
46. MADISON, D. V., A. P. FOX & R. W. TSIEN. 1987. Adenosine reduces an inactivating component of calcium current in hippocampal CA3 neurons (abstract). Biophys. J. 51: 30a.
47. ECKERT, R. & D. L. TILLOTSON. 1981. Calcium-mediated inactivation of the calcium conductance in caesium-loaded giant neurones of Aplysia californica. J. Physiol. (London) 314: 265–280.
48. BROWN, D. A. & W. H. GRIFFITH. 1983. Persistent slow inward calcium current in voltage-clamped hippocampal neurones of the guinea pig. J. Physiol. (London) 337: 303–320.
49. PITLER, T. A. & P. W. LANDFIELD. 1984. Inactivation of Ca^{2+}-dependent K^{+} conductance during repetitive intracellular stimulation of hippocampal neurons. Soc. Neurosci. Abstr. 10: 1075.
50. PITLER, T. A. & P. W. LANDFIELD. 1985. Ca^{2+}-mediated inactivation of Ca^{2+} spikes in hippocampal neurons. Soc. Neurosci. Abstr. 11: 519.
51. PITLER, T. A. & P. W. LANDFIELD. 1987. Probable Ca^{2+}-mediated inactivation of Ca^{2+} currents in mammalian brain neurons. Brain Res. 410: 147–153.
52. CAMPBELL, L. W., S-Y. HAO & P. W. LANDFIELD. 1988. Calcium-dependent inactivation of

calcium currents in hippocampal neurons: Effects of tetraethylammonium and nimodipine. Soc. Neurosci. Abstr. **14:** 138.
53. STORM, J. F. 1987. Action potential repolarization and a fast afterhyperpolarization in rat hippocampal pyramidal cells. J. Physiol. (London) **385:** 733–759.
54. THACH, W. T. 1975. Timing of activity in cerebellar dentate nucleus and cerebral motor cortex during prompt volitional movement. Brain Res. **88:** 237–247.
55. FOX, S. E. & J. B. RANCK, JR. 1981. Electrophysiological characteristics of hippocampal complex-spike cells and theta cells. Exp. Brain Res. **41:** 399–410.
56. WEST, M. O., E. P. CHRISTIAN, J. H. ROBINSON & S. A. DEADWYLER. 1981. Dentate granule cell discharge during conditioning. Relation to movement and theta rhythm. Exp. Brain Res. **44:** 287–294.

Evidence for Increased Calcium Buffering in Motor-Nerve Terminals of Aged Rats[a]

DEAN O. SMITH

Department of Physiology
University of Wisconsin
1300 University Avenue
Madison, Wisconsin 53706

INTRODUCTION

Depolarization of nerve terminals by invading action potentials opens voltage-sensitive calcium channels, causing a corresponding inward Ca^{2+} current that initiates vesicular fusion with the terminal membrane and, consequently, release of neurotransmitter. Release is then terminated as free calcium at the release sites returns to resting concentrations. This involves diffusion,[1] buffering,[2] and extrusion *via* a Na^+- or Ca^{2+}-Ca^{2+} exchange mechanism[3] and a Ca^{2+} ATPase.[4]

Calcium regulation during synaptic transmission may change during aging. Calcium uptake by depolarized synaptosomes decreases with age.[5] Moreover, synaptosomal membrane calcium transporters, specifically the Na^+-Ca^{2+} exchange pathway and the Ca^{2+}-Mg^{2+} ATPase, are demonstrably less efficient in tissue from aged preparations.[6] There is also indirect evidence of less efficient calcium clearance from the vicinity of the membrane in CA1 cells of hippocampal slices.[7]

These changes in calcium regulation may underlie age-related changes in neuronal function.[8,9] Therefore, we examined calcium regulation during aging in a well-defined physiological system, namely, the neuromuscular junction of diaphragm, soleus, and extensor digitorum longus (EDL) muscles of male Fischer 344 rats aged 10 (mature adult) and 25 (aged) months. Technical details of these experiments are presented in Hamilton and Smith[10] and Smith.[11]

Steady-State Intracellular Calcium

Changes in steady-state resting Ca^{2+} levels might be manifest as corresponding changes in m.e.p.p. frequency under resting, nonstimulated conditions. Thus, resting m.e.p.p. frequencies were determined.[11] M.e.p.p. frequencies normalized for the number of nerve terminals per end plate were similar between the two age groups for diaphragm and soleus.[11,12] However, the calculated rate per terminal in EDL increased by 79%. Assuming a fourth-power relationship between intracellular calcium, $[Ca^{2+}]_i$, and quantal transmitter release, this result indicates that there may be at least a 16% increase in steady-state, resting Ca^{2+} levels in the synaptic region of the EDL muscle.

[a]This work was supported by National Institutes of Health Grant AG01572 and the Muscular Dystrophy Foundation.

Calcium Dependence of Evoked Transmitter Release

The relationship between extracellular Ca^{2+} concentration, $[Ca^{2+}]_e$, and transmitter release was determined by measuring quantal content while increasing extracellular Ca^{2+} progressively from 0.5 to 1.2 mM. The results are illustrated in FIGURE 1. There are no apparent age-related differences in data obtained from diaphragm and soleus muscle, but there is a clear change in EDL (FIG. 1, bottom).

This relationship may be described simplisticly by a power function, $m = k[Ca^{2+}]_e^n$ with coefficient k and power n over this range of $[Ca^{2+}]_e$. The resulting estimates of n are presented in TABLE 1. The power coefficient, n, in the aged EDL muscles decreased by 22%, while the values in the other two muscles did not change significantly with age.

The theoretical relationship between quantal content and $[Ca^{2+}]_e$ was analyzed using a model in which transmitter release involves a transient increase in intracellular free Ca^{2+} which binds cooperatively to a saturable receptor.[13,14] This Ca^{2+}, which enters from the extracellular compartment through the voltage-dependent channels, adds to the resting, steady-state intracellular free Ca^{2+}, $[Ca^{2+}]_i$.

In general, this relationship may be formulated by the following equation:

$$m = k_{max} \left(\frac{k'[Ca^{2+}]_e + [Ca^{2+}]_i}{k'[Ca^{2+}]_e + [Ca^{2+}]_i + k_m} \right)^n \tag{1}$$

where k_{max} is the maximum quantal content, k_m is the concentration of intracellular Ca^{2+} needed to evoke half-maximal quantal release, and k' is a constant relating the magnitude of the Ca^{2+} influx to $[Ca^{2+}]_e$.

The relationship between $[Ca^{2+}]_e$ and the increase in intracellular free Ca^{2+} following an action potential, k', depends upon the amount of Ca^{2+} entering, the volume of the compartment within the cell membrane in which this Ca^{2+} is distributed, and the extent to which it is buffered. In motor-nerve terminals, peak Ca^{2+} currents, I_{Ca}, are about 10 pA over a 4-ms duration,[10] and it is assumed to distribute into a 100-nm wide shell inside the terminal membrane.[1] Assuming cylindrical terminals, the shell volume and the concentration of extracellularly derived Ca^{2+} within this shell, $[Ca^{2+}]_s$, were then calculated. Instantaneous buffering is further assumed to occur within this shell, and this is expressed as a bound/free ratio, B. The free Ca^{2+} concentration, $k'[Ca^{2+}]_e$, is thus equal to $[Ca^{2+}]_s/B$, from which k' may be calculated. Further details of the model are presented in Smith.[11]

The model for the 10-month data was evaluated initially by setting the power coefficient, n, to the value determined experimentally using the simplifying assumption of a power relationship (TABLE 1), k_m to 10 μM,[15] and k_{max} to 50000 quanta.[16] As shown in FIGURE 2A, a good fit was obtained. A similar evaluation was made using the value of n estimated from the 25-month data; however, k_{max}, the amount of Ca^{2+} inward current, and the total terminal volume were adjusted to account for the 36% decrease in the number of nerve terminals in the older preparations.[17] The result, also shown in FIGURE 2A, clearly indicates that some other factor must underlie the shift in the experimental data. Indeed, a much better fit of the 25-month data is obtained by using the 10-month parameters but reducing markedly the value of k_{max} to 2000 quanta (FIG. 2B).

By varying the parameters of this model (FIG. 3), further conclusions about the nature of the age-related changes in EDL may be drawn. For example, to produce changes similar to those observed during aging in EDL, $[Ca^{2+}]_i$ must increase at least 50-fold. This is much greater than the 16% increase deduced from the elevated m.e.p.p. frequency, and even then "basal" quantal content would be too high. However, increasing

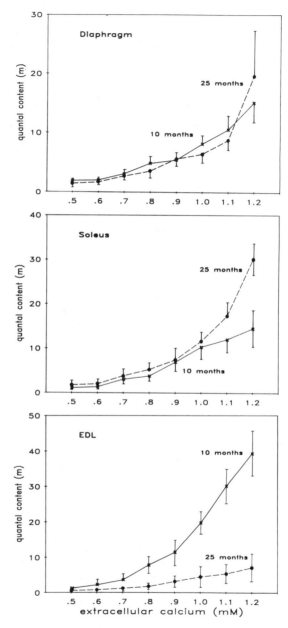

FIGURE 1. Relationship between quantal release and extracellular calcium. Each point represents the average (± SE) value of measurements obtained from at least 6 end plates from different animals aged 10 *(crosses, solid lines)* and 25 *(circles, dashed lines)* months. (From Smith.[11] Reprinted by permission from the *Journal of Neurophysiology.*)

either k_m (FIG. 3A) or B (FIG. 3B) or decreasing I_{Ca} (FIG. 3C) by about twofold in each case results in differences similar to those observed between 10- and 25-month animals. The magnitudes of these changes are fairly small, and the values are physiologically plausible in each case. Thus, to rule out age-related variation in any of these parameters as a possible cause for the observed shift observed in EDL from aged animals, further experimental evidence was obtained.

Calcium Currents

The magnitude of I_{Ca} was determined by direct recordings of Ca^{2+} currents entering the nerve terminals following depolarization by an action potential.[10] The results are summarized in FIGURE 4. Saturation was achieved at 2-mM $[Ca^{2+}]_e$, which is normal $[Ca^{2+}]_e$, in both age groups. In general, data obtained from the 10- and 25-month animals do not vary appreciably. Thus, age-related changes in I_{Ca} can be ruled out. By inference, significant differences in k_{max} and k_m also seem unlikely.

TABLE 1. Cooperative Action of Ca^{2+} on Transmitter Release[a]

| Age (mo) | Muscle | | |
	Diaphragm	Soleus	EDL
10	2.45 ± 0.20	3.25 ± 0.19	4.12 ± 0.13[b]
25	2.78 ± 0.30	3.26 ± 0.25	3.19 ± 0.15[b]

[a]Cooperativity, n, was determined by fitting a curve to the equation $m = k\ [Ca^{2+}]_e^n$ using least-squares techniques. The average (± SE) values of data obtained from 12 recording sites in 12 different animals are presented. (Data from Smith.[11])
[b]Corresponding values are different at the 0.05 level.

Calcium Clearance Following Action Potentials

An increase in Ca^{2+} buffering capacity, B, is expected to decrease Ca^{2+} clearance rates following an action potential. This should be manifest as prolonged decay rates of synaptic facilitation[18] and posttetanic augmentation.[19] Therefore, these two Ca^{2+}-dependent features of synaptic transmission were examined.

The extent of synaptic facilitation, which is primarily due to uncleared residual Ca^{2+}, provides an indirect measure of intracellular free Ca^{2+} in the immediate vicinity of the release sites.[18] Therefore, to determine whether the clearance rate of this residual Ca^{2+} was affected by age, paired pulses were delivered to the motor nerves at interpulse intervals ranging from 10 to 80 ms.[11] The degree of facilitation, F, was then calculated as the fractional increase in the amplitude of the e.p.p. in response to the second stimulus of the pair relative to the first. As the interimpulse period, t, increases, the facilitation becomes less, and over the range of short intervals used in this study, this decay may be described by a single exponential expression.[20] Therefore, the time constants of decay, τ, were estimated by fitting the data with the exponential expression $F = ke^{-t/\tau}$, where t represents interimpulse interval, using least-squares techniques.

The results, summarized in TABLE 2, indicate a significant (164%) age-related increase in the time constant for facilitation decay in EDL muscles only. This implies a corresponding increase in the time required to clear residual Ca^{2+} in this preparation and in the value of B.

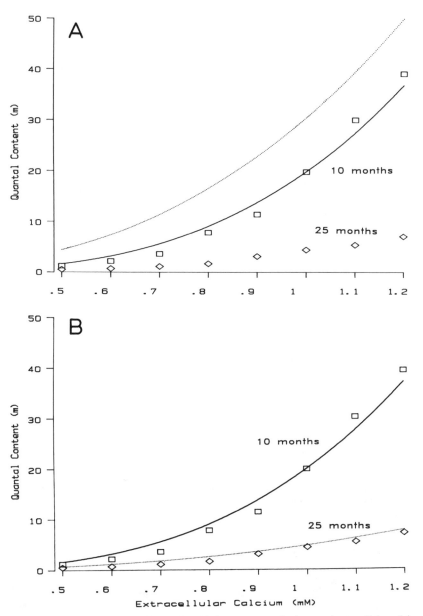

FIGURE 2. Observed and predicted relationship between quantal release and extracellular calcium in EDL muscle. The observed data are the same as shown in FIGURE 1. The curves were fit using EQUATION 1. **(A)** The 10-month data *(squares, solid curve)* were fit initially assuming $k_{max} = 50000$ quanta, $k_m = 10 \, \mu M$, $I_{Ca} = 10 \, pA$, $[Ca^{2+}]_i = 20 \, nM$, $n = 4.12$, and cylindrical terminals of length $5 \, \mu m$ and diameter $2 \, \mu m$. To account for a 36% decrease in the extent of terminal arborization, the 25-month data *(diamonds, dotted curve)* were fit after decreasing each of these parameter values by 36%; furthermore, $n = 3.19$ and $[Ca^{2+}]_i = 23.4 \, nM$ to reflect the apparent age-related increase. The resulting curve clearly does not fit the data. **(B)** The 10-month data were fit as in (A). Unlike (A), though, the 25-month data were fit assuming parameter values identical to those for the 10-month data, except $k_{max} = 2000$ and $n = 3.19$. (From Smith.[11] Reprinted by permission from the *Journal of Neurophysiology.*)

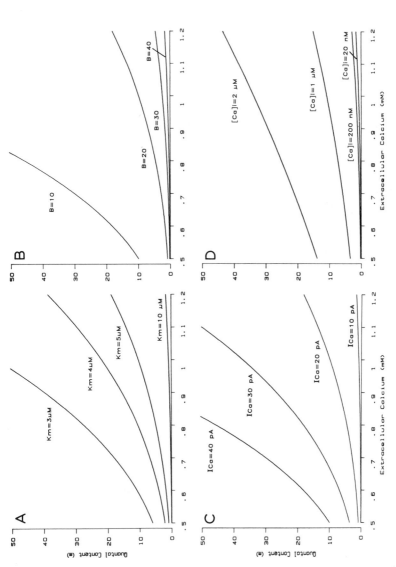

FIGURE 3. Predicted relationships between quantal release and extracellular calcium. In each case, n = 4, k$_{max}$ = 20000 quanta, k$_m$ = 10 μM, I$_{Ca}$ = 10 pA, B = 40, [Ca^{2+}]$_i$ = 20 nM, and terminal length and diameter = 5 and 2 μm, respectively, except for the noted changes. Current duration was assumed to be 2 ms. Families of curves were prepared to illustrate the effects of varying k$_m$ **(A)**, B **(B)**, I$_{Ca}$ **(C)**, and [Ca^{2+}]$_i$ **(D)**. (From Smith.[11] Reprinted by permission from the *Journal of Neurophysiology*.)

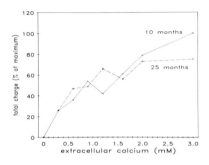

FIGURE 4. Nerve-terminal calcium current. After blocking end-plate currents and terminal K^+ currents, the inward calcium current following depolarization by an action potential was recorded. "Total charge" refers to the accumulated charge carried across the membrane of Ca^{2+}; it was determined by integrating the current. Each point represents the average value obtained from 7 animals aged 10 months *(crosses, solid lines)* and 3 animals aged 25 months *(circles, dashed lines);* the entire concentration range was tested for each recording site.

Following high-frequency stimulation, m.e.p.p. frequency is elevated above resting values for several minutes.[19] This posttetanic process can be resolved into four distinct phases: fast and slow facilitation, augmentation, and potentiation. The decay time course for each phase can be described by an exponential expression.[21] Moreover, the augmentation component, which decays with a time constant of about 10 s, is also quite likely due to residual calcium[19,22] as is facilitation. Thus, in related experiments, m.e.p.p. rate was recorded initially, the nerve was stimulated tetanically (500 impulses at 25 Hz), and then m.e.p.p. rate during the ensuing 20 s was measured at 2-s intervals.

Decay time constants were calculated from these data, and as in the facilitation analysis, there was a significant increase in augmentation decay rate only in the EDL muscles from aged animals (TABLE 2).[11] This provides further support for the notion that Ca^{2+} clearance is less efficient in the EDL muscle from aged animals. This is consistent with an age-related increase in the value of B.

CONCLUSIONS

These data provide direct evidence for muscle-specific, age-related changes in calcium regulation at intracellular synaptic release sites. Specifically in EDL, resting levels of intracellular Ca^{2+} apparently increase, and the power, n, of the relationship between $[Ca^{2+}]_e$ and transmitter release following an action potential also decreases with age.

The power relationship, n, between quantal release and $[Ca^{2+}]_e$ is quite dependent

TABLE 2. Time-Constants of Decay^a

	Muscle		
Age (mos)	Diaphragm	Soleus	EDL
Synaptic facilitation	35.9 ± 11.3	40.6 ± 17.1	13.5 ± 5.3^b
10			
25	39.3 ± 5.8	45.6 ± 16.7	35.6 ± 8.8^b
Posttetanic augmentation	9.3 ± 2.9	12.5 ± 2.6	10.3 ± 3.1^b
10			
25	13.8 ± 2.6	10.3 ± 2.6	23.4 ± 5.2^b

^{a}Time constants, τ, were estimated by fitting the equation $F = k\, e^{-t/\tau}$ to the data relating facilitation to interimpulse interval or the m.e.p.p. frequency to the time after tetanic stimulation. Averages values ($\pm$ SE) of data obtained from at least 8 different recording sites of each age group are presented. (Adapted from Smith.[11])

^{b}Corresponding values are different at the 0.05 level.

upon several parameters which need change only slightly to causes a shift resembling that observed experimentally between the ages of 10 and 25 months. Direct measurements, though, rule out changes in I_{Ca}, and theoretical considerations (*e.g.*, FIG. 3) demonstrate that the calculated changes in $[Ca^{2+}]_i$ are too small to affect this relationship meaningfully. In contrast, the prolonged decay times for synaptic facilitation and augmentation provide sound evidence for reduced Ca^{2+} clearance rates from synaptic release sites and, therefore, the capacity to buffer Ca^{2+} entering during the action potential, B. Indeed, age-related increase in the value of B is the parsimonious conclusion to be drawn from the results of this study.

These age-related phenomena have been observed only in EDL and not in diaphragm or soleus. Unlike the diaphragm, a fast-twitch vital muscle, and the soleus, a slow-twitch postural muscle, the EDL is a fast-twitch limb muscle which is seldom recruited.[23] Over the 25-month lifespan of the older animals, the occurrences of depolarization-dependent Ca^{2+} influx into the nerve terminals may conceivably become insufficient to preclude changes in the expression of proteins regulating this Ca^{2+}. Further disparities between the diaphragm and EDL include notable differences in the effects of age on terminal architecture[17,24] and on acetylcholine metabolism.[25] The relative extent of motoneuron recruitment may thus play a major role in determining the synaptic responses to aging, including those that are Ca^{2+}-dependent.

REFERENCES

1. ZUCKER, R. S. & N. STOCKBRIDGE. 1983. Presynaptic calcium diffusion and the time courses of transmitter release and synaptic facilitation at the squid giant synapse. J. Neurosci. **3**: 1263–1269.
2. BLAUSTEIN, M., R. W. RATZLAFF & E. SCHWEITZER. 1977. Calcium buffering in presynaptic nerve terminals. II. Kinetic properties of the nonmitochondrial Ca sequestration mechanism. J. Gen. Physiol. **72**: 43–66.
3. BLAUSTEIN, M. & A. C. ECTOR. 1975. Carrier-mediated sodium-dependent and calcium-dependent calcium efflux from pinched-off presynaptic nerve terminals (synaptosomes) *in vitro*. Biochim. Biophys. Acta **419**: 295–308.
4. SORENSEN, R. G. & H. R. MAHLER. 1981. Calcium-stimulated adenosine triphosphates in synaptic membranes. J. Neurochem. **37**: 1407–1418.
5. PETERSON, C. & G. E. GIBSON. 1983. Aging and 3,4-diaminopyridine alter synaptosomal calcium uptake. J. Biol. Chem. **258**: 11482–11485.
6. MICHAELIS, M. L., K. JOHE & T. E. KITOS. 1984. Age-dependent alterations in synaptic membrane systems for Ca^{2+} regulation. Mech. Ageing Dev. **25**: 215–225.
7. LANDFIELD, P. & T. PITLER. 1984. Prolonged Ca^{2+}-dependent afterhyperpolarizations in hippocampal neurons of aged rats. Science **226**: 1089–1092.
8. GIBSON, G. E. & C. PETERSON. 1987. Calcium and the aging nervous system. Neurobiol. Aging **8**: 329–343.
9. KHATCHATURIAN, Z. 1982. Towards theories of brain aging. *In* Handbook of Studies on Psychiatry and Old Age. D.W. Kay & G.D. Burrows, Eds.:7–30. Elsevier. Amsterdam.
10. HAMILTON, B. R. & D. O. SMITH 1987. Calcium currents in mammalian motor nerve terminals. Soc. Neurosci. Abstr. **13**: 312.
11. SMITH, D. O. 1988. Muscle-specific decrease in presynaptic calcium dependence and clearance during neuromuscular transmission in aged rats. J. Neurophysiol. **59**: 1069–1082.
12. SMITH, D. O. 1984. Acetylcholine storage, release and leakage at the neuromuscular junction of mature adult and aged rats. J. Physiol. (Lond.) **347**: 161–176.
13. COHEN, I. & W. VAN DER KLOOT. 1985. Calcium and transmitter release. Int. Rev. Neurobiol. **27**: 299–336.
14. DODGE, F. A., JR. & R. RAHAMIMOFF. 1967. Cooperative action of calcium ions in transmitter release at the neuromuscular junction. J. Physiol. (Lond.) **193**: 419–432.
15. KELLY, R. B., J. W. DEUTSCH, S. S. CARLSON & J. A. WAGNER. 1979. Biochemistry of neurotransmitter release. Annu. Rev. Neurosci. **2**: 399–446.

16. KATZ, B. & R. MILEDI. 1979. Estimates of quantal content during "chemical potentiation" of transmitter release. Proc. R. Soc. London B **205:** 369–378.
17. ROSENHEIMER, J. L. & D. O. SMITH. 1985. Differential changes in the end-plate architecture of functionally diverse muscles during aging. J. Neurophysiol. **53:** 1567–1581.
18. CHARLTON, M. P., S. J. SMITH & R. S. ZUCKER. 1982. Role of presynaptic calcium ions and channels in synaptic facilitation and depression at the squid giant synapse. J. Physiol. (Lond.) **323:** 173–193.
19. ERULKAR, S. D. & R. RAHAMIMOFF. 1978. The role of calcium ions in tetanic and post-tetanic increase of miniature end-plate potential frequency. J. Physiol. (Lond.) **278:** 501–511.
20. MALLART, A. & A. R. MARTIN. 1967. An analysis of facilitation of transmitter release at the neuromuscular junction of the frog. J. Physiol. (Lond.) **193:** 679–694.
21. ZENGEL, J. E. & K. L. MAGLEBY. 1982. Augmentation and facilitation of transmitter release. A quantitative description at the frog neuromuscular junction. J. Gen. Physiol. **80:** 583–611.
22. FOGELSON, A. L. & R. S. ZUCKER. 1985. Presynaptic calcium diffusion from various arrays of single channels. Implications for transmitter release and synaptic facilitation. Biophys. J. **48:** 1003–1017.
23. HENNIG, R. & T. LOMO. 1985. Firing patterns of motor units in normal rats. Nature **314:** 164–166.
24. ROSENHEIMER, J. L. 1985. Effects of chronic stress and exercise on age-related changes in end-plate architecture. J. Neurophysiol. **53:** 1582–1589.
25. SMITH, D. O. 1989. Acetylcholine synthesis and release in the infrequently used EDL muscle of mature and aged rats. J. Neurochem. In press.

Macromolecular Organization of the Neuromuscular Postsynaptic Membrane[a]

STANLEY C. FROEHNER

Department of Biochemistry
Dartmouth Medical School
Hanover, New Hampshire 03756

One of the hallmarks of a chemical synapse is the precise distribution of ion channels necessary for rapid signal transmission. Voltage-dependent calcium channels that mediate transmitter exocytosis are thought to be located at active zones, presynaptic sites specialized for vesicle fusion with the membrane. On the postsynaptic side, both acetylcholine receptors[1] and voltage-sensitive sodium channels[2] are highly concentrated at motor endplates. ACh receptors are restricted to the crests of the postjunctional folds and are present at these sites at remarkably high concentrations—approximately 8,000 receptors per square micron of membrane surface area.[1,3] Since ACh receptors are transmembrane proteins that are inherently capable of diffusion within the plane of the membrane, the muscle cell must elaborate a specialized macromolecular structure that anchors receptors at synaptic sites. By analogy with specialized membrane domains in other cells, it seems likely that a membrane-associated cytoskeleton will play a critical role in this process.[4] In this paper, I shall review the evidence that the postsynaptic cytoskeleton is involved in ACh receptor organization, describe the common cytoskeletal proteins that are concentrated at the neuromuscular synapse, and discuss other proteins that may serve to link ACh receptors to this cytoskeletal structure.

The involvement of the cytoskeleton in neuromuscular postsynaptic structure gained considerable support from experiments demonstrating that several proteins typically associated with an actin-based cytoskeleton are found at high concentrations in association with the endplate membrane. Along with a form of actin distinct from that of the myofibrils,[5] α-actinin,[6] filamin,[6] vinculin,[6] and talin[7] are major components of the postsynaptic cytoskeleton. Although these proteins are thought to be involved in establishing and/or maintaining high concentrations of receptors at the endplate, their interaction with the receptor is not understood.

Proteins of the postsynaptic cytoskeleton that may play a role in anchoring receptors have been identified in purified postsynaptic membrane preparations from *Torpedo* electric organ. These membrane vesicles are highly enriched in ACh receptor, which comprises as much as 50% of the total protein. In addition to the receptor, several other proteins are also found in these membrane preparations. Some of these proteins can be removed by procedures that extract peripheral membrane proteins, such as treatment with alkaline pH[8] or low concentrations of lithium diiodosalicylate.[9] Functional properties of the receptor, such as ligand binding and the conformational transitions necessary for channel opening and desensitization, are not affected by these treatments.[8] However, the extraction of peripheral membrane proteins with alkaline pH is accompanied by a dra-

[a]Work in the author's laboratory is supported by research grants from the National Institutes of Health (NS14871) and the Muscular Dystrophy Association. Dr. Froehner is an Established Investigator of the American Heart Association.

115

matic increase in the translational[10-12] and rotational[13,14] mobility of the receptor. These results suggest that peripheral membrane proteins are important in anchoring receptors at synaptic sites.

We have characterized three peripheral membrane proteins that can be extracted from *Torpedo* postsynaptic membranes by alkaline treatment. The most abundant of these, a protein of approximately 43,000 molecular weight, has been studied in considerable detail by our laboratory and by others. The 43kDa protein (also called 43K, ν_1, RAPsyn) is clearly distinct from actin and creatine kinase,[9] two major proteins in the electrocyte cell. We have prepared monoclonal antibodies to the 43kDa protein[15] and used them to determine the location of this protein in electrocytes and in mammalian skeletal muscle. In the electrocyte, mabs to the 43kDa protein label the cytoplasmic side of the postsynaptic membrane[16] (FIG. 1). Dense labeling occurs nearest the nerve terminal while only rarely

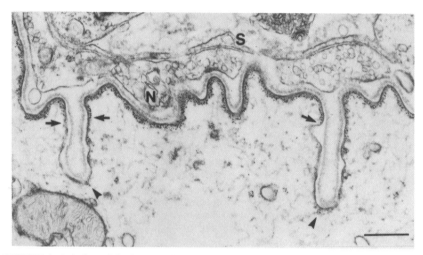

FIGURE 1. Labeling of the innervated face of *Torpedo* electrocyte with anti-43kDa monoclonal antibody. Colloidal gold particles can be seen on the cytoplasmic side of the postsynaptic membrane, and are most concentrated near the tops of the postjunctional folds. Dense labeling ends abruptly (*arrows*), although occasional patches can be seen in the deeper portions of the invaginations (*arrowheads*). Labeling in all other regions is at background levels, as assessed with control mouse IgG (not shown). N, nerve terminals; S. Schwann cells. Bar, 500 nm. (From Sealock *et al.*[16] Reprinted by permission from the *Journal of Cell Biology*.)

in the troughs of the membrane folds. This coincides with the distribution of the ACh receptor. Double label experiments, using a combination of 43kDa mabs and antibodies to the ACh receptor clearly demonstrate that these two proteins share a similar if not identical distribution in the postsynaptic membrane.[16] In addition, the 43kDa protein can be chemically cross-linked to the beta subunit of the receptor.[17] These results, along with the finding that the ACh receptor and the 43kDa protein occur in approximately equimolar concentrations,[18] suggest that the 43kDa protein may interact directly with cytoplasmic domains of the receptor.

The 43kDa protein appears to interact with other components of the postsynaptic membrane. Purified 43kDa protein binds tightly to liposomes devoid of other proteins.[19]

This interaction is relatively independent of the lipid composition of the liposome and is reversed by alkaline pH. The recent demonstration that 43kDa protein in cultured mouse muscle cells are derivatized with myristic acid at the amino-terminus[20] suggests a possible structural feature responsible for this property. Other parts of the protein, however, may also be involved in this interaction (see below). Under oxidative conditions, the membrane-bound form of 43kDa protein readily polymerizes via disulfide bond formation to form dimers and oligomers.[19] Although these multimers do not occur naturally, at least in adult tissue, this result does indicate that 43kDa proteins are closely aligned with each other along the inner surface of the membrane. Finally, there is evidence to suggest that the 43kDa protein binds actin.[21] Although the physiological importance of this interaction remains to be demonstrated, this finding raises the intriguing possibility that the 43kDa protein may serve as a link between the ACh receptor and the underlying actin-based cytoskeleton.

Recently, the primary structure of the 43kDa protein from *Torpedo* has been determined by protein sequencing[22] and by cDNA cloning.[23] Full length cDNA clones for the mouse skeletal muscle protein have also been isolated and sequenced[24] (also Froehner, unpublished results). Approximately 70% of the amino acid residues are identical in the *Torpedo* and mouse protein. However, no significant homology with any other protein in the protein sequence data banks was found. Thus, the 43kDa protein appears to represent a member of a unique protein family. Two interesting structural features are conserved between species. Five cysteine residues are arranged with the following spacing: -Cys-2X-Cys-15X-Cys-7X-Cys-8X-Cys- (where X represents any amino acid residue). This motif has been found in two other proteins, phospholipase A_2 and protein kinase C, in the domains that interact with the lipid bilayer.[25] In the 43kDa protein, this part of the protein may also serve to anchor the protein to the lipid bilayer, along with the myristic acid on the amino terminus. A potential phosphorylation site occurs near the carboxy terminus of the protein. Since phosphorylation of the 43kDa protein has not been demonstrated *in vivo*, the role of this posttranslational modification in 43kDa protein function is unknown.

The close association of the ACh receptor with the 43kDa protein is illustrated in FIGURE 2. This working model accounts for several features of the interaction of the 43kDa protein with the bilayer, through both the fatty acyl group and the cysteine motif, and with the beta subunit of the receptor. Cysteine groups on adjacent 43kDa proteins are situated close to one another so that disulfide bonds may form under oxidative conditions. Since this model shows only the receptor and the 43kDa protein, it is clearly incomplete. Other proteins are also found in close association with this structure. Two proteins of 58,000 (58K) and 280,000 (280K), first identified in *Torpedo* membranes, appear to be part of the postsynaptic cytoskeleton.

The 58K protein is present in much smaller amounts than the receptor in *Torpedo* membrane preparations.[26] A protein of very similar biochemical and immunological characteristics has been identified in mammalian skeletal muscle. The 58K protein is highly concentrated at postsynaptic sites in muscle but it is also found associated with extrasynaptic membrane (FIG. 3). The high concentration at muscle endplates is not due entirely to the postjunctional folds since synapses in neonatal rat muscle, which lack folds, are also stained more intensely than extrasynaptic membrane. It is noteworthy that maintenance of the high postsynaptic concentration of the 58K protein requires the presence of the nerve. Denervation of muscle causes a dramatic decrease in 58K protein at the synaptic site. The 58K protein is found in other tissues, especially brain and kidney, and clearly has a role in cellular functions not restricted to the postsynaptic membrane.

The 280K protein has a distribution in muscle very similar to that of the 58K protein. It is located on the cytoplasmic side of the postsynaptic membrane and has the characteristics of a peripheral membrane protein.[27] In contrast to the 58K and 43kDa proteins, the 280K protein is not as easily removed from *Torpedo* vesicles by lithium diiodosalicylate. Under certain conditions, virtually all of the 43kDa and 58K proteins can be removed from the membrane, while approximately 50% of the 280K protein remains membrane-associated. Thus, the 43kDa and 58K proteins do not appear to be necessary for association of the 280K protein with the membrane cytoskeleton. Like the 58K protein, the 280K protein has a widespread tissue distribution.

The evidence that these proteins are involved in anchoring ACh receptors at synaptic sites is indirect. Clearly, a direct test of this hypothesis is needed. We envision two approaches to this problem. First, the availability of cDNA clones encoding the 43kDa protein will allow us to study the effect of 43kDa protein on the distribution of ACh receptors, either in *Xenopus* oocytes injected with mRNA for each of the components, or in stably transfected mammalian cell lines. A second approach is to use anti-sense RNA or anti-sense oligodeoxynucleotides to block expression of the 43kDa protein in muscle cell lines that organize their ACh receptors into clusters. Once cDNA clones for the 58K and 280K proteins have been isolated, the similar approaches could be used to determine if these proteins are important in anchoring receptors at synaptic sites.

Are these proteins associated only with the neuromuscular postsynaptic membrane or are they also found at neuronal synapses? The study of neuronal ACh receptors has revealed a family of related but distinct receptors expressed in different regions of the central nervous system.[28] The association of postsynaptic cytoskeleton proteins with neuronal ACh receptors has not yet been investigated. However, the glycine receptor, when purified from mammalian spinal cord, has associated with it a 93kDa peripheral membrane protein.[29] It is clearly important to determine if closely associated peripheral membrane proteins are a common feature of synaptic ligand-gated ion channels and, if so, what roles they play in synaptic function.

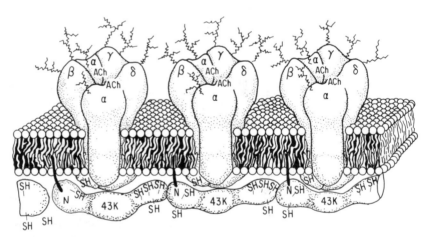

FIGURE 2. A model depicting possible interactions of the 43kDa protein with the ACh receptor and with the membrane. The receptor is a transmembrane glycoprotein pentamer of $\alpha_2\beta\gamma\delta$ subunit structure. The 43kDa protein is shown as a sulfhydryl (SH)-rich protein closely associated with the cytoplasmic domains of the receptor subunits. The *dark line* extending from the amino terminus (N) of the 43kDa protein represents myristate covalently attached to the terminal glycine residue and interacting with the lipid bilayer. A specific arrangement of cysteine residues near the carboxy terminus (see text) may also anchor the 43kDa protein to the membrane.

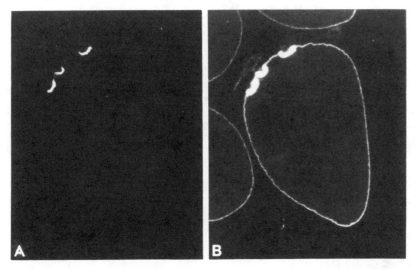

FIGURE 3. Sections of rat skeletal muscle stained **(A)** with rhodamine α-bungarotoxin to label acetylcholine receptors at neuromuscular synapses and **(B)** with anti-58K protein monoclonal antibodies and fluorescein-conjugated anti-mouse IgG. The 58K protein is highly concentrated at synaptic sites but is also found associated with the extrasynaptic membrane. (From Froehner *et al.*[26] Reprinted by permission from the *Journal of Cell Biology.*)

REFERENCES

1. FERTUCK, H. C. & M. M. SALPETER. 1974. Localization of acetylcholine receptor by ^{125}I-labeled alpha bungarotoxin binding at mouse motor endplates. Proc. Natl. Acad. Sci. USA **71:** 1376–1378.

2. BEAM, K. G., J. H. CALDWELL & J. T. CAMPBELL. 1985. Na channels in skeletal muscle concentrated near the neuromuscular junction. Nature **313:** 588–590.

3. MATTHEWS-BELLINGER, J. & M. M. SALPETER. 1978. Distribution of acetylcholine receptors at frog neuromuscular junctions with a discussion of some physiological implications. J. Physiol. **279:** 197–213.

4. FROEHNER, S. C. 1986. The role of the postsynaptic cytoskeleton in AChR organization. Trends Neurosci. **9:** 37–41.

5. HALL, Z. W., B. W. LUBIT & J. H. SCHWARTZ. 1981. Cytoplasmic actin in postsynaptic structures at the neuromuscular junction. J. Cell Biol. **90:** 789–792.

6. BLOCH, R. J. & Z. W. HALL. 1983. Cytoskeletal components of the vertebrate neuromuscular junction: Vinculin, a-actinin and filamin. J. Cell Biol. **97:** 217–223.

7. SEALOCK, R., B. PASCHAL, M. BECKERLE & K. BURRIDGE. 1986. Talin is a postsynaptic component of the rat neuromuscular junction. Exp. Cell Res. **163:** 143–150.

8. NEUBIG, R. R., E. K. KRODEL, N. D. BOYD & J. B. COHEN. 1979. Acetylcholine and local anesthetic binding to *Torpedo* nicotinic post-synaptic membranes after removal of non-receptor peptides. Proc. Natl. Acad. Sci. USA **76:** 690–694.

9. PORTER, S. & S. C. FROEHNER. 1983. Characterization and localization of the M_r 43,000 proteins associated with acetylcholine receptor-rich membranes. J. Biol. Chem. **258:** 10034–10040.

10. BARRANTES, F. J., D.-CH. NEUGEBAUER & H. P. ZINGSHEIM. 1980. Peptide extraction by alkaline treatment is accompanied by rearrangement of the membrane-bound acetylcholine receptor from *Torpedo marmorata*. FEBS Lett. **112:** 73–78.

11. BLOCH, R. J. & S. C. FROEHNER. 1986. The relationship of the postsynaptic 43K protein to

acetylcholine receptors in receptor clusters isolated from cultured rat myotubes. J. Cell Biol. **104**: 645–654.

12. CARTAUD, J., A. SOBEL, A. ROUSSELET, P. F. DEVAUX & J. P. CHANGEUX. 1981. Consequences of alkaline treatment for the ultrastructure of the acetylcholine-receptor rich membranes from *Torpedo marmorata* electric organ. J. Cell Biol. **90**: 418–426.

13. LO, M. M. S., P. B. GARLAND, J. LAMPRECHT & E. A. BARNARD. 1980. Rotational mobility of the membrane-bound acetylcholine receptor of *Torpedo* electric organ measured by phosphorescence depolarization. FEBS Lett. **111**: 407–412.

14. ROUSSELET, A., J. CARTAUD, P. F. DEVAUX & J.-P. CHANGEUX. 1982. The rotational diffusion of the acetylcholine receptor in *Torpedo marmorata* membrane fragments studied with a spin-labelled alpha-toxin: Importance of the 43,000 protein(s). EMBO J. **1**: 439–445.

15. FROEHNER, S. C. 1984. Peripheral proteins of postsynaptic membranes from *Torpedo* electric organ identified with monoclonal antibodies. J. Cell Biol. **99**: 88–96.

16. SEALOCK, R., B. E. WRAY & S. C. FROEHNER. 1984. Ultrastructural localization of the M_r 43,000 protein and the acetylcholine receptor in *Torpedo* postsynaptic membranes using monoclonal antibodies. J. Cell Biol. **98**: 2239–2244.

17. BURDEN, S. J., R. L. DEPALMA & G. S. GOTTESMAN. 1983. Crosslinking of proteins in acetylcholine receptor-rich membranes: Association between the beta-subunit and the 43kD subsynaptic protein. Cell **35**: 687–692.

18. LAROCHELLE, W. J. & S. C. FROEHNER. 1987. Comparison of the postsynaptic 43 kDa protein from muscle cells that differ in acetylcholine recognition clustering activity. J. Biol. Chem. **262**: 8190–8195.

19. PORTER, S. & S. C. FROEHNER. 1985. Interaction of the 43K protein with components of *Torpedo* postsynaptic membranes. Biochemistry **24**: 425–432.

20. MUSIL, L. S., C. CARR, J. B. COHEN & J. P. MERLIE. 1988. Acetylcholine receptor-associated 43K protein contains covalently-bound myristate. J. Cell Biol. **107**: 1113–1121.

21. WALKER, J. H., C. M. BONSTEAD & V. WITZEMANN. 1984. The 43K protein, ν_1, associated with acetylcholine receptor containing membrane fragments is an actin-binding protein. EMBO J. **3**: 2287–2296.

22. CARR, C., D. MCCOURT & J. B. COHEN. 1984. The 43kDa protein of *Torpedo* nicotinic postsynaptic membranes: Purification and determination of primary sequence. Biochemistry **26**: 7090–7102.

23. FRAIL, D. E., J. MUDD, V. SHAH, C. CARR, J. B. COHEN & J. P. MERLIE. 1987. cDNAs for the postsynaptic M_r 43,000 protein of *Torpedo* electric organ encode two proteins with different carboxy termini. Proc. Natl. Acad. Sci. USA **84**: 6302–6306.

24. FRAIL, D. E., L. L. MCLAUGHLIN, J. MUDD & J. P. MERLIE. 1988. Identification of the mouse muscle 43-dalton acetylcholine receptor-associated protein (RAPsyn) by cDNA cloning. J. Biol. Chem. **263**: 15602–15607.

25. MARAGANORE, J. M. 1987. Structural elements for protein-phospholipid interactions may be shared in protein kinase C and phospholipase A_2. Trends Biochem. Sci. **12**: 176–177.

26. FROEHNER, S. C., A. A. MURNANE, M. TOBLER, H. B. PENG & R. SEALOCK. 1987. A postsynaptic M_r 58,000 (58K) protein concentrated at acetylcholine receptor-rich sites in *Torpedo* electroplaques and skeletal muscle. J. Cell Biol. **104**: 1633–1646.

27. WOODRUFF, M. L., J. THERIOT & S. J. BURDEN. 1987. 300 kD subsynaptic protein copurifies with acetylcholine receptor-rich membranes and is concentrated at neuromuscular synapses. J. Cell Biol. **104**: 939–946.

28. GOLDMAN, D., E. DENERIS, A. KOCHBAR, J. BOULTER, J. PATRICK & S. HEINEMANN. 1986. Members of a nicotinic acetylcholine receptor gene family are expressed in different regions of the mammalian central nervous system. Cell **48**: 965–973.

29. SCHMITT, B., P. KNAUS, C.-M. BECKER & H. BETZ. 1987. The M_r 93,000 polypeptide of the postsynaptic glycine receptor is a peripheral membrane protein. Biochemistry **26**: 805–811.

Intracellular Ionic Calcium and the Cytoskeleton in Living Cells

MICHAEL L. SHELANSKI

Department of Pathology
and
The Center for Neurobiology and Behavior
College of Physicians and Surgeons
Columbia University
14th Floor, Room 436
630 West 168th Street
New York, New York 10032

The calcium ion is present in the cytoplasm of living cells at resting concentrations of between 50 and 150 nanomolar while total cellular calcium concentrations, primarily bound non-ionic calcium, are in the millimolar range. Over the past several decades this ion has been suggested to play a critical role in the control of a wide range of cellular functions including excitability, endocytosis, exocytosis, motility, and cell division. Attention has focused on calcium activation of various enzymes and on an expanding class of calcium-binding proteins. Except for the case of electrophysiological responses, most of our insight into calcium control mechanisms has been derived from studies on cell fractions rather than intact cells. The exceptions to this have been those which have used cells large enough to penetrate with calcium-sensitive ion selective electrodes or cells which have been injected with the calcium-sensitive photoprotein aequorin. Both of these approaches have serious limitations to their application. In the past five years the development and exploitation of calcium-sensitive fluorescent reporter dyes[1] has enabled the measurement of nanomolar concentrations of the cytoplasmic ionic calcium ($[Ca^{2+}]_i$) in living cells. In the studies reviewed here, I shall move away from the nervous system, which has been the focus of this symposium, to non-neural cells in culture in order to discuss the possible role of the calcium ion in the local control of cytoskeletal function in interphase cells, in mitosis, and finally in Alzheimer's disease.

Effects of Elevation of Intracellular Calcium and Calmodulin by Microinjection into Interphase Cells

Calcium mediation of microtubule disassembly is shifted from the millimolar to the micromolar range in the presence of calmodulin (CaM).[2] Therefore, it was of interest to know what effect, if any, the introduction of excess calmodulin into the cell might have on cytoskeletal integrity. To accomplish this we introduced excess calmodulin—in amounts adequate to approximately triple the basal CaM levels—into 3T3 cells by microinjection and followed the effect of this treatment on the cells by cytologic staining of microtubules and actin stress fibers. When CaM which had been preincubated with 0.1 mM EGTA was injected, no alterations whatsoever were seen in the actin or microtubular cytoskeletons. Similarly, when calcium alone (1 mM) was injected no changes were observed. However, when calmodulin is injected with sufficient calcium to saturate its calcium-binding sites the results are quite dramatic. The region of the cell radiating out from the site of injection shows complete absence of microtubules.[3] This region ends

121

quite abruptly and beyond it the microtubular cytoskeleton is normal in appearance. The microtubuler bundles appear to start at the end of this zone of "clearing" and then to course to the periphery as they would in a non-injected cell. A similar, but less extensive, rearrangement of the actin stress fibers is also seen though it is unclear whether this is a direct effect on the actin cables or secondary to the alterations in the microtubuler cytoskeleton. The sharp localization of microtubule disruption was unexpected and suggested a limitation of the mobility of the injected calmodulin in the cytoplasm. To test this hypothesis cells were injected either with EGTA-calmodulin or with Ca-calmodulin, then treated with the ionophore A-23187 to allow calcium to enter the cell. In this case, the cells injected with Ca-calmodulin still showed the limited area of microtubular breakdown while those injected with EGTA-calmodulin showed a global loss of microtubules in the cell. In either case microtubular dissolution was seen at external calcium concentration of 25 μM. In the absence of calmodulin injection calcium concentrations of 400 μM were necessary to obtain breakdown. These results are consistent with the idea that calcium, in addition to activating calmodulin, acts to bind it to cellular elements near the point of injection and limits its cytoplasmic mobility. The EGTA-calmodulin, on the other hand, is freely mobile in the cytoplasm and when subsequently activated by means of calcium entering through the ionophore causes microtubule breakdown throughout the cell. Direct immunological detection of injected Ca-calmodulin and EGTA-calmodulin shows the restriction of the former but not the latter. Subsequent studies using fluorescence photobleaching have confirmed this calcium-mediated restriction of the cytoplasmic mobility of calmodulin.[4]

Thus in the case of calmodulin, and possibly with other calcium-binding proteins, calcium is capable of controlling its cytoplasmic localization. This has important implications for the regulation of microtubule disassembly in living cells where the restriction of calmodulin to a specific region might underlie the localized disassembly of microtubules. This would, however, require a mechanism where the cell could locally control the cytoplasmic ionic calcium concentration.

Calcium and Mitotic Progression

The introduction of the calcium-sensitive report quin-2 and subsequently fura-2 enabled the measurement of $[Ca^{2+}]_i$ in living cells.[1,5] Our initial measurements were photometric and showed a modest fall in $[Ca^{2+}]_i$ in mitosis as compared to interphase.[6] To extend these observations we switched to the mitotic endosperm of the African Globe Lily, *Haemanthus Katherinae*, and repeated our studies using video image processing. In this approach, the video camera serves as a photometric array allowing intensity data to be taken simultaneously from over 250,000 points in the microscope field. The use of these very large plant cells allowed us to follow local changes in $[Ca^{2+}]_i$ as the cell progressed through mitosis. To recapitulate briefly, the distribution of calcium ion is uniform throughout the cytoplasm in interphase cells. At the beginning of prophase, calcium is elevated in an annular pattern corresponding to the area of nuclear envelope breakdown. Metaphase shows a uniform distribution except for the area of the chromosomes on the metaphase plate which is lower. As the chromosomes start to separate in anaphase the calcium ion concentration increases at the mitotic spindle poles covering the area of shortening microtubules from the kinetochore to the pole.[7] Of particular interest is that the area of calcium localization in anaphase corresponds to the localization of calmodulin in these cells at the same point in mitosis.[8] Thus it appears, at least in the case of *Haemanthus*, that calcium and calmodulin are both situated in a manner concordant with their mediation of the disassembly of kinetochore-to-pole microtubules in anaphase.

Telophase is marked by a return to uniform calcium ion distribution in the daughter cells. About 50% of the telophase cells also show a "line" of high calcium along the phragmoplast where the new cell plate is being formed. While the mechanism of generation of these long-lasting gradients is unclear, the cells in these experiments are grown and studied in citrate buffer at pH 5.7 making it highly unlikely that the calcium is derived from extracellular sources.

Fura-2,[5] with its greater quantum efficiency, allowed us to extend the image processing approach to smaller mammalian PtK2 cells in culture. The results here are similar to those in *Haemanthus* though no prophase ring could be identified. In metaphase an area of lowered $[Ca^{2+}]_i$ develops parallel to the metaphase plate while in anaphase $[Ca^{2+}]_i$ increases at the region extending from the kinetochore to the pole. In about half of the cells observed this region of elevated calcium is limited to the area between the kinetochore and the pole, while in the other cells it extends circumferentially as a thin rim of high calcium. With telophase the calcium distribution once again becomes uniform. Thus the major changes are from late metaphase through anaphase and consist of a decrease of $[Ca^{2+}]_i$ between the separating chromosomes where the polar microtubules involved in the separation of the poles would be expected to lengthen and an elevation in the kinetochore-to-pole region where the kinetochore-to-pole microtubules would be expected to shorten. One potential puzzle is posed by the fact that both the polar and the kinetochore microtubules pass through the region of elevated calcium but only the latter is affected. Some insight into this is given by the fact that the kinetochore microtubules are cold stable while the polar microtubules are not. Cold stability can be conferred on microtubules *in vitro* by the STOP protein which also is capable of binding calmodulin. Thus the sensitivity to cellular levels of ionic calcium may be due to the selective binding of calmodulin to these tubules via the STOP proteins.[9] Preliminary experiments in our laboratory indicate that a part of the calcium in these cells may come from extracellular sources.

In addition to these slow changes in calcium gradients, cells in mitosis show rapid (approximately 20-second duration) calcium transients.[10] Careful investigation of these transients in PtK2 cells shows that while a substantial percentage of these cells shows at least one transient within the three minutes before chromosome separation (anaphase onset), other cells spike much earlier or not at all.[11] Thus it is unlikely that these calcium transients serve as the "trigger" for the metaphase-anaphase transition. In contrast, there is a close correlation between spiking and the separation of the daughter cells during cytokinesis. In addition, approximately one third of the interphase cells show periodic spikes of calcium reminiscent of pacemaker activity in electrophysiological systems, though with a much slower time course. The activity at cytokinesis suggests a role for calcium in the mediation of the actomyosin system involved in the formation of the cleavage furrow in these cells. The periodic pacemaker-like spikes have also been seen in neutrophils,[12] macrophages,[13] and lymphocytes.[14]

The relative contributions of the slower changes in the calcium gradient such as are seen at anaphase and the more rapid transients to the regulation of the cytoskeleton are unclear. At the moment, the evidence for the relationship of either is circumstantial— temporal proximity, localization, and reasonable hypotheses for mechanism. Firmer links need to be sought. The periodic spiking could be the correlate of mechanisms similar to those invoked by Dr. Rasmussen earlier in this symposium as "calcium cycling."[15] However, the data on *Haemanthus* also underline the fact that significant regulation of ionic calcium in the cytoplasm can occur without use of extracellular calcium.

Important related information derives from studies on calcium in cell spreading of phagocytosis. In the case where polymorphonuclear neutrophils (PMNs) are allowed to settle and spread on an opsonized substrate, spreading is always preceded by a spike of intracellular calcium.[13] Similar increases are seen in the human neutrophil as it migrates toward and eventually engulfs an opsonized red blood cell.[16]

The Calcium Ion and Cell Function in Aging and Alzheimer's Disease

When human PMNs from humans with Alzheimer's disease are compared to control PMNs, they exhibit an inability to migrate normally in thermal gradients.[17] The defective cells closely resemble normal cells which have been treated with colchicine, a drug which leads to the breakdown of cytoplasmic microtubules. One possible reason for this failure could be a defect in the basal levels of $[Ca^{2+}]_i$ or the inability of the cell to regulate this ion properly. Such a defect would be in accord with the observation that as rodent leukocytes age, they show a decreased mitogenic response which is restored by a combination of calcium ionophore and PMA, a phorbol ester.[18] Modest decrements of radiocalcium uptake into unfractionated Alzheimer lymphocytes have also been observed.[19]

Studies from our laboratory on cell lines derived from patients with the familial form of Alzheimer's disease have shown reductions both in the resting levels of calcium ion and in the dynamic response of this ion to stimuli.[20,21] These experiments are discussed in detail by Dr. Peterson in this volume.

Whether the alterations in calcium are central to the development of Alzheimer's disease remains to be determined. The central role of this ion in a diverse range of biological actions makes it a promising subject for further investigation.

REFERENCES

1. TSIEN, R. Y., T. POZZAN & T. J. RINK. 1982. J. Cell Biol. **94:** 325–334.
2. WELSH, M. J., J. R. DEDMAN, B. R. BRINKLEY & A. J. MEANS. 1979. J. Cell Biol. **81:** 624.
3. KEITH, C., M. DIPAOLA, F. R. MAXFIELD & M. L. SHELANSKI. 1983. J. Cell Biol. **97:** 1918.
4. LUBY-PHELPS, K., F. LANNI & D. L. TAYLOR. 1985. J. Cell Biol. **101:** 1245.
5. GRYNKIEWICZ, G., M. POENIE & R. Y. TSIEN. 1985. J. Biol. Chem. **260:** 3440–3450.
6. KEITH, C. H., F. R. MAXFIELD & M. L. SHELANSKI. 1985. Proc. Natl. Acad. Sci. USA **82:** 800.
7. KEITH, C. H., R. R. RATAN, F. R. MAXFIELD, A. BAJER & M. L. SHELANSKI. 1985. Nature **316:** 848–850.
8. LAMBERT, A. M., M. VANTARD, L. VAN ELDIK & J. DEMEY. 1983. J. Cell Bio. **97:** 40.
9. MARGOLIS, R. L., C. T. RAUCH & D. JOB. 1986. Proc. Natl. Acad. Sci. USA **83:** 639.
10. POENIE, M., J. ALDERTON, R. STEINHARDT & R. Y. TSIEN. 1986. Science **233:** 886–889.
11. RATAN, R. R., F. R. MAXFIELD & M. L. SHELANSKI. 1988. J. Cell Biol. **107:** 993–999.
12. KRUSKAL, B. A. & F. R. MAXFIELD. 1987. J. Cell Biol. **105:** 2685–2693.
13. KRUSKAL, B. A., S. SHAK & F. R. MAXFIELD. 1986. Proc. Natl. Acad. Sci. USA **83:** 2919–2923.
14. WILSON, A. H., D. GREENBLATT, M. POENIE, F. D. FINKLEMAN & R. Y. TSIEN. 1987. J. Exp. Med. **166:** 601–666.
15. RASMUSSEN, H. This volume.
16. MARKS, P. & F. R. MAXFIELD. Cell Calcium. In press.
17. FU, T. -K., S. S. MATSUYAMA, J. D. KESSLER & L. F. JARVIK. 1986. Neurobiol. Aging **7:** 41–43.
18. MILLER, R. A., B. JACOBSON, G. WEIL & E. R. SIMONS. 1987. J. Cell Physiol. **132:** 337.
19. GIBSON, G. E., P. NIELSEN, K. SHERMAN & J. P. BLASS. 1987. Biol. Psychiatry **22:** 1079–1086.
20. PETERSON, C., R. R. RATAN, M. L. SHELANSKI & J. E. GOLDMAN. 1986. Proc. Natl. Acad. Sci. USA **83:** 7994.
21. PETERSON, C., R. R. RATAN, M. L. SHELANSKI & J. E. GOLDMAN. 1988. Neurobiol. Aging **9:** 261–266.

Pyramidal Cell Topography of Microtubule-Associated Proteins and Their Precipitation into Paired Helical Filaments

KENNETH S. KOSIK

Center for Neurologic Diseases
Department of Medicine (Division of Neurology)
Brigham and Women's Hospital
and
Department of Neurology (Neuroscience)
Harvard Medical School
Boston, Massachusetts 02115

It is well known that one of the defining features of Alzheimer's disease is the presence of insoluble polymers within neurons termed paired helical filaments (PHF). The occurrence of these structures is not confined to Alzheimer's disease, however, and they have been observed in other disease entities including subacute sclerosing panencephalitis, dementia pugilistica, and Guamanian Parkinson dementia complex among others.[1] In Alzheimer's disease a typical neocortical neurofibrillary tangle (NFT) has a flame-shaped appearance by light microscopy because it is delineated by the somatodendritic region of the pyramidal cell. A problem of considerable interest in Alzheimer's research has been the identification of the components of paired helical filaments and how these components are macromolecularly ordered within the polymer. The most well-established proteins thought to be components of PHF are tau protein[2-8] and ubiquitin.[9,10] Tau protein is a microtubule-associated protein shown to promote polymerization of microtubules and stabilize formed microtubules. Its migration on one-dimensional SDS-PAGE reveals a complex heterogeneous array of isoforms that range in mass from 55 kDa to 62 kDa. PC12 cells and some tissues also contain a tau isoform that migrates at 120 kDa.[11] Tau has the remarkable biophysical property of retaining its solubility after boiling, in 0.75 mM NaCl, and in 2.5% perchloric acid. Ubiquitin is a 8565 M_r protein found in many neural and nonneural intracellular inclusions.[12,13] One putative function of ubiquitin is to target a protein for degradation via conjugation with the amino terminus or an internal lysine of the substrate.

A number of highly specific polyclonal and monoclonal antibodies to tau[2-8] and ubiquitin[9,10] label neurofibrillary tangles by light microscopy, label PHF with immunogold particles, and label SDS-extracted PHF. For these reasons, tau and ubiquitin meet one set of criteria to consider them integral PHF components. A second criterion has also been achieved for these molecules; their sequence has been obtained from partially purified, partially solubilized PHF fractions.[9,14,15] Neither of these criteria is ideal due to the pitfalls of antibody cross-reactivities in the former case and to impurities in PHF preparations in the latter case. In particular the immunoreactivity of PHF with some neurofilament antibodies may be due to a phosphorylation site shared with tau protein,[16,17] and the immunoreactivity of PHF with some MAP2 antibodies may be due to a shared microtubule-binding site.[18,19] Sequences obtained from PHF fractions thought to derive from co-purifying contaminants are collagen, ferritin, and the β-amyloid protein. Among these proteins the most controversial is the β-amyloid protein which has been sequenced on PHF fractions from Guamanian Parkinson-dementia complex.[20] While this illness is not known to have senile plaques, the recent reports of a more widespread deposition of the β-amyloid using formic acid-treated sections[21] raises the distinct possibility of β-amyloid protein contamination in these PHF fractions.

High-resolution electron micrographs of PHF show a ''fuzzy'' material surrounding the helically wound filaments.[22] When the fuzzy material is removed with pronase one obtains a ''core'' fraction from which sequences derived from the carboxy half of tau have been obtained.[23] Tau epitopes from the amino half of tau can be released from the PHF core by treatment with trypsin.[24] Therefore, any model which seeks to solve the structure of PHF must describe how a carboxy portion of tau can self-assemble or co-assemble with an as yet unknown molecule into highly regular insoluble filaments, while maintaining a covalent linkage to the amino half of tau and to ubiquitin which may project out from the filament. Wischik and co-workers have reported that sequences they have obtained from PHF can only account for 10% of the mass of PHF.[22] Thus tau binding proteins may account for the as yet unidentified residual mass of these structures. Two proteins that bind to tau, tubulin and calmodulin, have been excluded from PHF on immunocytochemical grounds and because of the failure to find their sequences in partially solubilized PHF fractions.

While it is clear that the identification of the components of the PHF is an important goal, it seems increasingly clear that PHF represent only a small fraction of the more widespread neuritic response which occurs on the Alzheimer brain. And while microtubule-associated proteins are prominently involved in this response only one of the multifaceted features of the response is PHF. Perhaps most numerous are the widespread curly fibers or threads seen throughout the neuropil, particularly in brain regions affected by the presence of neurofibrillary tangles.[25,26] These dystrophic neurites are best visualized with tau antibodies and are more abundantly present than either senile plaques or neurofibrillary tangles. Their distribution extends well beyond the similar appearing neurites that surround the amyloid core of senile plaques. While it is not yet known what the ultrastructural counterpart of the curly fibers is, it seems unlikely based upon numerous previous electronmicroscopic surveys of Alzheimer brain, that such an abundantly present structure contains many PHF. Curly fibers may, however, contain the straight filaments also observed in Alzheimer brain.[27-31] Thus, the question is open as to why two polymerization processes occur in the Alzheimer brain, both involving tau protein, that result in two distinct ultrastructures—straight filaments and PHF.

It has been somewhat problematic to ascertain the relationship of the curly fibers to the surrounding neuronal architecture. Where it is possible to make the requisite observations, it appears that curly fibers tend to occur in dendritic fields. This feature is best illustrated in the dendritic-rich region which immediately surrounds the layer II star-cell clusters of the entorhinal cortex, that are characteristically affected by neurofibrillary pathology. Dividing the dendritic surrounds of these clusters are axonal bundles. In normal brain the distinctive anatomy of this region is well illustrated by MAP2 and tau immunocytochemistry, which label the somatodendritic elements and axonal elements, respectively. In Alzheimer's disease these dendritic fields contain abundant curly fibers, suggesting a dendritic origin of these fibers. However, the curly fibers are stained by the tau antibody, which, in normal brain exclusively labels axons. This dichotomy is also observed in the neurofibrillary tangle which is also tau-reactive, but is present in the cell body and proximal dendrite. A second curious feature of the tau-immunocytochemical pattern in Alzheimer brain is the progressive decrement in the ability of tau to stain axonal populations as the number of NFT and curly fibers increases. Whether these axonal populations are completely lost or whether they only lose their tau-reactivity, but retain other structural and functional capacities is unknown. This pattern in which a resorting of microtubule-associated proteins occurs is reminiscent of the neuronal regenerative response and indeed, it seems possible that curly fibers cannot be defined as either axons or dendrites, but instead represent aberrant or luxuriant sprouts.[32,33] Some of the regenerative principles which neurons undergo that may be relevant to sprouting in Alzheimer's disease include the transient co-localization of MAP2 and tau in newly formed elongating axons and dendrites,[34] the plasticity of dendrites sufficient to convert an established dendrite to an axon following close-axotomy,[35] the intrinsic axonal character of sprouts following close-axotomy,[36,37] and the great potential of differentiated vertebrate central neurons for new dendritic growth.[38–42] All these phenomena may be part of the growth response observed in the Alzheimer brain. Evidence that curly fibers do represent new growths include their association with structures that have the light microscopic features of filopodia and the fact that they are lined with microspikes. The presence of growth cone markers such as GAP43 immunoreactivity, associated with curly fibers has so far been negative, but is ongoing (Kowall, unpublished observations). The exploratory, rather than directed growth of these fibers is suggested by their course, which carries them in and out of the plane of section and thus creates the meshlike dystrophy in the neuropil. Normally pyramidal cell apical dendrites are radially deployed and, therefore, can often be seen to extend over several microns in a single section. While this proliferative response is abundantly present no evidence is available to suggest whether any fraction of these neurites is making synaptic connections and if so, whether any of these connections are correct ones. Given the chaotic appearance of the neuronal dystrophy the latter possibility seems unlikely.

The neuritic growth response has several additional morphological features. Supernumerary basilar dendrites on hippocampal and cerebrocortical pyramidal cells are apparent with MAP2 antibodies. Secondly, numerous contorted processes extend from the cell bodies of tangle-bearing neurons and thus confer bizarre morphologies upon these cells. Often these processes have the configuration of lamellipodia and filopodia. Although reactive with tau antibodies they often taper in a typical dendritic fashion. The exuberance of the response points to the vitality which tangle-bearing neurons retain. We have recently confirmed the viability of these neurons by hybridizing a tau cRNA to thioflavin S stained neurons containing NFT.[43] The response can, perhaps, be viewed as a postmitotic malignancy manifest not as cell division, but as uncontrolled neuritic growth of established cells. The utility of this view is that it begs the investigation of oncogene expression in Alzheimer's disease as a source of growth factor. The presence of an AP-1/fos promoter consensus sequence upstream of the β-amyloid gene is of interest in this regard.[44]

The presence of microtubule-associated protein often unassociated with microtubules appears to be a common denominator of the neuritic growth response. The tau-reactive NFT are unassociated with either microtubules or with tubulin-immunoreactivity. Only a subset of the curly fibers contain tubulin-immunoreactivity.[32] MAPs in the growth cone are thought to be either unassociated with microtubules or less tightly associated with microtubules.[45] Thus, a source of free tau, putatively capable of self-assembly or co-assembly into filaments, either straight or paired and helical, could be provided by the proliferating neurites. Indeed, neurite extension in PC12 cells is associated with a dramatic rise in tau protein[46,47] in excess of the rise in tubulin. Any rise in tau protein within tangle-bearing neurons is not capable of driving tubulin assembly since there is a paucity of microtubules in these neurons. One could postulate many explanations for the failure of tau to drive microtubule assembly in Alzheimer's disease. Some modification of tau, which might include a phosphorylation event, is one parsimonius explanation given the body of data[6,48,49] concerning aberrant phosphorylation of Alzheimer tau protein.

REFERENCES

1. WISNIEWSKI, K., G. A. JERVIS, R. C. MORETZ & H. M. WISNIEWSKI. 1979. Alzheimer neurofibrillary tangles in diseases other than senile and presenile dementia. Ann. Neurol. 5: 288–294.

2. BRION, J. P., E. PASSAREIRO, J. NUNEZ & J. FLAMENT-DURAND. 1985. Mise en evidence immunologique de la proteine tau au niveau des lesions de degenerescence neurofibrillaire de la maladie d'Alzheimer. Arch. Biol. (Bruxelles) 95: 229–235.

3. GRUNDKE-IQBAL, I., K. IQBAL, M. QUINLAN, Y.-C. TUNG, M. S. ZAIDI & H. M. WISNIEWSKI. 1986. Microtubule-associated protein τ: A component of Alzheimer paired helical filaments. J. Biol. Chem. 261: 6084–6089.

4. JOACHIM, C. L., J. H. MORRIS, D. J. SELKOE & K. S. KOSIK. 1987. Tau epitopes are incorporated into a range of lesions in Alzheimer's disease. J. Neuropathol. Exp. Neurol. 46: 611–622.

5. KOSIK, K. S., C. L. JOACHIM & D. J. SELKOE. 1986. Microtubule-associated protein tau (τ) is a major antigenic component of paired helical filaments in Alzheimer's disease. Proc. Natl. Acad. Sci. USA 83: 4044–4048.

6. WOOD, J. G., S. MIRRA, N. J. POLLOCK & L. I. BINDER. 1986. Neurofibrillary tangles of Alzheimer disease share antigenic determinants with the axonal microtubule-associated protein tau. Proc. Natl. Acad. Sci. USA 83: 4040–4043.

7. NUKINA, N. & Y. IHARA. 1986. One of the antigenic determinants of paired helical filaments is related to tau protein. J. Biochem. 99: 1541–1544.

8. DELACOURTE, A. & A. DEFOSSEZ. 1986. Alzheimer's disease: Tau proteins, the promoting factors of microtubule assembly, are major components of paired helical filaments. J. Neurosci. 76: 173–186.

9. MORI, H., J. KONDO & Y. IHARA. 1987. Ubiquitin is a component of paired helical filament in Alzheimer's disease. Science 235: 1641–1644.

10. PERRY, G., R. FRIEDMAN, G. SHAW & V. CHAU. 1987. Ubiquitin is detected in neurofibrillary tangles and senile plaque neurites of Alzheimer disease brains. Proc. Natl. Acad. Sci. USA 184: 3033–3036.

11. DRUBIN, D. G. & M. W. KIRSCHNER. 1986. Tau protein function in living cells. J. Cell. Biol. 103: 2739–2746.

12. MANETTO, V., G. PERRY, F. ABDUL-KARIM, M. TABATON, L. AUTILIO-GAMETTI & P. GAMBETTI. 1988. Ubuquitin: Selective presence in neuronal and non-neuronal inclusions. J. Neuropathol. Exp. Neurol. Abstracts of the 64th Annual Meeting. No. 107.

13. LOWE, J., A. BLANCHARD, K. MORRELL, G. LENNOX, L. REYNOLDS, M. BILLETT, M. LANDON & R. J. MAYER. 1988. Ubiquitin is a common factor in intermediate filament inclusion bodies of diverse type in man, including those of Parkinson's disease, Pick's disease, and Alzheimer's disease, as well as Rosenthal fibres in cerebellar astrocytomas, cytoplasmic bodies in muscle, and Mallory bodies in alcoholic liver disease. J. Pathol. 155: 9.

14. WISCHIK, C. M., M. NOVAK, H. C. THOGERSEN, P. C. EDWARDS, M. J. RUNSWICK, R. JAKES, J. E. WALKER, C. MILSTEIN, M. ROTH & A. KLUG. 1988. Isolation of a fragment of tau derived from the core of the paired helical filament of Alzheimer disease. Proc. Natl. Acad. Sci. USA 85: 4506–4510.

15. KONDO, J., T. HONDA, H. MORI, Y. HARRODA, R. MIURA, M. OGAWARA & Y. IHARA. 1988. The carboxyl third of tau is tightly bound to paired helical filaments. Neuron 1: 827–834.

16. NUKINA, N., K. S. KOSIK & D. J. SELKOE. 1987. Recognition of Alzheimer paired helical filaments by monoclonal neurofilaments antibodies is due to crossreaction with tau protein. Proc. Natl. Acad. Sci. USA 84: 3415–3419.

17. KSIEZAK-REDING, H., D. W. DICKSON, P. DAVIES & S.-Y. YEN. 1987. Recognition of tau epitopes by anti-neurofilaments antibodies that bind to Alzheimer neurofibrillary tangles. Proc. Natl. Acad. Sci. USA 84: 3410–3414.

18. YEN, S.-H., D. W. DICKSON, A. CROWE, M. BUTLER & M. L. SHELANSKI. 1987. Alzheimer's neurofibrillary tangles contain unique epitopes and epitopes in common with the heat-stable microtubule-associated proteins, tau and MAP2. Am. J. Pathol. 126: 81–91.

19. KOSIK, K. S., L. D. ORECCHIO, L. BINDER, J. Q. TROJANOWSKI, V. M.-Y. LEE & G. LEE.

1988. Epitopes that span the tau molecule are shared with paired helical filaments. Neuron **1:** 817–825.

20. GUIROY, D. C., M. MIYAZAKI, G. MULTHAUP, P. FISCHER, R. M. GARRUTO, K. BEY-REUTHER, C. L. MASTERS, G. SIMMS, C. J. GIBBS JR. & D. C. GAJDUSEK. 1987. Amyloid of neurofibrillary tangles of Guamanian Parkinsonism-dementia and Alzheimer disease share identical amino acid sequence. Proc. Natl. Acad. Sci. USA **84:** 2073–2077.

21. KITAMOTO, T., K. OGOMORI, J. TATEISHI & S. B. PRUSINER. 1987. Formic acid pretreatment enhances immunostaining of cerebral and systemic amyloidosis. Lab. Invest. **57:** 230–236.

22. WISCHIK, C. M., M. NOVAK, P. C. EDWARDS, A. KLUG, W. TICHELAAR & R. A. Crowther. 1988. Structural characterization of the core of the paired helical filament of Alzheimer disease. Proc. Natl. Acad. Sci. USA **85:** 4884–4888.

23. GOEDERT, M., C. M. WISCHIK, R. A. CROWTHER, J. E. WALKER & A. KLUG. 1988. Cloning and sequencing of the cDNA encoding a core protein of the paired helical filaments of Alzheimer's disease: Identification as the microtubule-associate protein tau. Proc. Natl. Acad. Sci. USA **85:** 4051–4055.

24. MERZ, P. A., M. WRZOLEK, R. J. KASCSAK, I. GRUNDKE-IQBAL, K. IQBAL, L. BINDER, K. KOSIK & H. M. WISNIEWSKI. 1987. Immune electron microscopy of paired helical filaments (PHF) with tau antibodies (abstract). J. Neuropathol. Exp. Neurol. **46:** 376.

25. KOWALL, N. W. & K. S. KOSIK. 1987. Axonal disruption and aberrant localization of tau protein characterize the neuropil pathology of Alzheimer's disease. Ann. Neurol. **22:** 639–643.

26. BRAAK, H., E. BRAAK, I. GRUNDKE-IQBAL & K. IQBAL. Occurrence of neuropil threads in the senile human brain and in Alzheimer's disease: A third location of paired helical filaments outside of neurofibrillary tangles and neuritic plaques. Neurosci. Lett. **65:** 351–355.

27. PERRY, G., P. MULVIHILL, V. MANETTO, L. AUTILIO-GAMBETTI & P. GAMBETTI. 1987. Immunocytochemical properties of Alzheimer straight filaments. J. Neurosci. **7:** 3736–3738.

28. OKAMOTO, K., A. HIRANO, H. YAMAGUCHI & S. HIRAI. 1982. The fine structure of eosinophilic stages of Alzheimer's neurofibrillary tangles. Clin. Neurol. **22:** 840–846.

29. SHIBAYAMA, H. & J. KITOH. 1978. Electron microscopic structure of the Alzheimer's neurofibrillary changes in case of atypical senile dementia. Acta Neuropathol. (Berlin) **41:** 229–234.

30. YAGISHITA, S., Y. ITOH, N. AMANO & T. NAKANO. 1980. The fine structure of neurofibrillary tangles in a case of atypical presenile dementia. J. Neurol. Sci. **48:** 325–332.

31. YAGISHITA, S., Y. ITOH, W. NAN & N. AMANO. 1981. Reappraisal of the fine structure of Alzheimer's neurofibrillary tangles. Acta Neuropathol. (Berlin) **54:** 239–246.

32. MCKEE, A. C., N. W. KOWALL & K. S. KOSIK. Microtubular reorganization and growth response in Alzheimer's disease. Ann. Neurol. In press.

33. IHARA, Y. 1988. Massive somatodendritic sprouting of cortical neurons in Alzheimer's disease. Brain Res. **459:** 138–144.

34. KOSIK, K. S. & E. A. FINCH. 1987. MAP2 and tau segregate into axonal and dendritic domains after the elaboration of morphologically distinct neurites: An immunocytochemical study of cultured rat cerebrum. J. Neurosci. **7:** 3142–3153.

35. DOTTI, C. G. & G. A. BANKER. 1987. Experimentally induced alterations in the polarity of developing neurons. Nature **330:** 254–256.

36. HALL, G. F., A. POULOS & M. J. COHEN. 1989. Sprouts emerging from the dendrites of axotomized lamprey central neurons have axonlike ultrastructure. J. Neurosci. **9:** 588–599.

37. HALL, G. F. & M. J. COHEN. 1983. Extensive dendritic sprouting induced by close axotomy of central neurons in the lamprey. Science **222:** 518–520.

38. SCHEIBEL, A. B. & U. TOMIYASU. 1978. Dendritic sprouting in Alzheimer's presenile dementia. Exp. Neurol. **60:** 1–8.

39. PROBST, A., V. BASLER, B. BRON & J. ULRICH. 1983. Neuritic plaques in senile dementia of Alzheimer type: A Golgi analysis in the hippocampal region. Brain Res. **268:** 249–354.

40. ARENDT, T., H. G. ZVENGINTSEVA & T. A. LEONTOVICH. 1986. Dendritic changes in the basal nucleus of Meynert and in the diagonal band nucleus in Alzheimer's disease—A quantitative Golgi investigation. Neurosci. **19:** 1265–1278.

41. PAULA-BARBOSA, M. M., R. M. CARDOSO, M. L. GUIMARAES & C. CRUZ. 1980. Dendritic

degeneration and regrowth in the cerebral cortex of patients with Alzheimer's disease. J. Neurol Sci. **45**: 129–134.

42. FERRER, I., A. AYMAMI, A. ROVIRA & J. M. G. VECIANA. 1983. Growth of abnormal neurites in atypical Alzheimer's Disease. A study with the Golgi method. Acta Neuropathol. (Berlin) **59**: 167–170.

43. KOSIK, K. S., J. E. CRANDALL, E. J. MUFSON & R. L. NEVE. Tau *in situ* hybridization in normal and Alzheimer brain: A predominant localization in the neuronal somatodendritic compartment. Ann. Neurol. In press.

44. SALBAUM, J. M., A. WEIDEMANN, H.-G. LEMAIRE, C. L. MASTER & K. BEYREUTHER. 1988. The promoter of Alzheimer's disease A4 precursor gene. EMBO J. **7**: 2807–2813.

45. BAMBURG, J. R., D. BRAY & K. CHAPMAN. 1986. Assembly of microtubules at the tip of growing axons. Nature **321**: 788–790.

46. DRUBIN, D., M. KIRSCHNER & S. FEINSTEIN. 1984. Microtubule-associated tau protein induction by nerve growth factor during neurite outgrowth in PC12 cells. *In* Molecular Biology of the Cytoskeleton. 343–355. Cold Spring Harbor Laboratory. Cold Spring Harbor, NY.

47. GREENE, L. A., R. K. H. LIEM & M. L. SHELANSKI. 1983. Regulation of a high molecular weight microtubule-associated protein in PC12 cells by nerve growth factor. J. Cell Biol. **96**: 87–93.

48. GRUNKE-IQBAL, I., K. IQBAL, Y.-C. TUNG, M. QUINLAN, H. M. WISNIEWSKI & L. I. BINDER. 1986. Abnormal phosphorylation of the microtubule-associated protein τ (tau) in Alzheimer cytoskeletal pathology. Proc. Natl. Acad. Sci. USA **83**: 4913–4917.

49. IHARA, Y., N. NUKINA, R. MIURA & M. OGAWARA. 1986. Phosphorylated tau protein is integrated into paired helical filaments in Alzheimer's disease. J. Biochem. **99**: 1807–1810.

Calcium-Regulated Contractile and Cytoskeletal Proteins in Dendritic Spines May Control Synaptic Plasticity[a]

E. FIFKOVÁ AND M. MORALES

Department of Psychology
Neuroscience Program
University of Colorado
Campus Box 345
Boulder, Colorado 80309

INTRODUCTION

Synaptic plasticity pertains to changes induced by various interventions or previous stimulus history at any level of the nervous system. Such changes may be implicated in the biological mechanism of the highest brain functions. Therefore, synaptic plasticity is potentially the most important property of the nervous system. Dendritic spines represent a major postsynaptic site of a number of neurons. They appear to be endowed with a considerable degree of plasticity as their dimensions are changed by electrical and environmental stimuli. Previously, we have demonstrated that dendritic spines are domains of the neuron with the highest concentration of actin filaments which are organized in a dense network and determine the particular shape of a spine.[1] In eukaryotic cells, actin networks were shown to have a stable structural organization but a dynamic composition of filaments which is regulated by Ca^{2+}-dependent, actin-associated proteins. These proteins thus control properties of actin systems. Various stimuli can rapidly alter the architecture of an actin network as was demonstrated in stimulated platelets,[2,3] and the induced changes may persist over varying periods of time.[4] Thus, the actin network in dendritic spines may be implicated in the spine plastic properties. Because actin-associated proteins determine the function of actin networks, detailed knowledge on their distribution within the spine and their relation to actin filaments is important. Therefore, we have studied two actin-associated proteins, myosin and MAP 2, in dendritic spines with immunogold electronmicroscopy.

Actin, Myosin, and MAP 2 in Dendritic Spines

Isolation of brain myosin and MAP 2 was done according to Burridge and Bray[5] and Tsuyama *et al.*,[6] respectively, and was described previously.[7,8] Primary antibodies used were mouse monoclonal antiactin antibody (IgM, Amersham, Orange, NJ), polyclonal antibody against human platelet myosin (kindly provided by Dr. J. Scholey, National Jewish Center for Immunology and Respiratory Medicine, Denver, CO) and anti-MAP 2 antibodies (IgG, clone AP14, kindly provided by Drs. L.I. Binder, University of Alabama and A. Frankfurter, University of Virginia). Protocol for immunoelectron microscopy was described in detail elsewhere.[7,8] Secondary antibodies used were goat antimouse IgM bound to 5 or 10 nm colloidal gold to detect actin, goat antimouse IgG bound to 5 nm

[a]Supported by National Institute on Aging Grant AG-04804-03 to E.F.

colloidal gold to detect MAP 2, and goat antirabbit IgG bound to 5 nm colloidal gold to detect myosin. The specificity of the primary antibodies used was tested on rat brain homogenate which was electrophoretically separated and transferred to nitrocellulose paper (FIG. 1).

Distribution of actin (FIG. 2A), MAP 2 (FIGS. 2B, 2C, 3) and myosin (FIGS. 4A, 4B) were studied in dendritic spines of the dentate fascia, hippocampus, and visual and cerebellar cortices. Both proteins, myosin and MAP 2, were found in dendritic spines associated with actin filaments. Together with actin, they were detected on membranes of the spine apparatus (SA [FIGS. 2A, 2C, 3, 4A, 4B]) and the smooth endoplasmic reticulum (SER). Actin and myosin were associated with the spine plasma membrane, and actin and MAP 2 were prominent in the postsynaptic density (PSD [FIG. 2C]). In dendrites, MAP 2 was associated with microtubules (FIG. 2B). Results concerning MAP 2 in dendritic spines extend those of Cáceres et al.[9] and Binder et al.[10] Myosin is demonstrated in dendritic spines for the first time.

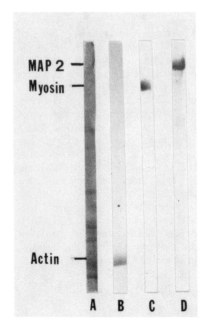

MAP 2 —
Myosin —

Actin —

A   B   C   D

FIGURE 1. Rat brain homogenate was electrophoretically separated and transferred to nitrocellulose paper. Adjacent strips were cut and treated with: A, Amido black; B, antiactin antibody; C, antimyosin antibody; or D, anti-MAP 2 antibody.

Synaptic Plasticity and Actin-Associated Proteins

It is widely accepted that synaptic plasticity involves both, the pre-[11] and postsynaptic component of a synapse. However, very often the postsynaptic membrane and PSD alone are investigated rather than the entire postsynaptic element, e.g., the dendritic spine. This may be caused by limited information available on the composition of such structures. Therefore, the reported experiments are part of our long-range effort to study the composition of the spine cytoplasm, most notably cytoskeletal and contractile proteins. These proteins have properties that may be important for mediating synaptic plasticity. The demonstration of actin in dendritic spines was the first step in this direction.[1] A logical continuation of this effort is the present paper where we have demonstrated that myosin and MAP 2, two actin-binding proteins, are colocalized with actin filaments in dendritic

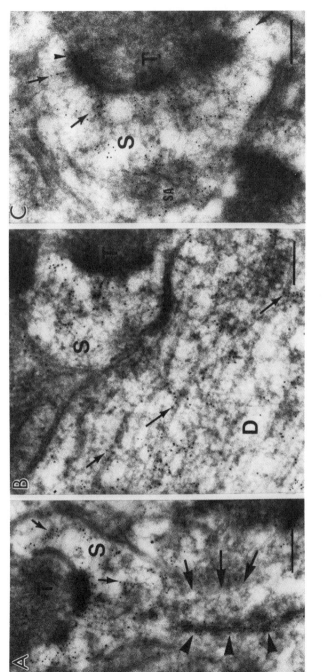

FIGURE 2. Distribution of actin (**A**) and MAP 2 (**B,C**) in dendritic spines (S) and axon terminals (T). (**A**) Actin is associated with filaments (*small arrows* on the spinal head) and the spine apparatus (*arrowheads*) on the neck of the spine (*large arrows*). (**B**) MAP 2 is present on the spine and dendrite (D). *Arrows* indicate the association of MAP 2 with microtubules. (**C**) MAP 2 is associated with filaments (*arrows*), PSD (*arrowhead*), and the spine apparatus (SA). Bars = 0.250 μm.

spines. *Myosin* is of particular significance as it can induce diverse types of cellular movements by its interaction with actin filaments. *MAP 2* binds microtubules and it cross-links actin filaments.[12]

The actin-MAP 2 and actin-myosin interactions are regulated by their respective levels of phosphorylation and by the concentration of cytoplasmic Ca^{2+}. Studies *in vitro* have shown that MAP 2 is a substrate for protein kinase C (PKC) and Ca^{2+}/calmodulin-activated protein kinase II (CAMKII). It was also shown that the regulatory portion of myosin, the myosin light chain (MLC) is a substrate for Ca^{2+}/calmodulin-activated myosin light chain kinase (MLCK) as well as PKC. Since calmodulin, PKC, and CAMKII are present in dendritic spines,[13-15] MAP 2 and MLC may constitute postsynaptic substrates of PKC and Ca^{2+}/calmodulin-dependent kinases.

Postsynaptic Mechanism of LTP

Depolarization of the postsynaptic membrane,[16,17] and increased Ca^{2+} entry into dendritic spines via NMDA receptors induced by repetitive activation of excitatory hippocampal synapses (which release glutamate[18]) has been shown to be necessary to induce LTP and sufficient to potentiate synaptic transmission.[19,20] While increased intraspinal Ca^{2+} seems to be critical for the induction of LTP, it seems to be also important in maintaining this phenomenon. The long-term increase in internal Ca^{2+}, induced by excitatory amino acids like glutamate, may be to a significant extent derived from smooth endoplasmic reticulum (SER), as this Ca^{2+} increase is suppressed by drugs like Dantrolene that block Ca^{2+} release from sarcoplasmic reticulum.[21] Since Dantrolene blocks LTP as long as two hours after it was induced, maintenance of LTP seems to depend also

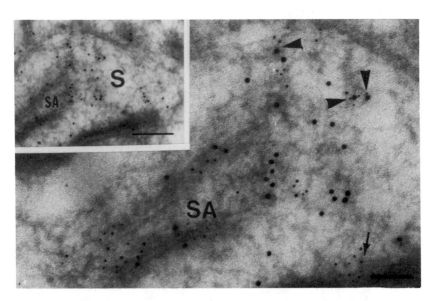

FIGURE 3. Colocalization of MAP 2 (5 nm colloidal gold) and actin (10 nm colloidal gold) on the dendritic spine. MAP 2 and actin are colocalized on the spine apparatus (SA) and filaments (*arrowheads*). *Small arrow* indicates a filament associated with PSD and containing MAP 2. Bar = 1 µm. *Inset,* low magnification of the dendritic spine. Bar = 0.250 µm.

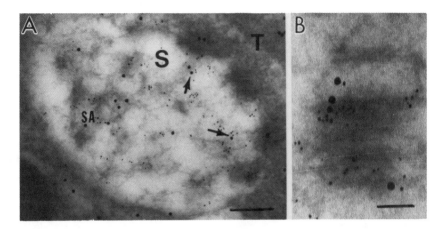

FIGURE 4. Colocalization of myosin (5 nm colloidal gold) and actin (15 nm colloidal gold) on the dendritic spine. (**A**) Myosin and actin are colocalized on the spine apparatus (SA) and filaments (*arrows*). Bar = 0.250 μm. (**B**) Colocalization of myosin and actin on the spine apparatus. Bar = 1 μm.

on Ca^{2+} derived from internal reservoirs.[22] In hippocampal dendritic spines, such a reservoir is likely to be the spine apparatus.[23] Additional influx of Ca^{2+} during the later phases of LTP may come through voltage-dependent Ca^{2+} channels.[24] The long-term effect of Ca^{2+} on LTP could be mediated through activation of protein kinases. Two of them, the Ca^{2+}/calmodulin-activated protein kinase II (CAMKII) and protein kinase C (PKC) are found in the postsynaptic density (PSD). The role that these kinases play in the maintenance of LTP is not clear. However, recent evidence indicates that persistent protein kinase activity is necessary for the maintenance of this phenomenon.[25] Substrate proteins of PKC with relevance to LTP are the Ca^{2+} ATPase, MAP 2, myosin light chain (MLC), and its kinase (MLCK).[26,24] Phosphorylated Ca^{2+} ATPase may be the additional source of continuous high levels of internal Ca^{2+}.[24] Increased Ca^{2+} concentration and activated kinases during LTP may affect properties of actin, myosin, and MAP 2 in the activated spines. Phosphorylated MAP 2 is not capable of cross-linking actin filaments. Consequently, changes in the architecture of the spine network may occur. The elevated Ca^{2+}, together with calmodulin, will also activate MLCK which by phosphorylating MLC will make myosin-actin interactions possible. This could induce movement of the cytoplasm within the spine with consequent morphometric changes which were observed in the stimulated spines during LTP[27–30] (also Dr. Per Andersen, personal communication). Modification of the architecture of the spine actin network will also affect its interaction with endomembranes of the SA which are suspended within the actin network of the spine head and stalk.[28,31] Since these membranes contribute to the electrical resistance of the narrow spine stalk, their redistribution may affect synaptic potentials generated by the spine.

CONCLUSION

The presence of actin in dendritic spines at densities exceeding those of other neuronal compartments and its colocalization with MAP 2 and myosin make the spine a unique compartment of the neuron. Given its actin cross-linking properties, MAP 2 may organize actin filaments into a network whose association with myosin may endow dendritic spines with dynamic properties necessary for synaptic plasticity in general and long-term potentiation in particular.

ACKNOWLEDGMENTS

The authors express their sincere thanks to Drs. L.I. Binder, University of Alabama and A. Frankfurter, University of Virginia for the generous gift of monoclonal anti-MAP 2 antibody (IgG, clone AP14) and to Dr. J. Scholey, National Jewish Center for Immunology and Respiratory Medicine, Denver, CO for the generous gift of the polyclonal human antiplatelet myosin antibodies.

REFERENCES

1. FIFKOVÁ, E. & R. J. DELAY. 1982. Cytoplasmic actin in neuronal processes as a possible mediator of synaptic plasticity. J. Cell Biol. **95:** 345–350.
2. ESCOLAR, C., M. KRUMWIDE & J. G. WHITE. 1986. Organization of the actin skeleton of resting and activated platelets in suspension. Am. J. Pathol. **123:** 86–94.
3. NAKATA, T. & N. HIROKAWA. 1987. Cytoskeletal reorganization of human platelets after stimulation revealed by the quick-freeze deep-etch technique. J. Cell Biol. **105:** 1771–1780.
4. STOSSEL, T. P. 1982. The spatial organization of cortical cytoplasm. Philos. Trans. R. Soc. Lond. B **299:** 275–289.
5. BURRIDGE, K. & D. BRAY. 1975. Purification and structural analysis of myosins from brain and other nonmuscle tissues. J. Mol. Biol. **99:** 1–13.
6. TSUYAMA, S., Y. TERAYAMA & S. MATSUYAMA. 1987. Numerous phosphates of microtubule-associated protein 2 in living rat brain. J. Biol. Chem. **262:** 10886–10892.
7. MORALES, M. & E. FIFKOVÁ. 1989. In situ localization of myosin and actin in dendritic spines with the immunogold technique. J. Comp. Neurol. **279:** 666–674.
8. MORALES, M. & E. FIFKOVÁ. 1989. Distribution of MAP 2 in dendritic spines and its colocalization with actin: An immunogold electron microscope study. Cell Tissue Res. **256:** 447–456.
9. CÁCERES, A., L. I. BINDER, M. R. PAYNE, P. BENDER, L. I. REBHUN & O. STEWARD. 1984. Differential subcellular localization of tubulin and the microtubule-associated protein MAP 2 in brain tissue as revealed by immunocytochemistry with monoclonal hybridoma antibodies. J. Neurosci. **4:** 394–410.
10. BINDER, L. I., A. FRANKFURTER & L. I. REBHUN. 1986. Differential localization of MAP 2 and tau in mammalian neurons in situ. Ann. N.Y. Acad. Sci. **466:** 145–166.
11. LOVINGER, D. M., K. L. WONG, K. MURAKAMI & A. ROUTTENBERG. 1987. Protein kinase C inhibitors eliminate hippocampal long-term potentiation. Brain Res. **436:** 177–183.
12. SATTILARO, R. F. 1986. Interaction of microtubule-associated protein 2 with actin filaments. Biochemistry **25:** 2003–2009.
13. CÁCERES, A., P. BENDER, L. SNAVELY, L. I. REBHUN & O. STEWARD. 1983. Distribution and subcellular localization of calmodulin in adult and developing brain tissue. Neuroscience **10:** 449–461.
14. OUIMET, C. C., T. L. MCGUINNESS & P. Greengard. 1984. Immunocytochemical localization of calcium/calmodulin-dependent protein kinase II in rat brain. Proc. Natl. Acad. Sci. USA **81:** 5604–5608.
15. SAITO, N., M. KIKKAWA, Y. NASHIZUKA & CH. TANAKA. 1988. Distribution of protein kinase C-like immunoreactive neurons in rat brain. J. Neurosci. **8:** 369–382.
16. GUSTAFSSON, B. & H. WIGSTRÖM. 1986. Hippocampal long-lasting potentiation produced by pairing single volley and brief conditioning tetani evoked in separate afferents. J. Neurosci. **6:** 1575–1582.

17. GUSTAFSSON, B., H. WIGSTRÖM, W. C. ABRAHAM & Y. -Y. Huang. 1987. Long-term poten-
     tiation in the hippocampus using depolarizing current pulses as the conditioning stimulus to
     single volley synaptic potentials. J. Neurosci. **7:** 774–780.
18. BLISS, T. V. P., M. P. CLEMENTS, M. L. ERRINGTON, M. A. LYNCH & J. H. WILLIAMS. 1988.
     Long-term potentiation is accompanied by an increase in extracellular release of arachidonic
     acid. Soc. Neurosci. Abstr. **14**(Part 1): 564.
19. KAUER, J. A., R. C. MALENKA & R. A. NICOLL. 1988. NMDA application potentiates
     synaptic transmission in the hippocampus. Nature **334:** 250–252.
20. MALENKA, R. C., J. A. KAUER, R. S. ZUCKER & R. A. NICOLL. 1988. Postsynaptic calcium
     is sufficient for potentiation of hippocampal synaptic transmission. Science **242:** 81–84.
21. MODY, T., J. F. MACDONALD & K. G. BAIMBRIDGE. 1988. Release of intracellular calcium
     following activation of excitatory amino acid receptors in cultured hippocampal neurons. Soc.
     Neurosci. Abstr. **14**(Part 1): 94.
22. OBENAUS, A., T. MODY & K. G. BAIMBRIDGE. 1988. Dantrolene blockade of long-term
     potentiation (LTP) in the rat hippocampal CA1 region. Soc. Neurosci. Abstr. **14**(Part 1): 567.
23. FIFKOVÁ, E., J. A. MARKHAM & R. J. DELAY. 1983. Calcium in the spine apparatus of spines
     in the dentate molecular layer. Brain Res. **266:** 163–168.
24. NISHIZUKA, Y. 1986. Studies and perspectives of protein kinase C. Science **233:** 305–311.
25. MALINOW, R., D. V. MADISON & R. W. TSIEN. 1988. Persistent protein kinase activity
     underlying long-term potentiation. Nature **335:** 820–824.
26. KIKKAWA, U., K. OGITA, M. GO, H. NOMURA, T. KITANO, T. HASHIMOTO, K. ASE, K.
     SEKIGUSHI, J. KOUMOTO, Y. NISHIZUKA, N. SAITO & CH. TANAKA. 1988. Protein kinase C
     in transmembrane signaling. *In* Advances in Second Messenger and Phosphoprotein Re-
     search, Vol. 21. R. Adelstein, C. Klee & M. Rodbell, Eds.: 67–74. Raven Press. New York,
     NY.
27. DESMOND, N. L. & W. B. Levy. 1983. Synaptic correlates of associative potentiation/depres-
     sion: An ultrastructural study in the hippocampus. Brain Res. **265:** 21–30.
28. FIFKOVÁ, E. 1985. Actin in the nervous system. Brain Res. Rev. **9:** 187–215.
29. FIFKOVÁ, E. & C. L. Anderson. 1981. Stimulation-induced changes in dimensions of stalks of
     dendritic spines in the dentate molecular layer. Exp. Neurol. **74:** 621–627.
30. FIFKOVÁ, E. & A. VAN HARREVELD. 1977. Long-lasting morphological changes in dendritic
     spines of dentate granule cells following stimulation of the entorhinal area. J. Neurocytol.
     6: 211–230.
31. MARKHAM, J. A. & E. FIFKOVÁ. 1986. Actin filament organization within dendrites and
     dendritic spines during development. Dev. Brain Res. **27:** 263–269.

Multiple Excitatory Amino Acid Receptor Regulation of Intracellular Ca²⁺

Implications for Aging and Alzheimer's Disease

CARL COTMAN AND DANIEL MONAGHAN

Department of Psychobiology
University of California, Irvine
Irvine, California 92717

INTRODUCTION

The overall state of the aged brain depends to a large degree on the proper operation of its synapses. The vast majority of synapses in the brain are excitatory and appear to employ an acidic amino acid (L-glutamate or L-aspartate) as their neurotransmitter. Only recently has the necessary information been gained in order to understand the specific receptors that mediate the responses at these synapses.

There are at least five distinct types of excitatory amino acid receptors all with distinct functions (TABLE 1). Three have been defined by the depolarizing actions of the specific agonists: N-methyl-D-aspartate (NMDA), kainate, quisqualate (or AMPA) and their differential blockage by selective antagonists. A fourth, the AP4 receptor, appears to represent an inhibitory autoreceptor. The fifth, the ACPD receptor, appears to modify inositol phosphate metabolism and is activated by quisqualate or more specifically by *trans*-1-aminocyclopentyl-1,3-dicarboxylic acid (*trans*-ACPD). Two of these receptors can regulate intracellular Ca²⁺. NMDA-receptor-coupled ion channels allow the entry of extracellular Ca²⁺ while the ACPD receptor can activate inositol phosphate metabolism which in turn can lead to mobilization of internal Ca²⁺ stores. An unexpected concept to emerge from this separation of classes is that these receptors often appear to work in one or more combinations.

At many of the central synapses, neurotransmission is mediated by combinations of receptors. For example, the most frequently observed receptor combination involves NMDA and non-NMDA receptors working in concert within the same synapse. Non-NMDA receptors open Na⁺/K⁺ channels causing rapid depolarization. The NMDA receptor channel, in contrast, is unusual in that it is activated (or gated) by its neurotransmitter in a voltage-dependent manner and is permeable to Ca²⁺. Whereas with non-NMDA receptors, the number of channels opened by the neurotransmitter is independent of the membrane potential, with the NMDA receptor complex, the number of channels opened increases as the neuron is depolarized (*i.e.*, when the membrane potential moves in a positive direction from −70 mV toward 0 mV). Thus, as the neuron is depolarized, increasing amounts of K⁺, Na⁺, and Ca²⁺ ions can diffuse through the channel.

The voltage-dependence of the NMDA receptor provides a mechanism to link receptor activation with the use-dependent regulation of intracellular Ca²⁺. In this way, the end response produced by the neurotransmitter is "conditional" upon the level of concurrent postsynaptic depolarization as controlled by any variety of transmitters or neuromodulatory actions. The requirement for a conditional or "associative" response to account for learning has been postulated for many years (for example, Hebbian synapses); the NMDA receptor appears to be well suited for such a role. Indeed, NMDA receptors have been shown to participate not only in synaptic transmission, but also in the use-dependent modification of synapses both in development and in adult learning. Ironically, however, NMDA receptors may also contribute to pathological processes such as epilepsy and the neuronal loss that follows ischemia and hypoglycemia.

In this paper we shall first briefly review the role of the NMDA receptor complex in synaptic transmission in relationship to other excitatory amino acid receptors and its organization in brain. Then we shall discuss the properties of the inositol phosphate coupled receptor, its possible role in synaptic modification, and the novel manner in

TABLE 1. Excitatory Amino Acid Receptors

Receptor Classes	Agonists	Antagonists
NMDA	NMDA	D-AP5
NMDA site	L-Glu	CPP
	L-Asp	CGS-19755
Gly site	Gly	HA-966
	D-Ser	KY
		7-Cl-KYN
		cyclo-Leu
Channel		PCP
		TCP
		MK-801
		ketamine
Kainate	kainate	CNQX
	L-Glu	DNQX
	quisqualate	
QA/AMPA	AMPA	CNQX
	quisqualate	DNQX
	L-Glu	
L-AP4	L-AP4	?
	L-Glu	
ACPD	trans-ACPD	?
	L-Glu	
	quisqualate	

which it may operate in relationship to the NMDA receptor. Finally, we shall present the hypothesis that various short-term compensation mechanisms in the course of Alzheimer's disease (AD) may contribute eventually to pathology. This discussion will focus on the hippocampus, a particular brain region known to be vulnerable to aging and AD.

There are several reasons for examining the NMDA receptor with respect to aging and AD. Since these receptors appear to participate in long-term potentiation, a synaptic analog of memory, a decline in this receptor might underlie possible age-related deficits in learning or memory. An increase in NMDA receptors, on the other hand, might make neurons more vulnerable to excitotoxic injury associated with excessive surges in extracellular glutamate or related agonists. The precise state of the NMDA receptor complex then must be precisely set to mediate plasticity and absorb toxic insults. It is the Ca^{2+} signal regulated by the NMDA receptor that appears to mediate the synaptic stabilization, plasticity, and also excitotoxic injury.[1,2]

The NMDA Receptor Complex and Its Anatomical Organization in Brain

The NMDA receptor is so named because it appears to be selectively activated by the ligand NMDA. It should be pointed out, however, that NMDA is not present in the mammalian brain. The natural ligand may be glutamate, aspartate or a related derivative which binds with high affinity to the recognition site. The NMDA receptor represents a complex of multiple functional components each with a discrete ligand site. These include: 1) a transmitter recognition site which binds to glutamate or NMDA, 2) a regulatory site which binds glycine, 3) a channel site which binds phencylidine or related compounds, 4) a voltage-dependent Mg^{2+} binding site, and 5) an inhibitory site which binds Zn^{2+}. It also appears as if there are also two distinct binding sites (or states) associated with the transmitter recognition site for NMDA (for a recent review on the properties of these sites see REF. 3).

It is now possible to study the organization and specific regional properties of excitatory amino acid receptors in brain using quantitative radioligand binding autoradiography. Under carefully optimized conditions NMDA sites can be specifically labelled with L-[^{3}H]glutamate. In ligand binding studies, it is important to verify that the binding site shows the kinetic and pharmacological properties of the receptor. Area CA1 of the hippocampus is an excellent system to perform such an analysis because it consists largely of one cell type and is amenable to detailed electrophysiological and biochemical analysis. As shown in FIGURE 1, there is close agreement between the dose-response curves for the antagonist D-AP5 and L-[^{3}H]glutamate binding, depolarization by NMDA applied to CA1 neurons, and the inhibition of long-term potentiation. These and other data support the hypothesis that these methods can be used to measure the NMDA recognition site associated with the receptor complex.

NMDA binding sites are organized heterogeneously in the brain. Highest densities of NMDA-displacable L-[^{3}H]glutamate binding sites are found in the hippocampus, cerebral cortex, and striatum.[4,5] Specifically, strata radiatum and oriens of the hippocampal CA1 region display the highest density of sites in the brain while binding levels in the dentate gyrus molecular layer are slightly lower. Binding within the CA3 region is moderately high except within the stratum lucidum which has low binding levels, a finding predictive of the properties of synaptic transmission in this region.[6] Cerebral cortex exhibits both regional and laminar variations in NMDA site density. Among cortical regions, frontal, insular, pyriform, perirhinal, and anterior cingulate display higher binding levels than temporal, occipital, parietal, and posterior cingulate.

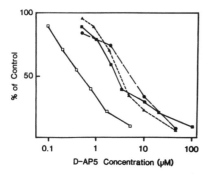

FIGURE 1. The different potencies of D-AP5 displacement of agonist (L-[^{3}H]glutamate) and antagonist (D-[^{3}H]AP5) binding are compared with D-AP5 potency at blocking NMDA-induced focal depolarizations and NMDA-receptor-mediated long-term potentiation. All measurements were made in the stratum radiatum of CA1 hippocampus. ▲, long-term potentiation; ●, NMDA depolarization; ■, L-[^{3}H]glutamate binding; and □, D-[^{3}H]AP5.

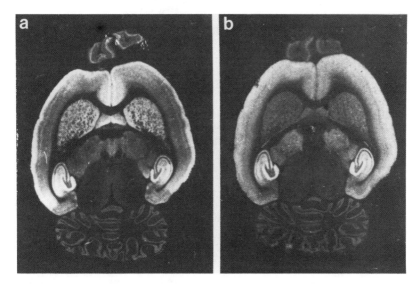

FIGURE 2. Autoradiograms of the rodent brain showing the distribution of the NMDA recognition site labeled by (**a**) [^{3}H]glutamate (agonist-preferring sites) compared to that labeled by (**b**) [^{3}H]CPP (antagonist-preferring sites).

Multiple Classes of NMDA Binding Sites: Agonist-Preferring and Antagonist-Preferring Subtypes

The widespread anatomical distribution and diverse function of NMDA receptors would suggest that specific subtypes or special properties exist for NMDA receptors in different brain areas. Since NMDA receptors are critical to brain function it is also of considerable relevance to answer such a question in order to develop regionally selective therapeutics.

Anatomical evidence suggests that L-[^{3}H]glutamate and radiolabelled antagonists bind to distinctly localized sites which may suggest multiple receptor types. NMDA receptors can be labelled with tritiated 3-((+)-2-carboxypiperazin-4-yl)-propyl-l-phosphonate ([^{3}H]CPP).[3,7–9] When the distribution of NMDA receptors is determined with the radiolabelled NMDA antagonist [^{3}H]CPP, the receptor distribution differs from that obtained with L-[^{3}H]glutamate[3] (see FIG. 2). In the parietal cortex the density of [^{3}H]CPP binding sites is greater than NMDA-sensitive L-[^{3}H]glutamate binding sites in layers IV and VI, resulting in a nearly uniform appearance in the cortex. Among other brain regions, NMDA-sensitive L-[^{3}H]glutamate binding site levels are relatively higher in the medial striatum and the lateral septum; [^{3}H]CPP binding sites are relatively high in the hippocampus, cerebral cortex, and various lateral thalamic nuclei (*e.g.,* ventral basal complex).[3] The radiolabelled NMDA antagonist D-[^{3}H]AP5 shows a pattern similar to that of [^{3}H]CPP binding sites (*e.g.,* high in hippocampus and ventral basal complex, and of uniform lamination in the cerebral cortex).[10]

Although two distinct NMDA binding sites have not been previously identified, their existence could account for different pharmacological actions obtained of the various NMDA receptor radioligands. For example,

- Antagonists generally display greater displacement potencies in [^{3}H]antagonist binding assays than in [^{3}H]agonist binding assays.[5,9,11–16] Conversely, agonists are relatively more potent in [^{3}H]agonist binding assays. As shown in FIGURE 1, for example, the NMDA antagonist D-AP5 is a much better displacer of antagonist than agonist binding.

• NMDA agonists and antagonists display complementary regional variations in their ability to displace L-[^{3}H]glutamate binding.[3] For example, NMDA antagonists selectively displace binding in the cerebral cortex and thalamus while agonists preferentially displace binding in the medial striatum.

• Glycine binding to the NMDA receptor can modify the relative affinities for agonists and antagonists at the glutamate recognition site.[3,17] Under certain conditions, glycine increases L-[^{3}H]glutamate binding but decreases [^{3}H]CPP binding.[3] This affect appears to be due to a glycine-induced change in the affinity of the NMDA receptor for L-[^{3}H]glutamate.[17] Although glycine appears to shift the NMDA receptor from a state which prefers antagonists to a state which prefers agonists, this glycine-induced shift in NMDA receptor properties does not appear to account for the observed regional variation in agonist and antagonist affinities at the NMDA receptor since in the presence of 5 μM glycine or 100 μM HA-966 (a glycine antagonist), NMDA-displacable L-[^{3}H]gluatmate and [^{3}H]CPP binding sites retain their differing regional distributions.

Thus, the two NMDA binding sites preferentially labelled by L-[^{3}H]glutamate and [^{3}H]CPP retain their differing distributions independent of the occupation state of the glycine subunit. These results indicate that the NMDA receptor has distinct forms corresponding to agonist-preferring (as found in the medial striatum) and antagonist-preferring (as found in the lateral portions of the thalamus). Presumably then, these two anatomically distinct sites can each display two differing states depending upon interactions with the glycine subunit.[3]

Several other lines of evidence are consistent with NMDA receptor heterogeneity. Molecular weights of the NMDA receptors estimated by radiation inactivation analysis differ for L-[^{3}H]glutamate and [^{3}H]CPP binding sites. The estimated size of the agonist-associated protein is 125 K whereas that of the anatagonist site is 209 K.[18] Further, it has been possible to affinity label the NMDA-gated ion channel with the PCP analogue azido-PCP.[19,20] The proteins which appear related to the NMDA receptor complex appear to be heterogeneously distributed among brain regions.[19,21] Other data also suggest a regional variation in sensitivity to NMDA agonists such as quinolinate. Binding studies show a 40% greater density of agonist binding sites in area CA1 than in dentate gyrus whereas antagonist sites are essentially equally distributed.[3] This is consistent with ex-citotoxicity and electrophysiological studies. For example, quinolinate is preferentially excitotoxic to hippocampal pyramidal cells relative to dentate gyrus granule cells while NMDA is equally toxic.[22] Electrophysiological studies show that pyramidal neurons are more sensitive to quinolinate-induced depolarizations than granule cells while NMDA is more similar.[23] Taken together, these data suggest a distinct heterogeneity of the NMDA system.

The apparent NMDA receptor heterogeneity in radioligand binding sites could be due to a variety of posttranslational factors, such as lipid environment, phosphorylation, oxidation/reduction, glycosylation, etc. It is also possible that these two subtypes are genetically distinct isoforms. With recent advances in molecular neurobiology, it is now apparent that the general rule for receptors is that a receptor exists as a family of closely related isoforms. Perhaps the closest analogy is with the GABA-A receptor system.

Comparisons between the Heterogeneity in NMDA and GABA Receptor Systems

Recently it was suggested that NMDA receptors are genetically related to GABA-A receptors.[24] GABA-A receptors, like NMDA receptors, show anatomically distinct agonist-preferring and antagonist-preferring receptor forms.[25] They also show regional allosteric regulation.[26] The GABA-A system shows at least three distinct α-subunit proteins[27] which have regional[28] and developmental[29] patterns of expression. Regional variations observed for multiple cDNA clones for GABA-A receptors appear to correspond to regional variations for agonist-preferring and antagonist-preferring GABA-A receptors.[30] Thus genetically determined forms may account for some of the structure activity profiles and allosteric interactions in both systems.

Possible Functional Significance of Different Forms of the NMDA Receptor Complex

The multiple forms of GABA-A receptor appear to display differences in receptor activation rate and desensitization rate.[31,32] These and related distinctions between NMDA receptors could greatly alter NMDA receptor activity during repetitive firing and seizures and influence the vulnerability of populations of neurons to excitotoxic insult. Genetically distinct GABA-A receptors also display differing sensitivities to GABA.[33] Such properties could play important and selective roles in NMDA-mediated functions such as LTP and the fine tuning of synaptic connections during development. Shifts in

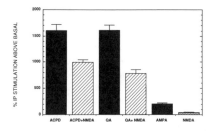

FIGURE 3. Stimulation of PI metabolism by excitatory amino acids and the inhibition by NMDA. Hippocampal slices from 10–13-day-old rats were incubated with ACPD (100 μM), QA (10 μM), AMPA (100 μM) or NMDA (100 μM). In the inhibition experiments NMDA (100 μM) was added 5 min before the addition of the agonist.

gene expression in the course of development, aging, and disease would enhance or compromise such functions. Perhaps the allosteric or voltage-dependent requirements differ for those of synaptic stabilization vs adult learning.

The Trans-ACPD Receptor: a Phosphoinositide Coupled Excitatory Amino Acid Receptor

As discussed above, glutamate's action at most brain synapses involves the interaction with at least two different glutamate receptors (NMDA and a non-NMDA (AMPA) receptor). In addition, however, glutamate will activate a receptor coupled to phosphoinositide (PI) turnover which provides additional control over intracellular free Ca^{2+}. Is this receptor interactive with other EAA receptors? Experiments were carried out to examine the activation of NMDA and non-NMDA receptors and combinations of these receptors on PI metabolism. This approach has yielded some surprising and interesting results.

Activation of this PI-coupled receptor with quisqualic acid (QA) or ACPD (non-NMDA agonists) stimulates PI formation 15-fold (see FIG. 3). NMDA has little effect on PI metabolism. Activation of the NMDA receptor, however, inhibits QA-induced stimulation by 70%. This inhibition is dependent on extracellular calcium and is reversible. That is, QA can reactivate PI metabolism after NMDA is removed from the media. The QA stimulation does not act via AMPA receptors, which electrophysiological evidence suggests is responsible for the initial fast depolarizations at dual receptor synapses in the hippocampus.[34]

The "Three Receptor" Model of the Excitatory Synapse and Its Possible Functional Significance

Synapses with two receptors may continuously regulate intracellular Ca^{2+} by ongoing synaptic activity. Thus, if the PI-coupled receptor site is also present near or in the synapses of young and adult animals, then glutamate neurotransmission could modulate intracellular Ca^{2+} through both the novel PI-coupled ACPD receptor and the NMDA receptor. By inhibiting PI metabolism, mobilization of Ca^{2+} is inhibited and Ca^{2+} entry would thus become dependent upon NMDA-associated channels. This system may adjust metabolism or it may even act to protect the cells against excessive intracellular free Ca^{2+}.

This "three receptor synapse" (i.e., NMDA, PI-coupled (ACPD), and non-NMDA (AMPA) receptors) may have significant value in potential plasticity mechanisms. A formal theory of synaptic modifications developed by Cooper and his associates[35] depends on a "modification threshold." The ability to set or adjust such a threshold might be related to the findings of an interaction between the NMDA and PI-coupled receptor. In this model, the efficacy of active synapses increases when the postsynaptic target is concurrently depolarized beyond the "modification threshold" which itself is continuously adjustable. At a mechanistic level, the critical question raised by Cooper et al. can be stated simply: When excitatory amino acid receptors are activated, what distinguishes the response of depolarized membrane potentials from the response at the resting potential? The NMDA-mediated influx of Ca^{2+} and the coupling to PI metabolism are certainly excellent candidates. This push-pull or bidirectional mechanism may be adjusted to stimulate, stabilize, or retract the response related to intracellular Ca^{2+}. Changes in the ratios of receptors and coupling mechanisms in the course of development and aging would change the threshold. Such a mechanism could be envisioned to play a key role in such processes at critical periods, adjust function with age, and perhaps participate in certain types of learning.

Recent data support the notion that the PI-coupled receptor may in fact participate in synaptic modifications. A late component of LTP, for example, is blocked by 2-amino-4-phosphonobutyrate which is thought to be an antagonist of glutamate activation of inositol phosphate formation.[36] This suggests that a third glutamate receptor, coupled to inositol phosphate formation (possibly the ACPD receptor, TABLE 1), may play a role in the initiation of the late component of LTP. Consistent with such a mechanism, a recent report has suggested the possible involvement of internal Ca^{2+} release in LTP.[37] This may provide a new mechanism for the multiple transmitter regulation of LTP, as a variety of neurotransmitters (e.g., norepinephrine, acetylcholine, etc.) may also influence inositol phosphate turnover and thereby regulate neuronal "set point."

NMDA Binding Sites and Normal Aging

At present a detailed analysis of the normal aged brain has not been carried out for the receptor types involved in multireceptor synapses. However, examination of the properties of binding sites in the normal rodent brain, of normal human aged brain, and of the brain of patients who have died from Alzheimer's disease was recently begun.

In the brain of aged Fischer 344 rats, NMDA receptors appear to show a regionally selective decline with age (FIG. 4). It appears that agonist and antagonist sites display anatomical differences in the course of development and aging. Agonist-preferring sites develop rather homogeneously over the cerebral cortex whereas antagonist sites are concentrated in entorhinal and parietal cortex. With age, selected regions such as the hippocampus and perirhinal cortex show a minor 15–20% loss of NMDA binding sites for both agonists and antagonists. In the striatum, these receptors show up to a 50% decline.[38] In the medial striatum the loss of antagonist sites exceeds the loss of agonist sites. In the rodent brain, the inner layers of the entorhinal cortex have a 30–40% loss of receptors with a greater loss of NMDA agonist sites. These data indicate that major as well as subtle changes occur in the NMDA receptor system with normal aging.

a

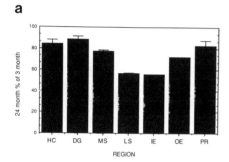

b

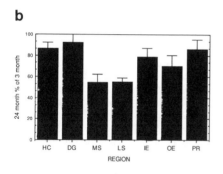

FIGURE 4. Histograms of NMDA binding site density in various brain regions of 2-year-old Fischer 344 rats labelled by **(a)** L-[^{3}H]glutamate and **(b)** [^{3}H]CPP. Data is expressed as % of the density in 3-month-old Fischer 344 rats. HC, hippocampus; DG, dentate gyrus; MS, medial striatum; LS, lateral striatum; IE, inner layers of entorhinal cortex; OE, outer layers of entorhinal cortex; PR, perirhinal cortex.

The status of NMDA receptors in the aged brain, however, may be even more complex. It appears that two different mouse strains show unique age-dependent changes in NMDA receptor properties,[39] suggesting a possible genetic component to age-dependent NMDA regulation. In Balb/c mice the Bmax values decline 16% by 10 months and 45% by 30 months of age when compared to 3 months. the K_D increased more by 10 months (+ 29%) than by 30 months (+ 14%). In the C57BI strain the Bmax was unaltered by 10 months but decreased 17% by 30 months. The K_D values, however, increased 121% by 10 months and 283% by 30 months of age. These changes may be due to properties of the receptor itself, its synthesis and/or its membrane environment.

Taken together it appears that NMDA receptors will show high individual variation both between brain areas, and perhaps over the life span.

NMDA Receptors in Alzheimer's Disease

A decline in L-[^{3}H]glutamate binding in the hippocampus of individuals with Alzheimer's disease of up to 80% was originally proposed by Young and colleagues in 1985[40,41] and attributed to an early loss in NMDA receptors. In contrast, other groups report that NMDA receptor number is, for the most part, maintained and only substantially declines in the later stages of the disease marked by severe cell loss.[42,43] Furthermore, these investigators have suggested that the conditions used in the earlier studies by Greenamyre and co-workers do not adequately resolve NMDA receptor binding from the binding of L-[^{3}H]glutamate to the other receptors and transport systems. If, as it now appears, NMDA receptor number is maintained to a large degree during the moderate cell loss in the early stages of Alzheimer's disease, it follows that the cellular density of receptors may actually increase slightly, possibly as a compensatory mechanism to maintain failing circuits. As the cell loss continues, however, a point is reached at which it is too great, and receptor number drops dramatically. It is tempting to speculate that the increased density of NMDA receptors, which was at first compensatory, may actually contribute to the subsequent cell death by an increased susceptibility to excitotoxic damage, particularly if associated with a decreased ability to handle intracellular Ca^{2+}. Overall, however, some caution needs to be exercised with respect to a final conclusion on the state of the NMDA receptor complex in the normal human and AD brain. Kinetic data is as yet incomplete and there is significant variation between individual cases.[4,42]

CONCLUSION

Excitatory synaptic transmission at many central synapses appears to operate via multiple receptors; non-NMDA, NMDA, and possibly one coupled to PI metabolism. The NMDA receptor provides these synapses with unusual properties as it is the only known neurotransmitter receptor that regulates a voltage-dependent channel directly linked to the second messenger Ca^{2+}. Thus, NMDA receptors are uniquely suited to activate Ca^{2+}-dependent synaptic modifications when there is concurrent presynaptic and postsynaptic activity, and to activate Ca^{2+}-dependent pathological events when there is excessive NMDA receptor activation. Further, a novel excitatory amino acid receptor has been described which activates phopshoinositide metabolism which itself is regulated by NMDA receptor activation. The central role played by the NMDA receptor in all of these processes suggests that other modulators known to control neuronal responsiveness may now need to be viewed in relation to the NMDA receptor system. This places NMDA receptors in a critical role in relationship to aging and age-related diseases where the substrates of learning must be maintained and excitotoxic injury avoided. In the aged brain NMDA receptors can be altered with respect to subtype or number depending on the species and brain region. These changes with age may depend on specific receptor properties and/or membrane environment which probably impact directly and indirectly on the activity-dependent regulation of Ca^{2+}.

REFERENCES

1. COTMAN, C. W., D. T. MONAGHAN & A. H. GANONG. 1988. Excitatory amino acid neurotransmission: NMDA receptors and Hebb-type synaptic plasticity. Ann. Rev. Neurosci. **11:** 61–80.
2. COTMAN, C. W., R. J. BRIDGES, J. S. TAUBE, A. S. CLARK, J. W. GEDDES & D. T.

MONAGHAN. 1989. The role of NMDA receptor in central nervous system plasticity and pathology. The Journal of WIH Research 1: 65–74.

3. MONAGHAN, D. T., H. J. OLVERMAN, L. NGUYEN, J. C. WATKINS & C. W. COTMAN. 1988. Two classes of NMDA recognition sites: differential distribution and differential regulation by glycine. Proc. Natl. Acad. Sci. USA 85: 9836–9840.

4. MARAGOS, W. F., J. B. PENNEY & A. B. YOUNG. 1988. Anatomic correlation of NMDA and [^{3}H]TCP-labelled receptors in rat brain. J. Neurosci. 8: 493–501.

5. MONAGHAN, D. T. & C. W. COTMAN. 1986. Identification and properties of NMDA receptors in rat brain synaptic plasma membranes. Proc. Natl. Acad. Sci. USA 83: 7532–7536.

6. HARRIS, E. W. & C. W. COTMAN. 1986. Long-term potentiation of guinea pig mossy fiber projections is not blocked by NMDA antagonists. Neurosci. Lett. 70: 132–137.

7. JARVIS, M. F., D. E. MURPHY & M. WILLIAMS. 1987. Quantitative autoradiographic localization of NMDA receptors in rat brain using [^{3}H]CCP: comparison with [^{3}H]TCP binding sites. Eur. J. Pharmacol. 141: 149–152.

8. OLVERMAN, H. J., D. T. MONAGHAN, C. W. COTMAN & J. C. WATKINS. 1986. [^{3}H]CPP, a new competitive ligand for NMDA receptors. Eur. J. Pharmacol. 131: 161–162.

9. MURPHY, D. E., J. SCHNEIDER, C. BOEHM, J. LEHMANN & M. WILLIAMS. 1987. Binding of [^{3}H]3-(2-carboxypiperazin-4-yl)propyl-l-phosphonic acid to rat brain membranes: a selective high-affinity ligand for N-methyl-D-aspartate receptors. J. Pharmacol. Exp. Ther. 240: 778–784.

10. MONAGHAN, D. T., D. T. YAO, H. J. OLVERMAN, J. C. WATKINS & C. W. COTMAN. 1984. Autoradiography of 3H-D-2-amino-5-phosphonopentanoate binding sites in rat brain. Neurosci. 52: 253–258.

11. GREENAMYRE, J. T., J. B. PENNEY, A. B. YOUNG, C. D'AMATO, C. J. HICKS & C. J. SHOULSON. 1985. Alterations in L-glutamate binding in Alzheimer's and Huntington's disease. Science 227: 1496–1499.

12. MONAGHAN, D. T., V. R. HOLETS, D. W. TOY & C. W. COTMAN. 1983. anatomical distribution of four pharmacologically distinct ^{3}H-L-glutamate binding sites. Nature 306: 176–179.

13. MONAGHAN, D. T., D. YAO & C. W. COTMAN. 1985. ^{3}H-L-Glutamate binds to kainate, NMDA and AMPA-sensitive binding sites: an autoradiographic analysis. Brain Res. 340: 378–383.

14. FOSTER, A. C. & G. E. FAGG. 1987. Comparison of L-[^{3}H]glutamate, D-[^{3}H]aspartate, DL-[^{3}H]AP5 and [^{3}H]NMDA as ligands for NMDA receptors in crude postsynaptic densities from rat brain. Eur. J. Pharmacol. 133: 291–300.

15. OLVERMAN, H. J., A. W. JONES & J. C. WATKINS. 1988. [^{3}H]D-2-Amino-5-phosphonopentanoate as a ligand for N-methyl-D-aspartate receptors in the mammalian central nervous system. Neuroscience 26: 1–15.

16. MONAHAN, J. B. & J. MICHEL. 1987. Identification and characterization of an N-methyl-D-aspartate-specific L-[^{3}H]glutamate recognition site in synaptic plasma membranes. J. Neurochem. 48: 1699–1708.

17. FADDA, E., W. DANYSZ, J. T. WROBLEWSKI & E. COSTA. 1988. Glycine and D-serine increase the affinity of NMDA sensitive glutamate binding sites in rat brain synaptic membranes. Neuropharmacology 27: 1183–1185.

18. HONORÉ, T. & J. DREJER. 1988. Binding characteristics of non-NMDA receptors. In Excitatory Amino Acids in Health and Disease. D. Lodge, Ed. 91-106. John Wiley & Sons, Ltd. Chichester, UK.

19. HARING, R., Y. KLOOG, N. HARSHAK-FELIXBRODT & M. SOKOLOVSKY. 1987. Multiple mode of binding of phencyclidines: high affinity association between phencyclidine receptors in rat brain and a monovalent ion-sensitive polypeptide. Biochem. Biophys. Res. Commun. 142: 501–510.

20. HARING, R., R. S. ZUKIN & S. R. ZUKIN. 1988. Photoaffinity labeling and binding studies reveal two types of phencyclidine (PCP) receptors in the NCB-20 cell line. Soc. Neurosci. Abstr. 14: 484.

21. HARING, R., Y KLOOG & M. SOKOLOVSKY. 1985. Regional heterogeneity of rat brain phencyclidine (PCP) receptors revealed by photoaffinity labeling with [^{3}H]azido phencyclidine. Biochem. Biophys. Res. Commun. 131: 1117–1123.

22. SCHWARCZ, R., G. S. BRUSH, A. C. FOSTER & E. D. FRENCH. 1984. Seizure activity and lesions after intrahippocampal quinolinic acid injection. Exp. Neurol. **84:** 1–17.
23. STONE, T. W. 1985. Differences of neuronal sensitivity to amino acids and related compounds in the rat hippocampal slices. Neurosci. Lett. **59:** 313–317.
24. BARNARD, E. A., M. G. DARLISON & P. SEEBURG. 1987. Molecular biology of the GABA receptor: the receptor/channel superfamily. Trends Neurosci. **10:** 502–509.
25. OLSEN, R. W., E. W. SNOWHILL & J. K. WAMSLEY. 1984. Autoradiographic localization of low affinity GABA receptors with [³H]bicuculline methochloride. Eur. J. Pharmacol. **99:** 247–248.
26. UNNERSTALL, J. R., M. J. KUHAR, D. L. NIEHOFF & J. M. PALACIOS. 1981. Benzodiazepine receptors are coupled to a subpopulation of γ-aminobutyric (GABA) receptors: evidence from a quantitative autoradiographic study. J. Pharmacol Exp. Ther. **218:** 797–804.
27. FUCHS, K., H. MÖHLER & W. SIEGHART. 1988. Various proteins from rat brain, specifically and irreversibly labeled by [³H]flunitrazepam, are distinct alpha subunits of the GABA-benzodiazepine receptor complex. Neurosci. Lett. **90:** 314–319.
28. SIEGHART, W. & G. DREXLER. 1983. Irreversible binding of [³H]flunitrazepam to different proteins in various brain regions. J. Neurochem. **41:** 47–55.
29. EICHINGER, A. & W. SIEGHART. 1986. Postnatal development of proteins associated with different benzodiazepine receptors. J. Neurochem. **46:** 173–180.
30. OLSEN, R. W., R. T. MCCABE, J. P. YEZUITA & J. K. WAMSLEY. 1988. Analysis of the distribution and density of the GABA/benzodiazepine/convulsant complex in the rat CNS. Soc. Neurosci. Abstr. **14:** 780.
31. CASH, D. J. & K. SUBBARAO. 1987. Channel opening of γ-aminobutyric acid receptor from rat brain: molecular mechanisms of receptor responses. Biochemistry **26:** 7562–7570.
32. AKAIKE, N., M. INOUE & O. A. KRISHTAL. 1986. Concentration-clamp study of gamma-aminobutyric-acid-induced chloride current kinetics in frog sensory neurones. J. Physiol. **379:** 171–185.
33. LEVITAN, E. S., P. R. SCHOFIELD, D. R. BURT, L. M. RHEE, W. WISDEN, M. KOHLER, N. FUJITA, H. F. RODRIGUEZ, A. STEPHENSON, M. G. DARLISON, E. A. BARNARD & P. H. SEEBURG. 1988. Structural and functional basis for GABA-A receptor heterogeneity. Nature **335:** 76–79.
34. PALMER, E., D. T. MONAGHAN & C. W. COTMAN. 1988. Glutamate receptors and phospho-inositide metabolism: stimulation via quisqualate receptors is inhibited by NMDA receptor activation. Mol. Brain Res. **4:** 161–165.
35. BEAR, M. F., L. N. COOPER & F. F. EBNER. 1987. A physiological basis for a theory of synapse modification. Science **237:** 42–48.
36. REYMANN, K. G. & H. MATTHIES. 1989. 2-Amino-4-phosphonobutyrate selectively eliminated late phases of long-term potentiation in rat hippocampus. Neurosci. Lett. **98:** 166–171.
37. OBENAUS, A., I. MODY & K. J. BAIMBRIDGE. 1989. Dantrolene-Na (Dantrium) blocks induction of long-term potentiation in hippocampal slices. Neurosci. Lett. **98:** 172–178.
38. MONAGHAN, D. T., K. A. ANDERSON, C. PETERSON & C. W. COTMAN. 1988. Age-dependent loss of NMDA receptors in rodent brain. Soc. Neurosci. Abstr. **14:** 486.
39. PETERSON, C. P. & C. W. COTMAN. Strain dependent decrease in glutamate binding to the NMDA receptor during aging. Neurosci. Lett. In press.
40. GREENAMYRE, J. T., J. M. OLSON, J. B. PENNEY & A. B. YOUNG. 1985. Autoradiographic characterization of NMDA-, quisqualate-, and kainate-sensitive glutamate binding sites. J. Pharmacol. Exp. Ther. **233:** 254–263.
41. GREENAMYRE, J. T., J. B. PENNEY, C. D'AMATO & A. B. YOUNG. 1987. Dementia of the Alzheimer's type: changes in hippocampal L-[³H]glutamate binding. J. Neurochem. **48:** 543–551.
42. GEDDES, J. W., H. CHANG-CHUI, S. M. COOPER, I. T. LOTT & C. W. COTMAN. 1986. Density and distribution of NMDA receptors in the human hippocampus in Alzheimer's disease. Brain Res. **399:** 156–161.
43. SIMPSON, M. D. C., M. C. ROYSTON, J. F. W. DEAKIN, A. J. CROSS, D. M. A. MANN & P. SLATER. 1988. Regional changes in [³H]D-aspartate and [³H]TCP binding sites in Alzheimer's disease brains. Brain Res. **462:** 76–82.

Neuronal Ca^{2+} Channels and Their Regulation by Excitatory Amino Acids

RICHARD J. MILLER, SHAWN N. MURPHY, AND
STEVEN R. GLAUM

Department of Pharmacological and Physiological Sciences
University of Chicago
947 East 58th Street
Chicago, Illinois 60637

INTRODUCTION

It is generally accepted that glutamate is the most widely distributed excitatory neurotransmitter in the central nervous system.[1] Under normal conditions glutamate released at central synapses causes fast neuronal excitation and also produces long-term changes in synaptic efficacy which may underlie phenomena such as learning and memory. However, under pathological conditions, such as those prevailing during periods of ischemia/hypoxia, abnormally large quantities of glutamate or similar substances are released from neurons and probably from glial cells as well.[2,3] These excessive quantities of glutamate produce a greater-than-normal activation of glutamate receptors, and this has been found to produce neuronal toxicity ("excitotoxicity") in most parts of the brain.[2] It is now believed that glutamate exerts both its physiological and pathophysiological effects through actions at at least four distinct types of receptors.[4] A question of great current interest is how glutamate exerts these effects at a molecular level. It has recently become clear that Ca^{2+} probably plays a key role in mediating the ability of glutamate to increase synaptic strength and also its long-term toxic effects under abnormal circumstances.[2,5] Thus removal of external Ca^{2+} has been reported to block both glutamate-mediated "long-term potentiation," a manifestation of synaptic plasticity,[5] and also glutamate-mediated neurotoxicity.[6,7] The purpose of this paper is to review current knowledge of our understanding of the ways in which glutamate can influence Ca^{2+} metabolism and disposition in central neurons.

Before embarking on this particular task, it is important to review what we know at this time about the pharmacology of glutamate receptors. As discussed above, it is currently thought that at least four types of glutamate receptors exist. These are generally defined through the use of archetypal glutamate agonists and in some cases glutamate antagonists as well.[4] The first receptor type we shall consider is that defined through the action of the glutamate agonist kainic acid. The second receptor is defined through the actions of the glutamate agonist quisqualic acid and the third through the actions of the glutamate agonist N-methyl-D-aspartic acid (NMDA). Finally, we shall discuss a second type of quisqualic-acid-specific glutamate receptor which appears to work in quite a different manner from the other three just defined. A summary of the pharmacology of these four types of glutamate receptors is provided in TABLE 1.

Kainic Acid Receptors

Addition of kainic acid to neurons from the hippocampus or stiatum *in vitro* produces a characteristic rise in $[Ca^{2+}]_i$ as can be seen in FIGURE 1. The $[Ca^{2+}]_i$ rises and falls and

TABLE 1. Pharmacology of Glutamate Receptors

Receptor	Kainate	Quisqualate	N-Methyl-D-Aspartate	Quisqualate II
Agonists	kainate	quisqualate	NMDA	quisqualate
	domoic acid	AMPA		ibotenate
			Competitive	
Antagonists	GAMS	CNQX	CPP	
	CNQX	DNQX	AP5	?
	DNQX		Noncompetitive	
			PCP	
			MK-801	
			7-Cl-Kynurenate	

then, in a large number of cases, is seen to rise again before finally declining to a plateau level which is some hundreds of nM above the normal resting level. FIGURE 1 also illustrates the effect of perfusing one of these central neurons with a depolarizing high K^+ solution. Again the $[Ca^{2+}]_i$ is seen to rapidly rise and then decline to a new plateau level. Removal of Ca^{2+} from the external solution completely blocks the effects of both high K^+ solutions and of kainic acid, indicating that these responses are entirely due to Ca^{2+} influx from the external solution. FIGURE 2 illustrates the effect of blocking voltage-sensitive Ca^{2+} channels in these neurons on the responses to kainic acid or high K^+. Ca^{2+} channels are inhibited in this experiment by a combination of the L-type Ca^{2+} channel blocking drug nitrendipine and voltage-dependent inactivation. Under these conditions the responses to depolarizing concentrations of K^+ are completely inhibited. Moreover, it can also be seen that the response to a moderate concentration of kainic acid (100 μM) is also virtually completely inhibited. This suggests that kainic acid is producing its response through depolarization of the cell and activation of voltage-sensitive Ca^{2+} channels. Such an explanation is quite reasonable considering what we know about the

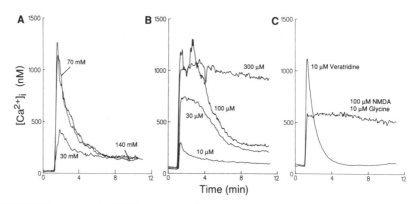

FIGURE 1. Increases in $[Ca^{2+}]_i$ recorded from single striatal neurons upon stimulation with various agents. Basal $[Ca^{2+}]_i$ was recorded for one minute after which the cells were perfused with the agents shown. (A) illustrates stimulation with various concentrations of $[K^+]_o$ containing 1 μM TTX and 100 μM AP5 to prevent glutaminergic synaptic transmission which is sometimes observed in these cultures. (B) shows application of various concentrations of kainate containing 1 μM TTX and 10 μM MK-801 to block activation of NMDA receptors. (C) shows depolarization with 10 μM veratridine and activation of NMDA receptors with 100 μM NMDA and 10 μM glycine in Mg^{2+}-free solution. Stimulating agents were present during entire period of the traces illustrated.

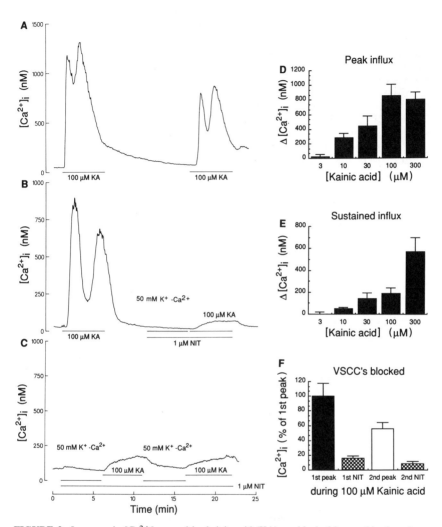

FIGURE 2. Increases in $[Ca^{2+}]_i$ caused by kainic acid (KA) are blocked by combined predepolarization and nitrendipine. The peak and sustained $[Ca^{2+}]_i$ rise elicted in striatal neurons by increasing concentrations of KA was quantified and illustrated in **(D)** and **(E)**. Two 100-μM KA applications to the same cell exhibited similar responses as shown in **(A)**. The effect of predepolarization with Ca^{2+}-free 50-mM K^+ solution which inactivates a portion of the voltage-sensitive Ca^{2+} channels in these neurons and 1 μM nitrendipine which blocks the rest was studied in comparison with the first response on the same cell **(B)** and also in comparison to the first response between cells (compare A and C). Combined results are shown in **(F)**. First two bars compare results between cells and second two bars compare results in the same cell (n ≥ 4).

mechanism of action of kainic acid. The kainate receptor is supposedly linked to a monovalent cation-specific ionophore.[8] Activation of this ionophore will therefore lead to depolarization of the neuron and subsequent opening of voltage-sensitive Ca^{2+} channels.[9] It is interesting to note, however, that at elevated concentrations of kainic acid (>100 μM) a response is still observed even following total blockade of voltage-sensitive Ca^{2+} channels.[9] It seems that this second type of response may be due to some limited permeability of the kainate receptor-gated ionophore to Ca^{2+}.[10] This limited influx may only become manifest at elevated kainate concentrations. The reason why kainate often produces the large oscillations illustrated in FIGURES 1 and 2 is not entirely clear. However, one possible explanation is as follows. During the response to kainate fairly large amounts of Na^+ will also enter the cell through the kainate-gated ionophore. This increase in the cytoplasmic $[Na^+]$ may cause Ca^{2+} to leave the mitochondria via Na^+/Ca^{2+} exchange, and this may therefore give rise to the secondary oscillation observed.[11] It is interesting to note that these secondary oscillations are never seen when all external Na^+ is replaced with Li^+ which does not participate in mitochondrial Na^+/Ca^{2+} exchange.[11]

Quisqualate Receptors

FIGURES 3 & 4 illustrate that quisqualic acid or its analog AMPA can produce similar effects to kainate. These effects are again dependent on Ca^{2+} influx and can be blocked when voltage-sensitive Ca^{2+} channels in the cells have been inhibited. Thus it is clear that

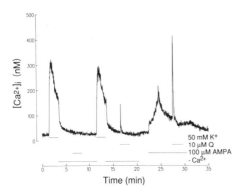

FIGURE 3. AMPA is unable to produce increases in $[Ca^{2+}]_i$ in Ca^{2+}-free medium although quisqualate (Q) is effective. However, in normal medium AMPA stimulated Ca^{2+} influx in this hippocampal neuron. Quisqualate then produced an additional response that was superimposed upon the AMPA response.

quisqualate or AMPA can also activate a glutamate receptor that results in cell depolarization and Ca^{2+} influx through voltage-sensitive Ca^{2+} channels. This is again consistent with our knowledge of the molecular basis of quisqualic acid action.[8] Stimulation of quisqualate receptors by quisqualate, glutamate, or AMPA is known to activate an ionophore with properties very similar to that activated by stimulation of kainic acid receptors. Specific competitive antagonists exist which can block the depolarizing effects of both kainic and quisqualic acids.[12] These are drugs such as CNQX and DNQX (TABLE 1). FIGURE 4 shows that the responses to quisqualate that result from depolarization and Ca^{2+} influx can be completely blocked by CNQX as would be expected.

NMDA Receptors

FIGURES 1 and 5 illustrate the consequences of the addition of the glutamate analog NMDA to central neurons. Here again $[Ca^{2+}]_i$ rises rapidly and remains at elevated levels for as long as the stimulus is present. FIGURE 5 illustrates two important properties of the

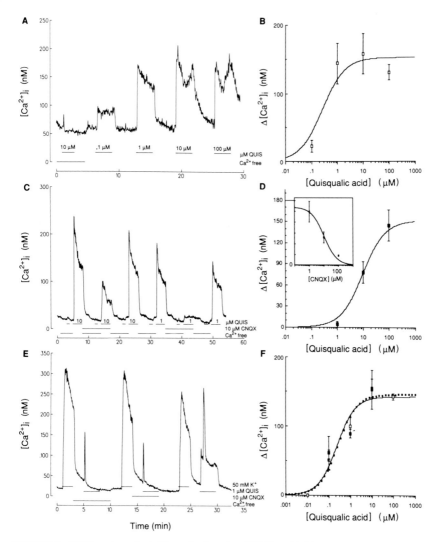

FIGURE 4. (A) Dose-dependent effects of quisqualate (QUIS) on $[Ca^{2+}]_i$ in hippocampal neurons. This is an example of a cell that showed little quisqualate-induced Ca^{2+} mobilization and therefore predominantly produced Ca^{2+} influx. The small contribution of Ca^{2+} mobilization can be observed during the initial addition of quisqualate (10 μM) in Ca^{2+}-free medium. (B) Dose-response relationship for quisqualate-induced Ca^{2+} influx. $EC_{50} = 263$ nM. (C) Effect of CNQX on quisqualate-induced Ca^{2+} influx. Trace shows the effect of 10 μM CNQX on Ca^{2+} influx induced by various concentrations of quisqualate. In this cell there was little contribution from quisqualate-induced Ca^{2+} mobilization. (D) Dose-response curve for quisqualate-induced Ca^{2+} influx in the presence of 10 μM CNQX. Dose-response curve is shifted to the right in a parallel manner ($EC_{50} = 9.77$ μM). Comparison of curves in B and D enabled us to calculate a K_i value of 250 nM for CNQX action at the quisqualate receptor responsible for activating Ca^{2+} influx. *Inset* in D illustrates the dose-dependent inhibition of Ca^{2+} influx induced by 10 μM quisqualate produced by CNQX ($IC_{50} = 10.3$ μM). (n = 3). (E) In this cell which exhibited both quisqualate-induced Ca^{2+} influx and Ca^{2+} mobilization, CNQX clearly blocked influx but not mobilization. (F) Dose-response curves for quisqualate-induced Ca^{2+} mobilization in the absence (*open squares*) and the presence (*filled squares*) of 10 μM CNQX. Theoretical curves in the absence of CNQX (*open dots*) and the presence (*continuous line*) of 10 μM CNQX show that there was no effect of this concentration of CNQX. In the absence of CNQX the EC_{50} for quisqualate-induced Ca^{2+} mobilization was 277 nM and in the presence of CNQX the EC_{50} was 239 nM.

response to NMDA which are again both consistent with our knowledge of the molecular actions of this particular glutamate analog. The left hand panel shows that the response to NMDA is much greater when all Mg^{2+} is removed from the extracellular bathing solution. When physiological concentrations of Mg^{2+} are added back the response to NMDA is considerably smaller. This is due to the well-known blocking action of Mg^{2+} on NMDA receptors.[13] It is known that NMDA receptor activation leads to the gating of an ion channel which has properties that are somewhat different from those gated by quisqualate or kainate receptors described above. For example, the NMDA receptor-gated ionophore is much more permeable to Ca^{2+} than monovalent cations ($P_{ca}/P_{na} \simeq 10$).[10] It is known that Mg^{2+} at physiological concentrations will block this ionophore in a highly voltage-dependent manner such that the block is relieved at depolarized membrane potentials.[13] A second characteristic of the response to NMDA is illustrated in the right hand panel. It can be seen that when cells are perfused rapidly so that there is no build up of any excreted materials in the extracellular medium, the response to NMDA is no longer apparent. However, if low concentrations of the amino acid glycine (>1 nM) are added back to the perfusing solution, the effects of NMDA are now quite apparent. This result

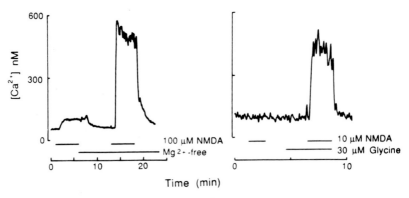

FIGURE 5. Effects of NMDA on $[Ca^{2+}]_i$ in hippocampal neurons *in vitro*.

is again consistent with the known interaction of glycine and glutamate at the NMDA receptor.[14] It is clear that the actions of NMDA or glutamate are greatly potentiated by glycine in the concentration range of 10 nM to 1 μM. A recent study has indicated that this may be due to the fact that under normal conditions NMDA receptors rapidly inactivate in response to agonists and that this inactivation is prevented by the action of glycine.[15] NMDA receptors also possess a characteristic pharmacology which is quite distinct from that associated with the kainate and quisqualate receptors already described. A variety of drugs have been described that interact with the NMDA receptor at various points (TABLE 1). Thus agonists and antagonists are known which act at the glutamate recognition site or the glycine regulatory site or bind within the gated ionophore.[4,16] A summary of some of these agents is provided in TABLE 1. It is thought likely that when large amounts of glutamate are released during hypoxic insult to the brain, much of the Ca^{2+} entry into neurons that occurs does so via the NMDA receptor-gated ionophore.[2] Consequently there is great interest at this time in compounds which can block this receptor in various ways as these may be potentially useful drugs for combating brain injury resulting from the Ca^{2+} overload associated with ischemia or epilepsy, for example.

"Quisqualate II" Receptors

It has recently become apparent that a fourth type of glutamate receptor exists that can regulate $[Ca^{2+}]_i$ through a mechanism that is completely distinct from those already described. Several laboratories have reported that activation of glutamate receptors, particularly by the glutamate analog quisqualic acid, can lead to a large stimulation of the production of inositol triphosphate (IP$_3$).[17,18] Presumably this is achieved through the activation of the enzyme phospholipase C. It is well established that in many types of cells IP$_3$ can stimulate the mobilization of Ca^{2+} from intracellular storage sites. These storage sites may be specific calsequestrin-containing vesicular entities which have recently been named "calciosomes."[19] FIGURES 3 and 6 illustrate the effect of quisqualic acid on hippocampal neurons in vitro in both Ca^{2+}-containing and Ca^{2+}-free medium. It is quite clear that in Ca^{2+}-free medium a rapid transient increase in $[Ca^{2+}]_i$ can be observed.[20,21] Indeed under some circumstances quisqualate will trigger a series of oscillations in $[Ca^{2+}]_i$ as also illustrated in FIGURE 6. Such rapid oscillations in $[Ca^{2+}]_i$ are quite typical of responses to agents which produce their effects through the stimulation of IP$_3$ production. The ability of quisqualic acid to mobilize intracellular Ca^{2+} is clearly mediated through a type of glutamate receptor which is quite distinct from that mediating quisqualate-dependent depolarization and Ca^{2+} influx as described above. Thus AMPA is not able to produce the rapid $[Ca^{2+}]_i$ transients produced by quisqualate under these circumstances (FIG. 3). Moreover, CNQX is unable to inhibit the Ca^{2+} mobilizing responses of quisqualate (FIG. 4). In addition to quisqualate, ibotenate and glutamate are also effective agonists at this type of receptor site, although aspartate is not. No specific antagonists are known at this time. In order to distinguish this receptor from that producing quisqualate-dependent depolarization we shall refer to it as the "quisqualate II" type of glutamate receptor (TABLE 1).

Further Effects of Glutamate on Intracellular Calcium Concentrations

Under normal conditions the response of $[Ca^{2+}]_i$ to stimulation by glutamate will result from a combination of the effects already described in this paper (FIG. 7). However, we have recently observed that further effects of glutamate on $[Ca^{2+}]_i$ homeostasis can also be demonstrated which are as yet not completely explained. These types of effects can be seen in the experiment illustrated in FIGURE 7, for example. In this case, the cell was first perfused with glutamate producing a transient response. The cell was then depolarized with K$^+$ and a rise and fall in $[Ca^{2+}]_i$ was observed. Glutamate was then added a second time under Ca^{2+}-free conditions and a transient rise in $[Ca^{2+}]_i$ of the type described above was also observed. This response was presumably mediated by quisqualate II receptors. However, when the glutamate was washed out of the medium and Ca^{2+} returned to the extracellular bathing solution, a large influx of Ca^{2+} into the cell occurred, and this resulted in $[Ca^{2+}]_i$ that remained elevated for very long periods of time. Thus it seems that the cell had been "primed" in some way by the initial addition of glutamate and that Ca^{2+} influx subsequently occurred even in the absence of the agonist. Similar effects have also been described by Connor et al.[22] It is clear that such processes may be of considerable significance when considering conditions such as ischemia, for example, where high concentrations of glutamate occur extracellularly in the brain. The mechanism which triggers these large Ca^{2+} influxes has not been completely determined. However, it has been reported that it can be reduced when the enzyme protein kinase C is inhibited pharmacologically with agents such as sphingosine, K525A or staurosporine (REF. 22 and unpublished observations). Moreover it is also quite clear that

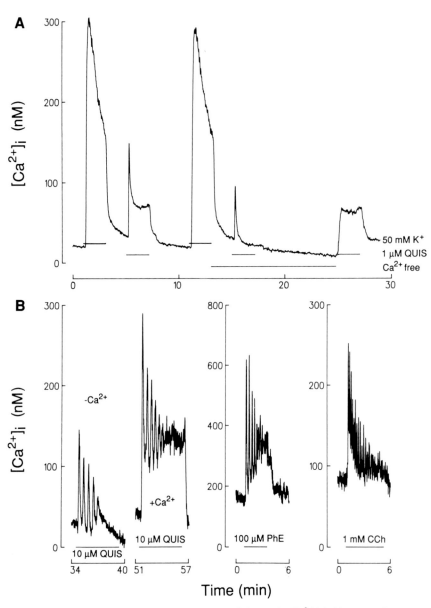

FIGURE 6. Quisqualate (QUIS) produces two types of changes in $[Ca^{2+}]_i$ in hippocampal neurons. (A) In normal medium quisqualate (1 μM) produced a response consisting of a transient spike superimposed on top of a more maintained plateau response. On removal of external Ca^{2+} only the transient spike was observed. The sustained plateau was produced once again when external Ca^{2+} was readmitted. (B) In Ca^{2+}-free medium quisqualate sometimes produced a series of oscillations rather than a single transient spike. When Ca^{2+} was present these could be observed on top of the more sustained plateau response as in (A). Also shown are oscillatory responses obtained with phenylephrine (PHE) and carbachol (CCh) in a hippocampal and striatal neuron, respectively.

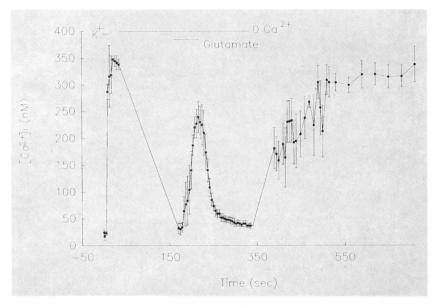

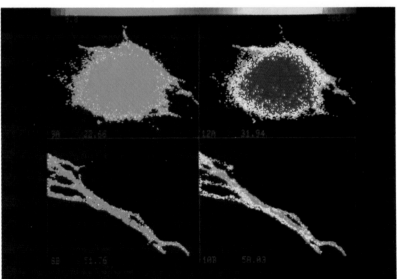

FIGURE 7. Changes in $[Ca^{2+}]_i$ in rat hippocampal neurones *in vitro* produced by glutamate (500 μM). Rat hippocampal neurons were cultured for 2 weeks and $[Ca^{2+}]_i$ recorded using a fura-2 based digital imaging technique. In the *upper trace* cells were first treated with 500 μM glutamate and a transient rise in $[Ca^{2+}]_i$ was obtained (see *lower panels*). Following this the experiment proceeded as shown. First cells were perfused with a depolarizing high-K^+ solution. External Ca^{2+} was then removed and the cells were again perfused with glutamate. A transient rise in $[Ca^{2+}]_i$ was observed, presumably due to the activation of ''quisqualate-II'' receptors. When glutamate was washed out and external Ca^{2+} added back a large sustained Ca^{2+} influx was observed. (n = 4). *Bottom images* illustrate the maximal response of a hippocampal pyramidal neuron to a single challenge with glutamate (500 μM). Images of the response in the cell soma and dendrites are illustrated separately. Scale illustrates range of $[Ca^{2+}]_i$ in this experiment.

the increase in $[Ca^{2+}]_i$ is the result of sustained Ca^{2+} influx as it can be reduced on removal of the extracellular Ca^{2+}. It will clearly be important to determine exactly how these effects are produced and the precise nature of the Ca^{2+} influx pathway involved. Drugs that specifically inhibit this pathway may also find therapeutic utility in a number of conditions.

REFERENCES

1. MAYER, M. L. & G. L. WESTBROOK. 1987. The physiology of excitatory amino acids in the vertebrate central nervous system. Prog. Neurobiol. **28:** 197–278.
2. CHOI, D. W. 1988. Glutamate neurotoxicity and diseases of the nervous system. Neuron **1:** 623–634.
3. SANCHEZ-PRIETO, J. & P. GONZALEZ. 1988. Occurrence of a large Ca^{2+} independent release of glutamate during anoxia in isolated nerve terminals (synaptosomes). J. Neurochem. **50:** 1322–1324.
4. WATKINS, J. H. & H. J. OLVERMAN. 1987. Agonists and antagonists for excitatory amino acid receptors. Trends Neurosci. **7:** 265–272.
5. NICOLL, R. A., J. A. KAUER & R. C. MALENKA. 1988. The current excitement in long term potentiation. Neuron **1:** 97–101.
6. CHOI, D. W. 1987. Ionic dependence of glutamate neurotoxicity. J. Neurosci. **7:** 369–379.
7. SIESJO, B. K. 1988. Calcium, ischaemia and the death of brain cells. Ann. N. Y. Acad. Sci. **522:** 638–662.
8. CULL-CANDY, S. & M. USOWICZ. 1987. Patch clamp recording from single glutamate receptor channels. Trends Pharmacol. Sci. **8:** 218–224.
9. MURPHY, S. N. & R. J. MILLER. 1989. Regulation of Ca^{2+} influx into striatal neurons by kainic acid. J. Pharmacol. Exp. Ther. **249:** 184–193.
10. VYCKLICKY, L., J. KRUSEK & C. EDWARDS. 1988. Difference in the pore size of the N-methyl-D-aspartate and kainate cation channels. Neurosci. Lett. **89:** 313–318.
11. NICHOLLS, D. G. 1987. A role for the mitochondrion in the protection of cells against calcium overload. Prog. Brain Res. **63:** 97–106.
12. HONORÉ, T., S. N. DAVIES, J. DREJER, E. J. FLETCHER, P. JACOBSEN, D. LODGE & F. E. NIELSEN. 1988. Quinoxalinediones: potent competitive non-NMDA glutamate receptor antagonists. Science **241:** 701–704.
13. MAYER, M. L., G. L. WESTBROOK & P. B. GUTHRIE. 1984. Voltage dependent block by Mg^{2+} of NMDA responses in spinal cord neurons. Nature **309:** 261–263.
14. JOHNSON, J. W. & P. ASCHER. 1987. Glycine potentiates the NMDA response in cultured mouse brain neurons. Nature **325:** 529–531.
15. MAYER, M. L., L. VYCKLICKY & J. CLEMENTS. 1989. Regulation of NMDA receptor desensitization in mouse hippocampal neurons by glycine. Nature **338:** 425–427.
16. REYNOLDS, I. J. & R. J. MILLER. 1989. Allosteric regulation of NMDA receptors. In Advances in Pharmacology 1989. D. Schumacher, Ed. Academic Press. New York, NY. In press.
17. SCHOEPP, D.D. & B. C. JOHNSON. 1988. Excitatory amino acid agonist/antagonist interactions at AP4 sensitive quisqualate receptors coupled to phosphoinositide hydrolysis in slices of rat hippocampus. J. Neurochem. **50:** 1605–1613.
18. NICOLETTI, F., M. J. IADAROLA, J. WROBLEWSKI & E. COSTA. 1986. Excitatory amino acid recognition sites coupled with inositol phospholipid hydrolysis: developmental changes and interaction with $\alpha 1$-receptors. Proc. Natl. Acad. Sci. USA **83:** 1931–1935.
19. MELDOLESI, J., P. VOLPE & T. POZZAN. 1988. The intracellular distribution of calcium. Trends Neurosci. **11:** 449–452.
20. MURPHY, S. N. & R. J. MILLER. 1988. A glutamate receptor regulates Ca^{2+} mobilization in hippocampal neurons. Proc. Natl. Acad. Sci. USA **85:** 8737–8741.
21. MURPHY, S. N. & R. J. MILLER. 1989. Two distinct quisqualate receptors regulate Ca^{2+} homeostasis in hippocampal neurons in vitro. Molec. Pharmacol. In Press.
22. CONNOR, J. A., W. J. WADMAN, P. E. HOCKBERGER & R. K. S. WONG. 1988. Sustained dendritic gradients of Ca^{2+} induced excitatory amino acids in CA1 hippocampal neurons. Science **240:** 649–653.

The Discovery of New Chemical Modulators of the N-Methyl-D-Aspartate Receptor

ALAN P. KOZIKOWSKI, WERNER TUECKMANTEL, AND
KEITH MALONEYHUSS

Departments of Chemistry and Behavioral Neuroscience
University of Pittsburgh
1101 Chevron Science Center
Pittsburgh, Pennsylvania 15260

The N-methyl-D-aspartate (NMDA) receptor is a glutamatergic receptor present in high density in the brain, and most abundant in the hippocampus, a primary conduit in memory processing. Glutamate itself represents the most abundant amino acid to be found in the brain, and as an excitatory amino acid it serves to activate at least three glutamate receptor subtypes, the NMDA, kainate, and quisqualate receptors.[1] Additionally, glutamate plays an important role in a host of other biological processes that take place in the brain, one of these being its precursor role to the major inhibitory neurotransmitter γ-aminobutyric acid or GABA (EQUATION 1). A model of the NMDA receptor-channel complex and its accompanying recognition sites are provided in FIGURE 1.

Glutamate   GABA

Over the past few years intimate links have been built between the NMDA receptor and the phenomenon of LTP (long-term potentiation), the best current substrate of memory.[2] In addition to this NMDA↔memory link, it has further been shown that improper regulation of the NMDA receptor-channel complex may be involved in certain neuropathological disorders such as the brain damage following ischemic episodes, the progressive mental deterioration accompanying Huntington's disease and Alzheimer's disease, the amyotrophic lateral sclerosis resulting from the ingestion of certain unusual amino acids, and, perhaps, even schizophrenia.[3]

The fact that this receptor system has such a "pinwheel" of clinical disorders associated with it (FIG. 2) makes it an exceedingly important target for drug discovery. Perhaps by learning to regulate this receptor complex through chemical agents not unlike those already present in the brain, one can create chemical mimics capable of facilitating memory processing, protecting the brain from the ischemic conditions accompanying a stroke or heart attack,[4] retarding the symptoms of Alzheimer's and Huntington's disease, or even alleviating the disordered throught processes of the schizophrenic patient.[5]

Given such a high payoff in terms of clinical benefits, it would appear a timely matter to learn more about the chemical regulation of the NMDA receptor system. An X-ray

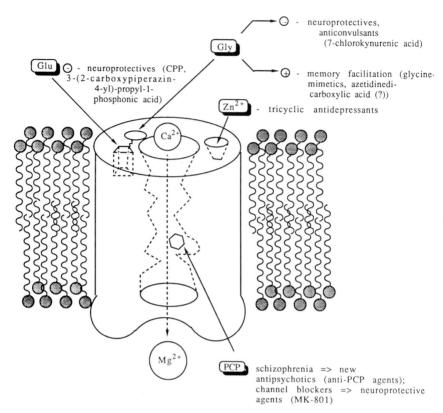

FIGURE 1. NMDA receptor-associated recognition sites. − indicates inhibitory or antagonistic effect; + indicates agonistic or potentiating effect.

structure of the NMDA receptor complex would certainly make the chemical design and synthesis task simpler for the medicinal chemist; however, at the moment no such X-ray data is available. The NMDA receptor has not yet been purified, and even if purification is accomplished, no adequate methods are currently available for obtaining a crystalline form of a membrane-bound glycoprotein. Such methods represent as yet but a deep yearning in the hearts of the X-ray crystallographer.

Nevertheless, given the present state of receptor technology, the sophisticated tools of the synthetic organic chemist, and certain structural leads provided by phytochemists, pharmacologists, and neuroscientists, the medicinal chemist can enter into a rational drug

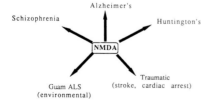

FIGURE 2. The NMDA receptor and neuro-pathologies.

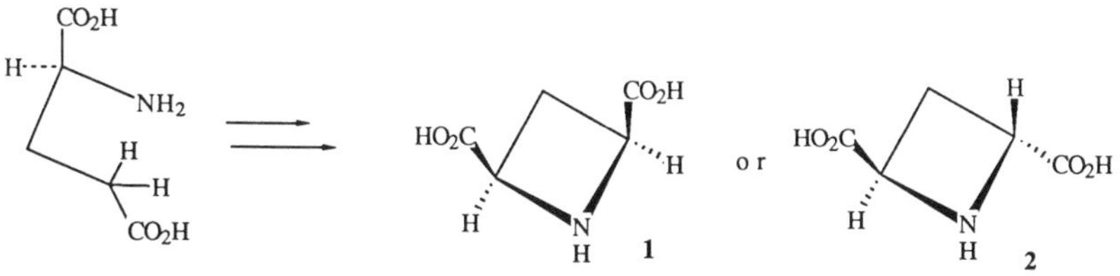

FIGURE 3. Generation of a conformationally constrained analogue of glutamate.

discovery program aimed at the generation of new ''unnatural'' modulators of the NMDA receptor. Not only may whole new classes of drugs be thus discovered, but through a properly integrated program one is certain to learn more about the relationships between this receptor and human behavior.

Over the past few years we have initiated a program of the type outlined above. We began our studies by reviewing the nature of the ligands which had been synthesized or isolated from nature and shown to act at the various glutamate receptor subtypes. In particular, we felt that the construction of certain conformationally constrained analogues of glutamate might provide a useful design component in learning more about the conformation(s) of glutamate which are relevant to the elicitation of an intracellular signal (PI turnover, cGMP synthesis, or Ca^{++} influx) upon receptor binding. In particular, we were captivated by the fact that the simplest of the conformationally constrained analogues of glutamate, the azetidine diacids **1** or **2**, represented virtually unknown compounds. In comparison with glutamate these rigid glutamate analogues contain no additional atoms; two hydrogen atoms have simply been removed from the glutamate molecule and a new C-N bond installed to arrive at **1/2** (FIG. 3).

Given the novelty of these structures, we accordingly undertook a synthesis of these molecules in order to examine their function in cerebellar granule cell cultures. The azetidines were readily prepared from glutaric anhydride as sketched in FIGURE 4.[6]

Much to our surprise, in $^{45}Ca^{++}$ influx experiments, neither of these compounds proved active. Only when exogenous glutamate was added to the culture dishes to stimulate $^{45}Ca^{++}$ influx was an effect observed. The *cis*-azetidine-2,4-dicarboxylic acid (**1**) was found capable of potentiating glutamate's ability to induce $^{45}Ca^{++}$ influx into neurons. This positive allosteric modulatory action was found only for the *cis*-isomer **1**, and not the *trans*-isomer **2** *(stereoselectivity of action)*. The potentiating effect of **1** was observed upon using glutamate, aspartate, or NMDA as the primary stimulating ligand. No effect was observed with kainate, and only a weak action with quisqualate.[6] Subsequent to our discovery of the allosteric modulatory action of **1**, Ascher and Johnson reported that glycine exhibited a similar modulatory role on glutamate, potentiating its ability to induce NMDA-associated ion channel openings.[7]

These electrophysiological findings for glycine led us to examine the ability of **1** to displace glycine binding from the strychnine insensitive glycine binding sites. While **1** bears some structural resemblance to glycine, no displacement of glycine binding was in

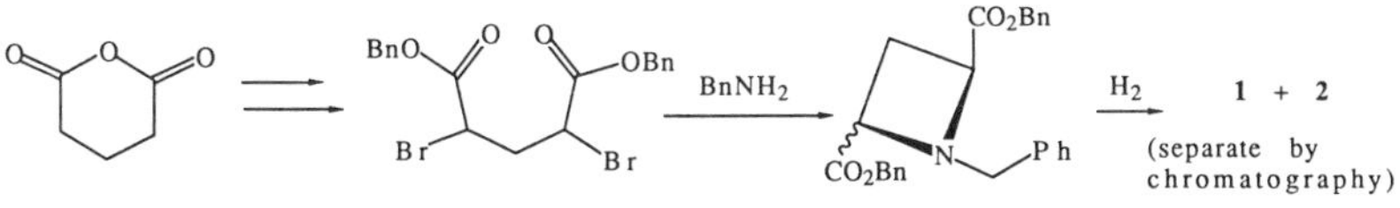

FIGURE 4. Synthesis of the azetidines **1** and **2**.

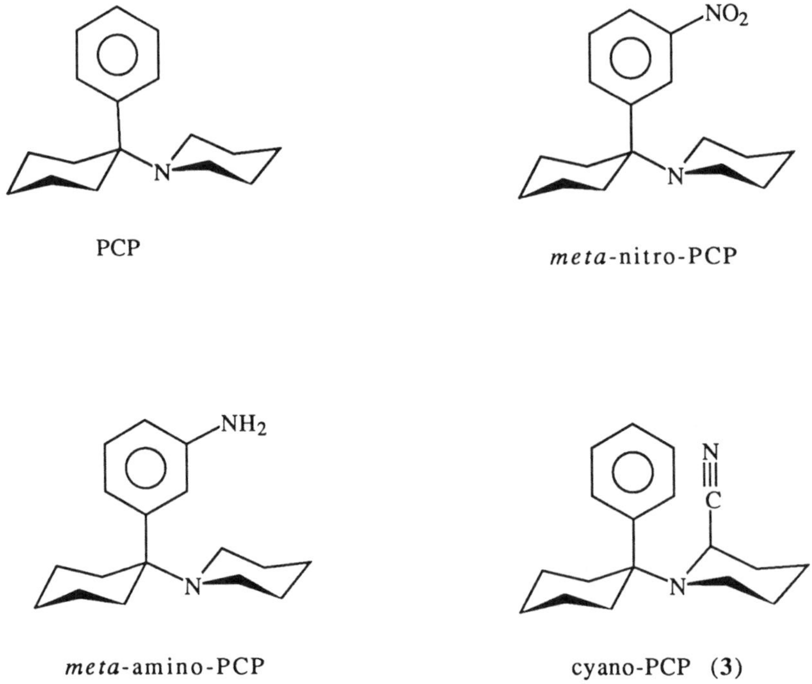

PCP   *meta*-nitro-PCP

meta-amino-PCP   cyano-PCP (3)

FIGURE 5. PCP analogues.

fact found. The azetidine **1** also failed to displace PCP binding, indicating that it was not working inside the NMDA-operated ion channel. Displacement of [^{3}H]-glutamate binding was observed only upon employing very high concentrations of the azetidine **1**.[8] Furthermore, the azetidine **1** was found to increase cGMP formation in rats when administered icv, and to possess considerably less neurotoxicity than NMDA itself. Curiously, in extracellular recordings from the CA1 region of the hippocampus, **1** was found to behave like NMDA, depressing the population EPSPs, an effect which could be blocked by the noncompetitive NMDA antagonist MK-801.[9] While the glycine-like modulatory action of **1** in the granule cell experiments and its NMDA-like activity in the hippocampal slice preparation might be explainable by the fact that the granule cells represent a more artificial neuronal population whereas the slice maintains many more of its intrinsic connections (it's almost a "whole animal," in the words of Phillip Landfield), we believe the explanation to be more complex than this. Perhaps the azetidine **1** is working as a presynaptic uptake inhibitor for glutamate, or somehow it helps to retard postsynaptic receptor desensitization. Clearly, more experiments are needed to arrive at a better understanding of the mechanism of action for the azetidines. The synthesis of other rigid glutamate analogues is also clearly needed.

As we leave the extracellular surface of the NMDA receptor and move into the channel, we arrive at the PCP (phencyclidine) recognition site.[10,11] Here too we have begun a program to build new agents capable of modulating the activity of the NMDA receptor through use of derivatives of PCP. First, to test the ability of the granule cell cultures to appropriately respond to any chemical modifications built into our PCP derivatives, we began our study by constructing and evaluating several known PCP derivatives (FIG. 5). The placement of a *m*-nitro group on the aromatic ring of PCP has been shown to decrease PCP-like behavioral effects in rats (rotarod assay), while a *m*-amino group was found to enhance PCP-like activity. When these two readily synthesized compounds were tested in cell culture experiments they were, in fact, found to exhibit the expected action; *m*-amino-PCP > PCP > *m*-nitro-PCP in blocking Ca^{++} influx.[12,13] A variety of other analogues were also made before we discovered that a very simple derivative, the cyano-PCP compound **3**, could modulate the NMDA receptor so as to enhance rather than to inhibit $^{45}Ca^{++}$ uptake. The pharmacological action is quite unusual, and, in fact, it represents a rather exciting finding if the effect observed is truely due to an action at the PCP recognition site. Such a finding would suggest that the

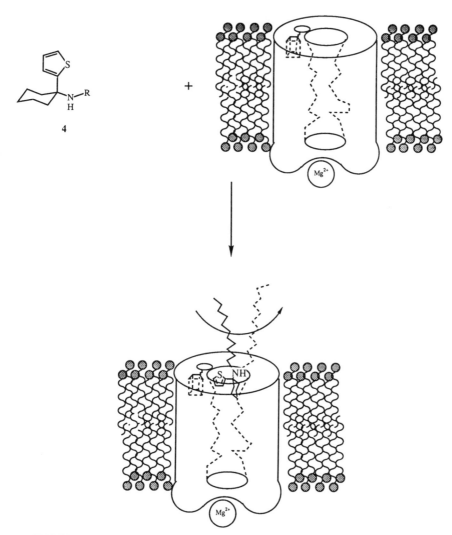

FIGURE 6. Interaction of a PCP analogue with the NMDA receptor-operated ion channel.

164 ANNALS NEW YORK ACADEMY OF SCIENCES

negative allosteric modulatory action of PCP or MK-801 is not simply a consequence of these molecules acting as steric plugs for the channel, but rather that these agents act to change the "breathing" motion of the ion channel. Thus, again a biologically intriguing compound has been discovered, one which requires both further modification and study in order to more precisely pinpoint its mechanism of action. While compound 3 does displace PCP binding, PCP is a multifarious agent, one which works at other receptors and channels. Such facts, of course, suggest that the inverse activity of cyano-PCP 3 could possibly be due to non-NMDA-mediated processes.

Some indirect chemical evidence for the location of the PCP binding site within the NMDA receptor channel was also gleaned from the PCP-based structure/activity studies. The thienylcyclohexylamine 4 was found to become a poorer displacer of [^{3}H]MK-801 binding on increasing the length of its carbon chain R [R = Et was the best with an IC_{50} $\cong$ 90 nM; R = $C_{10}H_{21}$, IC_{50} > 100 μM].[14] The conformational bulk of the long carbon chain presumably impedes the threading of the molecule into the ion channel (FIG. 6).

Lastly, the large number of PCP-analogues which were synthesized in order to establish these structure-activity relationships have led to the discovery of a ligand which binds with high specificity and *irreversibly* to the PCP site. *We believe that this agent now places isolation of the PCP receptor within our grasp.* If the Ca^{++} channel is an intimate part of a larger protein structure, then we should be able to isolate and purify the entire NMDA receptor-channel complex. By isolating, purifying, and sequencing the NMDA receptor, we will gain a better understanding of the nature of the amino acids, which make up the glutamate and glycine recognition sites, and the channel itself. Such information will ultimately provide a further element of rationality to the design of chemical modulators of the NMDA receptor.

We remain convinced that studies like the ones described above will help unveil the secret workings of the NMDA receptor and of the molecular mechanisms through which it transduces its signals. Moreover, given the fact that man's quest of the unknown has often led to discoveries which improve the human condition, one remains hopeful that new agents for the treatment of certain neuropathological disorders might be discovered along the way as well.

ACKNOWLEDGMENTS

Alan P. Kozikowski is indebted to Drs. Wroblewski, Danysz, Fadda, Krueger, and Nicoletti of the Fidia-Georgetown Institute for the Neurosciences and Dr. Barrionuevo of the University of Pittsburgh for their assistance in carrying out the biological studies pertaining to the compounds described in this article. Alan P. Kozikowski also thanks Dr. Erminio Costa for many stimulating lessons in neuroscience.

REFERENCES

1. WATKINS, J. C. & H. J. OLVERMAN. 1987. Trends Neurosci. **10:** 265–272.
2. COLLINGRIDGE, G. L. & T. V. P. BLISS. 1987. Trends Neurosci. **10:** 288–293.
3. ROTHMAN, S. M. & J. W. OLNEY. 1987. Trends Neurosci **10:** 299–302.
4. KEMP, J. A., A. C. FOSTER & E. H. F. WONG. 1987. Trends Neurosci. **10:** 294–298.
5. WINGER, G. 1987. Trends Pharmacol. Sci. **8:** 323.
6. KOZIKOWSKI, A. P., W. TUECKMANTEL & J. T. WROBLEWSKI. J. Med. Chem. Submitted.
7. JOHNSON, J. W. & P. ASCHER. 1987. Nature **325:** 529–531.
8. DANYSZ, W. Unpublished results from the Fidia-Georgetown Institute for the Neurosciences, Georgetown University, Washington, DC.

9. BARRIONUEVO, G. Unpublished results from the Department of Behavioral Neuroscience, University of Pittsburgh, Pittsburgh, PA.
10. JOHNSON, K. M. & L. D. SNELL. 1986. Phencyclidine: An Update. NIDA Research Monograph 64. D. H. Clouet, Ed.: 52–66. National Institute on Drug Abuse. Rockville, MD.
11. WROBLEWSKI, J. T., F. NICOLETTI, E. FADDA & E. COSTA. 1987. Proc. Natl. Acad. Sci. USA **84:** 5068–5072.
12. COSTA, E. E. FADDA, A. P. KOZIKOWSKI, F. NICOLETTI, J. T. WROBLEWSKI. 1988. Classification and allosteric modulation of excitatory amino acid signal transduction operative in brain slices and primary cultures of cerebellar neurons. *In* Neurobiology of Amino Acids, Peptides and Trophic Factors. J. Serendelli, R. Collins & E. Johnson, Eds.: 35–50. Nijhoff, Boston, MA.
13. WROBLEWSKI, J. T., F. NICOLETTI, E. FADDA, A. P. KOZIKOWSKI, J. W. LAZAREWICZ & E. COSTA. 1989. Modulation of glutamate signal transduction. *In* Allosteric Modulation of Amino Acid Receptors: Therapeutic Implications. E. A. Barnard & E. Costa, Eds.: 287–300. Raven Press. New York, NY.
14. KOZIKOWSKI, A. P. & K. KRUEGER. Unpublished results.

The Role of Calcium in Long-Term Potentiation

R. A. NICOLL, R. C. MALENKA, AND J. A. KAUER

Departments of Pharmacology and Physiology
University of California School of Medicine
San Francisco, California 94143

The mechanisms by which nervous systems store information remains perhaps the most fascinating and perplexing question in modern neuroscience. For over 75 years a number of prominent psychologists and neuroscientists have proposed that changes in the strength of synaptic transmission, due to usage, at certain critical synapses in the brain might underlie our ability to learn and remember. Although many attempts were made to experimentally verify this hypothesis, the first report in support of this hypothesis did not occur until 1973 when Bliss and Lømo[1] found that brief, repetitive activation of excitatory pathways in the hippocampus caused an increase in the strength of synaptic transmission that could last for many hours or even days to weeks in the intact animal. This was a particularly exciting finding in light of the role the hippocampus may play in information storage. This long-lasting synaptic enhancement, which has come to be known as long-term potentiation (LTP), has now been accepted as the prime candidate for a cellular mechanism related to learning and memory.

Use of the *in vitro* hippocampal slice preparation, which allows for stable intra- and extracellular electrophysiological recording and for complete control over the perfusing medium, has led to a detailed understanding of the mechanisms underlying the initiation of LTP. It is now clear that at least for the CA1 region of the hippocampus, depolarization of the postsynaptic membrane is a requirement for LTP induction. This was demonstrated in two ways. First, preventing the depolarization that normally occurs during a LTP-inducing tetanus, either by voltage-clamping the cell[2] or hyperpolarizing[3] it with current injection, blocks LTP induction. Second, pairing low-frequency stimuli, which alone do not cause LTP, and postsynaptic membrane depolarization induces LTP.[4]

This requirement for some critical level of postsynaptic depolarization can explain one of the most interesting properties of LTP known as associativity. Weak stimuli, even when given at high frequencies, are normally unable to induce LTP, presumably because they do not generate sufficient postsynaptic depolarization. However, when a weak input is activated concomitantly (within about 20–100 msec) with an independent strong input (one that is able to induce LTP), the weak input will now exhibit LTP. The association between the strong and weak inputs is due to the passive electrotonic spread of the depolarization generated by the strong input to the site of the weak input.

The elucidation of the novel biophysical properties of the N-methyl-D-aspartate (NMDA) receptor/ion channel have clarified how the postsynaptic membrane potential influences the initiation of LTP. Synaptically released glutamate, the excitatory transmitter used by these synapses, acts as two subtypes of glutamate receptors, NMDA receptors and quisqualate/kainate (Q/K) receptors.[5] D-2-amino-5-phosphovalerate (APV), a competitive antagonist of NMDA receptors, blocks LTP induction while having minimal effects on baseline synaptic transmission.[6] This indicates that during normal synaptic transmission Q/K receptors are primarily responsible for the generation of the excitatory postsynaptic potential (EPSP), yet NMDA receptor activation is required to elicit LTP. It turns out that physiological levels of extracellular magnesium (Mg^{++}) exert a voltage-dependent block of the NMDA channel such that at the resting membrane potential, the NMDA channel is unable to pass much current even when activated by synaptically released glutamate.[7,8] In contrast, when the cell is depolarized, the Mg^{++} is expelled from the channel allowing the channel to pass significant amounts of current when the NMDA receptor is activated. In addition to differing from Q/K receptors in their voltage-dependence, NMDA receptor/ion channels also differ dramatically in their permeability to ions. Q/K receptor/ion channels are permeable to K^+ and Na^+ but not to Ca^{++}, while NMDA receptor/ion channels have a high permeability to Ca^{++}[9-12] in addition to K^+ and Na^+.

Given the novel biophysical properties of the NMDA receptor/ion channel and the requirements for LTP induction of postsynaptic depolarization and NMDA receptor activation, a very specific model has been developed outlining the steps involved in LTP induction. When the postsynaptic cell is depolarized, either by a high frequency tetanus or artificially by current injection, the Mg^{++} block of the NMDA channel is relieved permitting Ca^{++} to enter the dendritic spine, the site of the synapse and a structure which may serve the function of isolating this Ca^{++} rise from other parts of the cell. This rise in Ca^{++} within the dendritic spine may be the trigger for LTP.

The first evidence in support of this Ca^{++} hypothesis for LTP induction came from the finding that loading the postsynaptic cell with EGTA, a calcium chelator which should buffer any rise in intracellular Ca^{++}, blocked the induction of LTP. More recently we have obtained additional and more direct evidence in support of a critical role for postsynaptic calcium in LTP.[13] In collaboration with Dr. R. Zucker (University of California, Berkeley) we injected the caged Ca^{++} compound, Nitr-5, into hippocampal pyramidal cells to examine the effect on synaptic transmission of directly raising postsynaptic Ca^{++}.[13] Photolysis of Nitr-5 that has been preloaded with Ca^{++} greatly decreases the affinity of Nitr-5 resulting in the release of Ca^{++} into the cell. This photolysis of Nitr-5 routinely produced a large increase in the amplitude and rise time of the intracellularly recorded EPSP while having no effect on the extracellularly recorded field EPSP (FIG. 1). The lack of change in the field EPSP demonstrates that the light flash used to photolyze the Nitr-5 had no effect on synaptic transmission in surrounding cells. As a further control we injected cells with Nitr-5 that had not been preloaded with Ca^{++}. Photolysis of this Nitr-5 had no lasting effect on synpatic transmission indicating that it was the release of Ca^{++} from the Nitr-5 that caused the change in the EPSP.

We also found that injection of Nitr-5 into cells blocked the occurrence of LTP. Like EGTA, this is presumably because of the ability of Nitr-5 to rapidly buffer any increases in Ca^{++} caused by LTP-inducing stimuli. However, buffering intracellular Ca^{++} with Nitr-5 or EGTA does not distinguish between rises in Ca^{++} due to influx of Ca^{++} across the membrane (as would be expected if the Ca^{++} were entering through the NMDA receptor/ion channel) from that due to release of Ca^{++} from intracellular stores. To distinguish these possibilities we took advantage of the finding that holding the membrane potential of cultured CNS neurons at values more positive than about $+20$ mV suppresses the influx of Ca^{++} through the NMDA channel.[10] Therefore, by strongly depolarizing the cell it should be possible to prevent LTP induction. In agreement with this proposal we found that pairing membrane depolarization to a level far beyond the EPSP reversal potential with low frequency stimuli had no effect on synaptic transmission, whereas in the same cell using the same pairing photocol with more modest membrane depolarization resulted in a robust increase in the EPSP.

Taken together, these experiments provide extremely strong support for the hypothesis that the rise in Ca^{++} in the dendritic spine due to the influx of Ca^{++} through the NMDA receptor/ion channel during membrane depolarization is a critical trigger for LTP.

An obvious question is which biochemical processes turned on by this rise in Ca^{++} are responsible for the enhancement in synaptic transmission. Several possibilities have

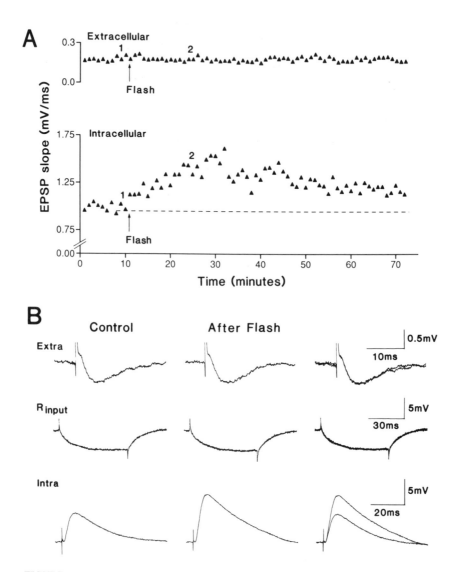

FIGURE 1. Photolysis of intracellularly injected Nitr-5 enhances synaptic transmission. (**A**) Graphs of the slope of the extracellular EPSP recorded in a stratum radiatum (*upper*) and the slope of the simultaneously recorded intracellular EPSP (*lower*). Each point represents the average of six slope measurements. The cell was penetrated 15 min before time 0 on the graph. At the time marked by the *arrow* (flash) the slice was exposed to ultraviolet light for 25 s. (**B**) Sample records obtained at the times indicated by the numbers 1 and 2 in (A) (records are the average of six sweeps). *Upper records:* the extracellularly recorded EPSP (Extra). *Middle records:* the response to a constant current hyperpolarizing pulse (0.11 nA) used to monitor the input resistance (R_{input}). *Bottom records:* the intracellularly recorded EPSP (Intra). The *right-hand column* shows superimposed records before and after the flash.

been proposed, three of which will be briefly mentioned here. The type II Ca^{++}/calmodulin-dependent kinase is found in high concentrations in the postsynaptic density[14,15] and may become constitutively active following autophosphorylation.[16] Protein kinase C is also found in high concentrations in brain, may be activated indirectly by increases in Ca^{++},[17,18] and has dramatic effects on synaptic transmission.[19] Finally, the calcium-activated protease, calpain has been proposed to play a critical role in the processes underlying LTP.[20] At the present time we are examining the effects on LTP of a variety of synthetic peptides[21,22] to help delineate which of these biochemical processes play a critical role in LTP.

During aging and perhaps with Alzheimer's disease, the intracellular handling of Ca^{++} and a variety of Ca^{++}-dependent processes are significantly altered. A more detailed understanding of the mechanisms underlying LTP will undoubtedly lay the foundation for further investigation of how these changes interact with the ability of aged or diseased nervous systems to exhibit plasticity.

REFERENCES

1. BLISS, T. V. P. & T. LØMO. 1973. Long-lasting potentiation of synaptic transmission in the dentate area of the anaesthetized rabbit following stimulation of the perforant path. J. Physiol. **232:** 331–356.
2. KELSO, S. R. A., A. H. GANONG & T. H. BROWN. 1986. Hebbian synapses in hippocampus. Proc. Natl. Acad. Sci. USA **83:** 5326–5330.
3. MALINOW, R. & J. P. MILLER. 1986. Postsynaptic hyperpolarization during conditioning reversibly blocks induction of long-term potentiation. Nature **320:** 529–530.
4. GUSTAFSSON, B., H. WIGSTRÖM, W. C. ABRAHAM & Y. -Y. HUANG. 1987. Long-term potentiation in the hippocampus using depolarizing current pulses as the conditioning stimulus to single volley synaptic potentials. J. Neurosci. **7:** 774–780.
5. WATKINS, J.C. & R. H. EVANS. 1981. Excitatory amino acid transmitters. Annu. Rev. Pharmacol. Toxicol. **21:** 165–204.
6. COLLINGRIDGE, G. L., S. J. KEHL & H. McCLENNAN. 1983. Excitatory amino acids in synaptic transmission in the Schaffer collateral-commissural pathway of the rat hippocampus. J. Physiol. **334:** 33–46.
7. NOWAK, L. M. & R. L. MACDONALD. 1982. Substance P: Ionic basis for depolarizing responses of mouse spinal cord neurons in cell culture. J. Neurosci. **2:** 1119–1128.
8. MAYER, M. L., G. L. WESTBROOK & P. B. GUTHRIE. 1984. Voltage-dependent block by Mg^{2+} of NMDA responses in spinal cord neurones. Nature **309:** 263.
9. JAHR, C. E. & C. F. STEVENS. 1987. Glutamate activates multiple single channel conductances in hippocampal neurones. Nature **325:** 522–525.
10. MAYER, M. L., A. B. MacDERMOTT, G. L. WESTBROOK, S. J. SMITH & J. L. BARKER. 1987. Agonist- and voltage-gated calcium entry in cultured mouse spinal cord neurons under voltage. J. Neurosci. **7:** 3230–3244.
11. ASCHER, P. & L NOWAK. 1988. The role of divalent cations in the N-methyl-D-aspartate responses of mouse central neurones in culture. J. Physiol. **399:** 247–266.
12. MAYER, M. L. & G. L. WESTBROOK. 1987. Permeation and block of N-methyl-D-aspartic acid receptor channels by divalent cations in mouse central neurones. J. Physiol. **394:** 501–527.
13. MALENKA, R.C., J. A. KAUER, R. J. ZUCKER & R. A. NICOLL. 1988. Postsynaptic calcium is sufficient for potentiation of hippocampal synaptic transmission. Science **242:** 81–84.
14. Kennedy, M.B., M. K. Bennett & N. E. ERONDU. 1983. Biochemical and immunochemical evidence that the "major postsynaptic density protein" is a subunit of a calmodulin-dependent protein kinase. Proc. Natl. Acad. Sci. USA **80:** 7357–7361.
15. OUIMENT, C. C., T. L. McGUINNESS & P. GREENGARD. 1984. Immunocytochemical localization of calcium/calmodulin-dependent protein kinase II in rat brain. Proc. Natl. Acad. Sci. USA **81:** 5604–5608.
16. MILLER, S. G. & M. B. KENNEDY. 1986. Regulation of brain type II Ca^{2+}/

calmodulin-dependent protein kinase by autophosphorylation: A Ca^{2+}-triggered molecular switch. Cell **44:** 861–870.

17. KACZMARCK, L. K. 1987. The role of protein kinase C in the regulation of ion channels and neurotransmitter release. Trends Neurosci. **10:** 30–34.

18. MILLER, R. J. 1986. Protein kinase C: A key regulatory of neuronal excitability? Trends Neurosci. **9:** 538–541.

19. MALENKA, R. C., D. V. MADISON & R. A. NICOLL. 1986. Potentiation of synpatic transmission in the hippocampus by phorbol esters. Nature **321:** 695–697.

20. LYNCH, G. & M. BAUDRY. 1984. The biochemistry of memory: A new and specific hypothesis. Science **224:** 1057–1063.

21. KELLY, P. T., R. P. WEINBERGER & M. N. WAXHAM. 1988. Proc. Natl. Acad. Sci. USA **85:** 4991–4995.

22. HOUSE, C. & B. E. KEMP. 1987. Protein kinase C contains a pseudosubstrate prototype in its regulatory domain. Science **238:** 1726–1728.

Links between Long-Term Potentiation and Neuropathology

An Hypothesis Involving Calcium-Activated Proteases

GARY LYNCH AND PETER SEUBERT

Center for the Neurobiology of Learning and Memory
University of California, Irvine
Irvine, California 92717

INTRODUCTION

Recent experimental work has reinforced the idea[1] that mechanisms responsible for promoting synaptic plasticity are also involved in some instances of neuropathology. Antagonists of one class of glutamate receptors that block the induction of long-term potentiation (LTP) and selectively disrupt certain forms of learning[2,3] also greatly ameliorate the neuronal damage that follows ischemia[4–8] or hypoglycemia.[9] Moreover, calcium, which has been implicated in a diverse array of cell pathologies,[10] has also been linked to the induction of LTP (see below).

Questions now arise about the nature of the chemical processes stimulated by transient increases in intracellular calcium and in particular if pathology results from an exaggeration of those events that modify synaptic strength. In this paper we will consider evidence that calcium-dependent proteases are activated by circumstances that trigger LTP as well as by a variety of pathogenic conditions.

Triggers and Substrates of Long-Term Potentiation

Glutamate Receptors, Calcium, and Structural Changes

The initial steps in triggering LTP are thought to involve i) stimulation of the N-methyl-D-aspartate (NMDA) receptors, a subclass of glutamate receptors and ii) a transient increase in calcium levels in the postsynaptic neuron. Thus, selective antagonists of the NMDA receptor have little effect on synaptic transmission in hippocampus but completely suppress the development of LTP,[11–13] results that are also obtained with intracellular injections of calcium buffering agents.[14,15] These observations are connected by an unusual property of the NMDA receptor ionophore, namely, that it is permeant to calcium.[16,17]

The NMDA receptor-calcium influx steps define the beginnings of a sequence that results in LTP, but what of the final changes that express it? The synapse specificity[18] and stability of LTP—it can persist without change for weeks[19,20]—greatly limit the possible substrates. The simplest modification that could satisfy the specificity-stability criteria would be a structural alteration in existing synapses, and electronmicroscopic studies have shown that effects of this kind accompany LTP. In field CA1 of hippocampus, there is an increase in the frequency of two types of relatively unusual synaptic profiles as well as a change in the shape of spines.[21–23] It seems likely then that morphology is changed to a significant degree in those spines experiencing LTP.

The Locus of LTP

There are any number of ways in which the transformation of a spine could result in a potentiated synapse, including a passive adjustment in its tightly apposed presynaptic terminal.[21] Recent work using newly introduced pharmacological agents has provided unexpected results that narrow the range of possible loci at which the substrates of LTP might be found.

As noted, synapses in hippocampus and elsewhere in forebrain possess two subtypes of glutamate receptors which are usually referred to as NMDA and non-NMDA ("quisqualate/kainate" or "AMPA") receptors. Increased release from a fixed population of terminals should increase the currents generated by both classes of receptors, and indeed there is experimental evidence for this. Thus, paired pulse and frequency facilitation, phenomena thought to reflect a transient increase in transmitter release (see REF. 24 for a recent review), augment those components of the postsynaptic response that are sensitive to antagonists of the NMDA receptor as well as those suppressed by non-NMDA receptor blockers. Surprisingly enough, this does not occur after induction of LTP. That is, the component of the evoked response that is blocked by an antagonist of the NMDA receptor changes little if at all after LTP, while that aspect of the same potential which is suppressed by a quisqualate receptor antagonist increases by about 50%.[25-27] These observations point to the conclusion that one class of glutamate receptors induces LTP while the other subgroup expresses it. The extent to which this is the case is illustrated in an experiment in which high frequency stimulation was applied to one of two collections of inputs to a common target while the quisqualate receptors were blocked, *i.e.,* under conditions in which only the NMDA receptors were functional. The LTP-inducing stimulation caused only a very small increase in the NMDA-dependent responses; however, the potentiation effect was fully evident on the pathway given the high frequency stimulation once the quisqualate receptors were unmasked.[25]

How is it possible that LTP can selectively modify the postsynaptic machinery generating the response to released transmitter? A number of possibilities exist, but two seem most likely. *First,* the structural changes in spines could modify surface chemistry such that the quisqualate but not NMDA receptors are rendered more potent. For example, it is possible that the number of quisqualate receptors or their association with ionophores is sensitive to the configuration of the spine or its internal cytoskeleton. *Second,* anatomical transformations could modify the resistance of the spine. Computer simulations of spines suggest that a reduction in the resistance of the spine neck would amplify fast synaptic ionic currents while having little effect on slower currents,[28] and it is the case that NMDA-mediated responses are somewhat slower than those associated with the quisqualate receptor.

The Connection between Calcium and Structural Changes

The above results lead to the hypothesis that a calcium-dependent process exists in spines which causes rapid and lasting changes in morphology. One candidate for a key component of this process is the partial breakdown of the membrane cytoskeleton by calcium-dependent proteases. *A priori,* it might be expected that pronounced anatomical transformations would require disassembly and reassembly of the cytoskeletal elements that maintain spines in particular configurations. Pertinent to this, there is evidence that anatomical changes in erythrocytes[29] and blood platelets[30] involve the digestion of several membrane structural proteins (including spectrin) by calpain, a calcium-activated neutral thiol protease. Calpain has also been implicated in the growth and retraction of axon terminal branches in the periphery.[31]

Neuronal spectrin (fodrin) is a heterodimer composed of subunits of nearly equivalent sizes which form a flexible, rod-like structure (for a review see REF. 32). The dimers consist of a calmodulin-binding 240-kDa subunit and a 235-kDa subunit which has at least two isoforms. Tetramers formed by the head-to-head attachment of dimers are thought to be the prevalent configuration. Spectrin forms a meshwork underlying the membrane which serves to indirectly link actin filaments to integral membrane proteins. Both subunits of spectrin are excellent substrates for calpain.[33] The functional consequences of calpain activation would then include the disruption of actin filament attachment. The association of calmodulin with the 240-kDa subunit accelerates the proteolysis of both subunits by calpain[34] with cleavage occurring near the calmodulin binding site.[35] Cleavage of the 240-kDa subunit occurs at only a small number of sites resulting in the formation of a 155-kDa breakdown product (BDP) which can be further cleaved to a 150-kDa species which resists further proteolysis.[36] These breakdown products provide a convenient marker for the occurrence of a calpain-spectrin interaction.

Since the calpain-spectrin interaction has been related to drastic morphological transformations in blood cells, it is not implausible that it produces comparable effects in spines and thus could account for the morphological changes seen after LTP induction. Studies on the localization of spectrin and calpain provide indirect support for this idea. Spectrin is a major cytoskeletal protein in postsynaptic densities (psds),[37] and calpain has been visualized by electronmicroscopic immunocytochemistry in spines and psds.[38] Indirect evidence that calpain attacks postsynaptic spectrin at least under some circumstances is found in a report that the 150-kD BDP is present in cortical psds purified to virtual homogeneity.[37] Two recent studies have provided more direct support for the hypothesis that the calpain-spectrin interaction is involved in LTP. *First,* application of NMDA to slices of hippocampus causes a rapid increase in the concentration of the 155-kDa spectrin BDP (see REF. 39 and below). *Second,* infusion of leupeptin, an antagonist of calpain, into the cerebral ventricles for two to four days (a treatment paradigm that suppresses calpain activity) inhibits the induction of LTP without obvious disturbances of baseline physiology.[40] Suppression of LTP by leupeptin has also been obtained in slices of hippocampus under conditions in which synaptic transmission appears normal (Oliver *et al.,* submitted).

In all, partial breakdown of the spine cytoskeleton by calpain provides a plausible explanation for the anatomical changes that accompany LTP and that are likely to be responsible for its expression. The mechanism is activated by the triggers for LTP (NMDA receptors, calcium), while the drastic modifications of structural proteins that it produces account for the synaptic specificity and extreme duration of the potentiation effect.

Calpain and Brain Pathology

Calpain degrades a number of structural proteins in addition to spectrin including microtubule-associated proteins,[41] tubulin,[42] and neurofilament protein.[43] Widespread activation of the protease would thus be expected to produce cytoskeletal disassembly and pathology. The idea that calpain is involved in atrophy and degeneration is supported by studies showing that inhibitors of calpain retard the breakdown of severed peripheral nerves[44] and the atrophic response of muscle to denervation.[45] The question thus arises as to whether the calpain-mediated degradation of spectrin, postulated above to produce the substrates of LTP, precedes and contributes to development of neuronal degeneration in brain. Recent experimental work suggests this is the case and thereby points to a specific relationship between plasticity and pathology.

Denervation-Induced Dendritic Atrophy

Lesions to the entorhinal cortex remove the great majority of extrinsic afferents to the dentate gyrus of hippocampus. The terminals generated by these inputs begin exhibiting morphological signs of degeneration at about 24 hours after destruction of their parent cell bodies. This is accompanied over 1–3 days by a loss of spines and a severe atrophy of the granule cell dendrites. At five days postlesion, a growth response by the undamaged residual afferents of the dentate gyrus (from septum and hippocampus) begins and proceeds rather slowly for weeks.[46,47] Glial responses in the denervated zone include a proliferation of microglia starting at about 20 hours postlesion[48] and an hypertrophic reaction of astrocytes starting shortly thereafter.[49]

The time course of the various events following denervation is sufficiently well defined to test if and when in the course of pathology spectrin proteolysis begins. Experiments directed at this revealed that a sizeable increase in BDP is evident *at 4 hours after denervation,* well in advance of any signs of degeneration. The concentration of BDP increased 50-fold between its first appearance and 2 days postlesion and then began to decline 2 days later.[36] These results provide the first evidence that spectrin proteolysis is a very early step in a sequence leading to spine loss and dendritic atrophy.

Colchicine-Induced Pathology

The microtubule-disrupting drug colchicine is preferentially toxic to granule cells of the dentate gyrus in the hippocampal formation[50] and is known to affect organelles thought to be involved in calcium sequestration.[51] Colchicine, like denervation, causes a pronounced increase in spectrin breakdown in the dentate gyrus region.[52] Subdissection of the molecular layer (the zone containing dendrites) from the granule cell body layer revealed that most of the breakdown occurred in the former area and thus increased proteolysis did not result from increased somatic turnover of newly synthesized spectrin which could not be transported to the distal processes due to disrupted miocrotubule systems.

Intraventricular infusion of chloroquine, a nonspecific lysosomal inhibitor, prior to colchicine injection caused a comparative increase in the levels of spectrin breakdown products suggesting that lysosomal proteases participate in the removal, but not the generation of these proteolytic fragments. On the other hand, infusion of the protease inhibitor leupeptin[53] markedly reduced spectrin breakdown.[52]

Excitatory Amino Acid Receptor-Mediated Proteolysis

The demonstrated proteolytic degradation of spectrin in dendritic fields under conditions of extensive restructuring or degeneration lent support to the original hypothesis, namely, that neuroplastic and/or pathological events involve the action of calcium-activated proteases in response to activation of NMDA receptors.[1] Administration of excitotoxic amounts of either kainate or NMDA has also been found to produce noticeable spectrin breakdown in hippocampus.[54] This study added the information that levels of calpain I decreased during this period, presumably through its degradation following autoproteolytic activation and that a nonneuronal calpain substrate protein (glial fibrillary acidic protein) was unaffected, supporting the idea that spectrin proteolysis is largely a neuronal process.

As noted above, a more direct demonstration of NMDA receptor activation coupled to spectrin proteolysis resulted from studies undertaken using hippocampal slices. In those experiments excitatory postsynaptic potentials (EPSPs) were monitored to assess neuronal viability. Application of NMDA was found to increase spectrin proteolysis within 15 minutes of its application.[39] This effect required extracellular calcium, was blocked by the NMDA receptor antagonist APV, and was not reproduced by potassium-induced depolarization. The concentration of NMDA used in these experiments reversibly depolarized the neurons, *i.e.*, EPSPs returned when the NMDA was removed by perfusion. While a pathological component cannot be ruled out, the data suggest that spectrin proteolysis can rapidly occur in viable neurons in response to activation of NMDA receptors and therefore is likely to be the cause rather than a consequence of neuropathology.

A Plasticity-Pathology Continuum—?

Requirements for a Memory Mechanism

If the "calpain" hypothesis for LTP and learning[55] proves to be correct, then it is of interest to ask why the brain would employ what seems to be a potentially lethal mechanism for what is presumably a recurring process.

One possible answer is that the requirements for recognition and associative memory as used in everyday life by mammals can only be accomplished by a very narrow range of cellular processes. The vast capacity and relative stability (*i.e.*, freedom from interference by subsequent storage events) of memory implies that the storage elements are present in enormous numbers and are reasonably independent from each other. This strongly suggests that highly localized chemistries are responsible for the encoding event and that synapses are the sites at which modifications occur. The extreme duration (years) of memory also imposes severe constraints on underlying chemistries in that it is difficult to imagine anything but a structural change as being involved. Modifications of proteins, for example, would not outlast the proteins themselves.

It would seem then that memory requires a process that rapidly and selectively reorganizes the anatomy of a subset of synapses on a neuron. This points to cellular chemistries which are located within the synaptic region and which produce significant changes in structural proteins that maintain the morphology of synapses. Partial digestion of the synaptic cytoskeleton may be a necessary step in satisfying these conditions.

It is of interest that a mechanism involving calcium, calpain, and spectrin has been implicated in shape changes in platelets,[30] red blood cells,[29] and perhaps heptocytes,[56] while the NMDA receptor-calcium influx appears to play a prominent role in timing activity patterns in lamprey spinal cord.[57] The postulated memory mechanism thus involves a combination of steps that may have evolved separately for very different purposes.

Regulatory Processes

Given its potential for causing cell damage, it can be expected that cellular devices exist for regulating the NMDA receptor-calcium-calpain-structural protein sequence. One intriguing possibility is the spine itself. Spine heads provide small volumes that lack endoplasmic reticulum and mitochondria, two critical elements for suppressing the buildup of calcium. At the same time, the very thin necks of spines may restrict the flow of calcium into the parent dendrite, making the spine a specialization wherein large fluctuations of calcium can occur without placing the central cytoskeleton of the neuron at risk.

Intracellular recording studies have shown that a number of physiological systems act to circumscribe the actions of the NMDA receptor and channel complex. The channel itself is blocked in a voltage-dependent fashion by magnesium and therefore is regulated by events that hyperpolarize the cell or shunt excitatory inputs. For example, the disynaptic feedforward IPSPs which accompany and truncate EPSPs in hippocampus and other forebrain regions normally prevent the occurrence of large NMDA receptor-mediated responses. However, the IPSPs exhibit a transient refractoriness such that once initiated they cannot be reactivated for several hundred milliseconds. Certain patterns of activity can thus obviate the IPSP and thereby stimulate large NMDA-dependent currents; that is, if afferents are stimulated with very short high frequency bursts with an interval of 200 ms between bursts, then the second and subsequent bursts will generate EPSPs that are largely free of IPSPs. Not surprisingly, these conditions (which occur during learning) are optimal for activating the NMDA system and for inducing LTP.[13,58,59]

Subsequent physiological experiments revealed that the intense depolarization needed to elicit large NMDA-mediated currents activates mechanisms that suppress further depolarization. This was evident in an experiment in which three inputs to a common cell were stimulated in rapid succession 200 msec after a priming burst of stimulation (*i.e.*, under conditions in which IPSPs were refractory). The earliest burst in the sequence was found to depress responses to the later bursts;[60] Grillner and colleagues[57] have shown that the NMDA receptor-mediated calcium influx in lamprey spinal cord triggers a hyperpolarizing current, and it is possible that similar events occur in hippocampus. Whatever its origins, the depression effect that develops while stimulating a target cell in a rapid sequence with several inputs may serve to protect the neuron from prolonged activation by currents carried through NMDA receptor channels and hence from extended periods of calcium influx.

It follows from the above arguments that disturbances of inhibitory interneurons (and perhaps hyperpolarizing afterpotentials) will open the way to excessive activation of NMDA receptor channels. The evidence that aging perturbs certain potassium- (*i.e.*, hyperpolarizing) mediated currents[61] is of interest in this regard.

In addition to the anatomical and physiological regulatory devices, there are a variety of cellular mechanisms that buffer intracellular calcium levels and thus presumably insure that any stimulation of calpain will be brief and spatially restricted. These include membrane-associated calcium pumps, the endoplasmic reticulum, and mitochondria. Brain mitochondria provide a high capacity sequestration system that in the presence of appropriate polyamine levels will buffer calcium into the low micromolar range,[62] *i.e.*, well below the level needed for calpain activation. Mitochondria are likely to contribute indirectly to calcium homeostasis by providing the ATP needed for pumps and more directly by sequestering calcium within dendritic and axonal branches. Denervation of the hippocampus disturbs the calcium sequestering activities of mitochondria[63] and the activity of ornithine decarboxylase,[64] the key regulatory enzyme in polyamine biosynthesis. It is thus possible that a type of trophic relationship exists between central neurons involving ornithine decarboxylase, polyamines, and calcium buffering by mitochondria.

Repair and Stability in Neurons

The causes of damage are only one aspect of neuropathology; repair or the absence thereof being the other. In the context of the above discussion we might ask why the damage produced to a dendritic cytoskeleton by proteolytic activity cannot be reversed or repaired and indeed why the neuron seems to be incapable of shedding and replacing its branches as very probably recurs in development. Striking age-related changes in axonal and dendritic growth have been documented for hippocampus of rat using the ''sprouting paradigm'' in which part of a dendritic field is denervated and the rate and degree of which remaining (intact) afferents invade and form synapses are measured. Early in the third week of life axons stop the rapid formation of long collateral branches and dendrites begin exhibiting permanent atropic reactions.[65,66] Growth, after this age consists of a meager collateralization of short branches and synaptogenesis. The latter effect declines in vigor from puberty into early adulthood (*i.e.*, up to about 3 months postnatal) whereupon it stabilizes prior to a further decline in old age.[67]

Two types of processes may be at work here. One, the degenerative-atrophic reaction itself is different. That is, degenerating axons and terminals persist much longer in the adult than in the young rat[67] suggesting that the cytoskeletal system is more stable and resistant to dissolution. Second, it is possible that the signals for growth are reduced in intensity or efficiency. We have elsewhere hypothesized that both effects reflect a slowing of turnover of the constituents of the cytoskeleton and membrane; according to this idea, "stability" of neuronal branches (*i.e.*, the half-life of a specific dendritic process or spine) increases during maturation, and this is achieved by slowing breakdown and synthesis.[1] A side effect of reducing synthesis and assembly processes is that the neuron is less capable of replacing elements affected by pathological events.

Stability is necessary for a long-term memory system. That is, if modifications of spines and synapses are to encode information that will persist for months or years then the branches upon which the altered contacts are found must themselves remain in place for that period of time. Here again we encounter the possibility that the demands for stable memory require cellular properties that affect the response of the neurons to pathogenic events.

REFERENCES

1. Lynch, G., J. Larson & M. Baudry. 1986. Proteases, neuronal stability, and brain aging: A hypothesis. *In* Treatment Development Strategies for Alzheimer's Disease. T. Crook, R. Bartus, S. Ferris & S. Gershon, Eds.: 119–139. Mark Powley. Madison, CT.
2. Morris, R. G. M., E. Anderson, G. Lynch & M. Baudry. 1986. Selective impairment of learning and blockade of long-term potentiation by an N-methyl-D-aspartate receptor antagonist, AP-5. Nature **319:** 774–776.
3. Staubli, U., O. Thibault, M. DiLorenzo & G. Lynch. 1989. Antagonism of NMDA receptors impairs acquisition but not retention of olfactory memory. Behav. Neurosci. **103:** 54–60.
4. Boast, C. A., S. C. Gerhardt, G. Pastor, J. Lehmann, P. E. Etienne & J. M. Liebman. 1988. The N-methyl-D-aspartate antagonists CGS 19755 and CPP reduce ischemia brain damage in gerbils. Brain Res. **442:** 345–348.
5. Gill, R., A. C. Foster & G. N. Woodruff. 1988. MK-801 is neuroprotective in gerbils when administered during the post-ischaemic period. Neuroscience **25:** 847–855.
6. Gill, R., A. C. Foster & G. N. Woodruff. 1987. Systemic administration of MK-801 protects against ischemia-induced hippocampal neurodegeneration in the gerbil. J. Neurosci. **7:** 3343–3349.
7. Rothman, S. 1984. Synaptic release of excitatory amino acid neurotransmitter mediates anoxic neuronal death. J. Neurosci. **4:** 1884–1891.
8. Simon, R. P., J. H. Swan, T. Griffiths & B. S. Meldrum. 1984. Blockade of N-methyl-D-aspartate receptors may protect against ischemic damage in the brain. Science **226:** 850–854.
9. Wieloch, T. 1985. Hypoglycemia-induced neuronal damage prevented by an N-methyl-D-aspartate antagonist. Science **230:** 681–683.
10. Farber, J. L. 1981. The role of calcium in cell death. Life Sci. **29:** 1289–1295.
11. Collingridge, G. L., S. J. Kehl & H. McLennan. 1983. Excitatory amino acids in synaptic transmission in the Schaffer-collatoral-commissural pathway of the rat hippocampus. J. Physiol. (Lond.) **334:** 33–46.
12. Harris, E. W., A. H. Ganong & C. W. Cotman. 1984. Long-term potentiation in the

hippocampus involves activation of N-methyl-D-aspartate receptors. Brain Res. **323:** 132–137.

13. LARSON, J. & G. LYNCH. 1988. Role of N-methyl-D-aspartate receptors in the induction of synaptic potentiation by burst stimulation patterned after the hippocampal theta rhythm. Brain Res. **441:** 111–118.

14. LYNCH, G., J. LARSON, S. KELSO, G. BARRIONUEVO & F. SCHOTTLER. 1983. Intracellular injections of EGTA block induction of hippocampal long-term potentiation. Nature **305:** 719–721.

15. MALENKA, R. C., J. A. KAUER, R. S. ZUCKER & R. A. NICOLL. 1988. Postsynaptic calcium is sufficient for potentiation of hippocampal synaptic transmission. Science **242:** 81–84.

16. MACDERMOTT, A. B., M. L. MAYER, G. L. WESTBROOK, S. J. SMITH & J. L. BARKER. 1986. NMDA-receptor activation increases cytoplasmic calcium concentration in cultured spinal cord neurones. Nature **321:** 519–522.

17. MAYER, M. L. & G. L. WESTBROOK. 1987. Permeation and block of N-methyl-D-aspartic acid receptor channels by divalent cations in mouse cultured central neurones. J. Physiol. (Lond.) **394:** 501–527.

18. DUNWIDDIE, T. & G. LYNCH. 1978. Long-term potentiation and depression of synaptic responses in the rat hippocampus: Localization and frequency dependency. J. Physiol. (Lond.) **276:** 353–367.

19. DOUGLAS, R. M. & G. V. GODDARD. 1975. Long-term potentiation of the perforant path-granule cell synapse in the rat hippocampus. Brain Res. **86:** 205–215.

20. STAUBLI, U. & G. LYNCH. 1987. Stable hippocampal long-term potentiation elicited by "theta" pattern stimulation. Brain Res. **435:** 227–234.

21. LEE, K., F. SCHOTTLER, M. OLIVER & G. LYNCH. 1980. Brief bursts of high-frequency stimulation produce two types of structural change in rat hippocampus. J. Neurophysiol. **44:** 247–258.

22. LEE, K., M. OLIVER, F. SCHOTTLER & G. LYNCH. 1981. Electron microscopic studies of brain slices: The effects of high frequency stimulation on dendritic ultrastructure. *In* Electrical activity in isolated mammalian CNS preparations. G. Kerkut & H. V. Wheal, Eds.: 189–212. Academic Press. New York, NY.

23. CHANG, F. L. F. & W. T. GREENOUGH. 1984. Transient and enduring morphological correlates of synaptic activity and efficacy change in the rat hippocampal slice. Brain Res. **309:** 35–46.

24. ZUCKER, R. S. 1988. Frequency dependent changes in excitatory synaptic efficacy. *In* Mechanisms of Epileptogenesis. M. A. Dichter, Ed.: 153–167. Plenum. New York, NY.

25. MULLER, D., M. JOLY & G. LYNCH. 1988. Contributions of quisqualate and NMDA receptors to the induction and expression of LTP. Science **242:** 1694–1697.

26. MULLER, D. & G. LYNCH. 1988. Long-term potentiation differentially affects two components of synaptic responses in hippocampus. Proc. Natl. Acad. Sci. USA **85:** 9346–9350.

27. MULLER, D., J. LARSON & G. LYNCH. 1988. The NMDA receptor mediated components of responses evoked by patterned stimulation are not increased by long-term potentiation. Brain Res. **477:** 396–399.

28. WILSON, C. J. 1984. Passive cable properties of dendritic spines and spiny neurons. J. Neurosci. **4:** 281–297.

29. SIMAN, R., M. BAUDRY & G. LYNCH. 1987. Calcium-activated proteases as possible mediators of synaptic plasticity. *In* Synaptic Function. G. M. Edelman, W. E. Gall & W. M. Cowan, Eds.: 519–548. John Wiley. New York, NY.

30. FOX, J. E. B., C. C. REYNOLDS, J. S. MORROW & D. R. PHILLIPS. 1987. Spectrin is associated with membrane-bound actin filaments in platelets and is hydrolyzed by the Ca^{2+}-dependent protease during platelet activation. Blood **69:** 537–545.

31. LASEK, R. J. & P. N. HOFFMAN. 1976. The neuronal cytoskeleton, axonal transport and axonal growth. *In* Cell Motility. R. Goldman, T. Pollard & J. Rosenbaum, Eds.: 1021–1049. Cold Spring Harbor Laboratory. Cold Spring Harbor, NY.

32. GOODMAN, S. & I. ZAGON. 1984. Brain spectrin: A review. Brain Res. Bull. **13:** 813–832.

33. SIMAN, R., M. BAUDRY & G. LYNCH. 1984. Brain fodrin: Substrate for the endogenous calcium-activated protease calpain I. Proc. Natl. Acad. Sci. USA **81:** 3276–3280.

34. SEUBERT, P., M. BAUDRY, S. DUDEK & G. LYNCH. 1987. Calmodulin stimulates the degra-
     dation of brain spectrin by calpain. Synapse **1:** 20–24.
35. HARRIS, A. S., D. E. CROALL & J. S. MORROW. 1988. The calmodulin-binding site in a-fodrin
     is near the calcium dependent protease-I cleavage site. J. Biol. Chem. **263:** 15754–15761.
36. SEUBERT, P., G. IVY, J. LARSON, J. LEE, K. SHAHI, M. BAUDRY & G. LYNCH. 1988. Lesions
     of entorhinal cortex produce a calpain-mediated degradation of brain spectrin in dentate gyrus.
     I. Biochemical studies. Brain Res. **459:** 226–232.
37. CARLIN, R. K., D. C. BARTELT & P. SIEKEVITZ. 1983. Identification of fodrin as a major
     calmodulin-binding protein in postsynaptic density preparations. J. Cell Biol. **96:** 443–448.
38. PERLMUTTER, L. S., R. SIMAN, C. GALL, P. SEUBERT, M. BAUDRY & G. LYNCH. 1988. The
     ultra-structural localization of calcium-activated protease "calpain" in rat brain. Synapse
     2: 79–88.
39. SEUBERT, P., J. LARSON, M. OLIVER, M. W. JUNG, M. BAUDRY & G. LYNCH. 1988. Stim-
     ulation of NMDA receptors in proteolysis of brain spectrin in hippocampal slices. Brain Res.
     460: 189–194.
40. STAUBLI, U., J. LARSON, O. THIBAULT, M. BAUDRY & G. LYNCH. 1988. Chronic adminis-
     tration of a thiol-proteinase inhibitor blocks long-term potentiation of synaptic responses.
     Brain Res. **444:** 153–158.
41. SANDOVAL, I. V. & K. WEBER. 1978. Calcium-induced inactivation of miocrotubule formation
     in brain extracts. Eur. J. Biochem. **92:** 463–470.
42. MALIK, M. N., L. A. MEYERS, K. IQBAL, A. M. SHEIKH, L. SCOTTO & H. M. WISNIEWSKI.
     1981. Calcium activated proteolysis of fibrous proteins in central nervous system. Life Sci.
     29: 795–802.
43. SCHLAEPFER, W. W. & M. B. HASLER. 1979. Characterization of the calcium-induced dis-
     ruption of neurofilaments in rat peripheral nerve. Brain Res. **168:** 299–309.
44. SCHLAEPFER, W. W. & R. P. BUNGE. 1973. Effects of calcium ion concentration on the
     degeneration of amputated axons in tissue culture. J. Cell Biol. **59:** 456–470.
45. LIBBY, P. & A. L. GOLDBERG. 1978. Leupeptin, a protease inhibitor, decreases protein deg-
     radation in normal and diseased muscles. Science **199:** 534–536.
46. LYNCH, G., D. MATTHEWS, S. MOSKO, T. PARKS & C. W. COTMAN. 1972. Induced acetyl-
     cholinesterase-rich layer in rat dentate gyrus following entorhinal lesions. Brain Res. **42:**
     311–318.
47. LYNCH, G., S. DEADWYLER & C. COTMAN. 1973. Postlesion axonal growth produces perma-
     nent functional connections. Science **180:** 1364–1366.
48. GALL, C., G. ROSE & G. LYNCH. 1978. Proliferative and migratory activity of glial cells in the
     deafferented hippocampus. J. Comp. Neurol. **183:** 539–550.
49. ROSE, G., C. W. COTMAN & G. S. LYNCH. 1976. Hypertrophy and redistribution of astrocytes
     in the deafferented hippocampus. Brain Res. Bull. **1:** 87–92.
50. GOLDSCHMIDT, R. B. & O. STEWARD. 1980. Preferential neurotoxicity of colchicine for
     granule cells of the dentate gyrus of the adult rat. Proc. Natl. Acad. Sci. USA **77:** 3047–3051.
51. HINDELANG-GERTNER, C., M. E. STOECKEL, A. PORTEL, F. STUTINSKY. 1976. Colchicine
     effects on neurosecretory neurons and other hypothalamic and hypophysial cells, with special
     reference to changes in the cytoplasmic membranes. Cell Tiss. Res. **170:** 17–41.
52. SEUBERT, P., Y. NAKAGAWA, G. IVY, P. VANDERKLISH, M. BAUDRY & G. LYNCH. Intra-
     hippocampal colchicine injection results in spectrin proteolysis. Neuroscience. In press.
53. TOYO-OKA, T., T. SHIMIZA & T. MASAK. 1978. Inhibition of proteolytic activity of calcium
     activated neutral protease by leupeptin and antipain. Biochem. Biophys. Res. Commun.
     82: 484–491.
54. SIMAN, R. & J. C. NOSZEK. 1988. Excitatory amino acids activate calpain I and induce
     structural protein breakdown in vivo. Neuron **1:** 279–287.
55. LYNCH, G. & M. BAUDRY. 1984. The biochemistry of memory: A new and specific hypoth-
     esis. Science **224:** 1057–1063.
56. NICOTERA, P., P. HARTZELL, G. DAVIS & S. ORRENIUS. 1986. The formation of plasma
     membrane blebs in hepatocytes exposed to agents that increase cytosolic Ca^{2+} is mediated by
     the activation of a non-lysosomal proteolytic system. FEBS Lett. **209:** 139–144.
57. WALLEN, P. & S. GRILLNER. 1987. N-methyl-D-aspartate receptor induced, inherent oscilla-

tory activity in neurons active during fictive locomotion in the lamprey. J. Neurosci. **7:** 2745–2755.

58. LARSON, J., D. WONG & G. LYNCH. 1986. Patterned stimulation at the theta frequency is optimal for induction of hippocampal long-term potentiation. Brain Res. **368:** 347–350.
59. LARSON, J. & G. LYNCH. 1986. Induction of synaptic potentiation in hippocampus by patterned stimulation involves two events. Science **232:** 985–988.
60. LARSON, J. & G. LYNCH. 1988. Theta pattern stimulation and the induction of LTP: The sequence in which synapses are stimulated determines the degree to which they potentiate. Brain Res. **489:** 49–58.
61. LANDFIELD, P. W. & T. A. PITLER. 1984. Prolonged Ca^{2+}-dependent afterhyperpolarizations in hippocampal neurons of aged rats. Science **226:** 1089–1091.
62. JENSEN, J. R., G. LYNCH & M. BAUDRY. 1987. Polyamines stimulate mitochondrial calcium transport in rat brain. J. Neurochem. **48:** 765–772.
63. BAUDRY, M., C. GALL, M. KESSLER, H. ALAPOUR & G. LYNCH. 1983. Denervation-induced decrease in mitochondrial calcium transport in rat hippocampus. J. Neurosci. **3:** 252–259.
64. BAUDRY, M., G. LYNCH & C. GALL. 1986. Induction of ornithine decarboxylase as a possible mediator of seizure-elicited changes in genomic expression in rat hippocampus. J. Neurosci. **6:** 3430–3435.
65. GALL, C. & G. LYNCH. 1978. Rapid axon sprouting in the neonatal rat hippocampus. Brain Res. **153:** 357–362.
66. GALL, C. & G. LYNCH. 1981. Fiber architecture of the dentate gyrus following ablation of the entorhinal cortex in rats of different ages: Evidence for two forms of axon sprouting in the immature brain. Neuroscience **6:** 903–910.
67. MCWILLIAMS, J. R. & G. LYNCH. 1983. Rate of synaptic replacement in denervated rat hippocampus declines precipitously from the juvenile period to adulthood. Science **221:** 572–574.

The Protein Kinase C Family in the Brain: Heterogeneity and Its Implications

AKIRA KISHIMOTO, USHIO KIKKAWA, KOUJI OGITA, MARK S. SHEARMAN, AND YASUTOMI NISHIZUKA

Department of Biochemistry
Kobe University School of Medicine
Kobe 650, Japan

Although once considered as a single entity,[1] molecular cloning and enzymological analysis have revealed the existence of multiple subspecies of protein kinase C (PKC) in the brain tissues. These subspecies have a common structure closely related to, yet distinctly different from one another. The relative distribution of several PKC subspecies varies markedly with the different areas of the brain and cell types examined. It is possible that each member of this family has a defined function in processing and modulating cellular responses to a wide variety of external stimuli. The integrated nomenclature used herein for the several PKC subspecies and its correspondence to those of other workers are as given elsewhere.[2]

Molecular Heterogeneity

Four cDNA clones which encode the α-, βI-, βII-, and γ-subspecies were initially found in the mammalian brain cDNA libraries (TABLE 1). The clones α, β(βI plus βII), and γ are encoded by separate genes located in different chromatosomes. The βI- and βII-subspecies are derived from a single mRNA transcript by alternative splicing; these two subspecies differ from each other only in a short range of ~50 amino acid residues in their carboxy-terminal end region. Recently, another group of cDNA clones encoding the δ, ε-, and ζ-sequence have been isolated from the rat brain library.[3]

The structures of these PKC subspecies are schematically shown in FIGURE 1. The group of the α-, βI-, βII-, and γ-subspecies each have four conserved (C_1–C_4) and five variable (V_1–V_5) regions, whereas the second group of the δ-, ε-, and ζ-subspecies lack the C_2 region, though the mass of the enzyme molecule is similar. In general, the enzyme is composed of two, regulatory and protein kinase, domains. The amino-terminal half of the molecule appears to be the regulatory domain. The conserved region C_1 contains a tandem repeat of a cysteine-rich sequence, except for the ζ-subspecies which has only one set of this sequence. The cysteine-rich sequence matches the consensus sequence of "cysteine-zinc DNA-binding finger," which is found in many DNA-binding proteins that are related to transcriptional regulation, but no obvious evidence is available indicating that PKC will bind to DNA. On the other hand, the carboxy-terminal half containing C_3 and C_4 regions appears to be the protein kinase domain, as it shows large clusters of sequence homology to many other protein kinases. The conserved region C_3 has an ATP-binding sequence, GXGXXG——K, where G, K and X represent glycine, lysine and any amino acid, respectively.

TABLE 1. Subspecies of Protein Kinase C from Mammalian Tissues

Subspecies[a]	α	βI	βII	γ	δ	ε	ζ
Amino acid residues	672	671	673	697	673	737	592
Calculated molecular weight	76,799	76,790	76,933	78,366	77,517	83,474	67,740
Chromatographic sub-fraction	type III		type II	type I	not identified	not identified	not identified
Activators[b]	PS+DG+Ca²⁺ AA+Ca²⁺	PS+DG+Ca²⁺		PS+DG+ Ca²⁺AA	PS+DG+(Ca²⁺)	PS+DG+ (Ca²⁺)	PS+ (DG+Ca²⁺)
Tissue expression	universal	some tissues & cells	many tissues & cells	brain & spinal cord only	many tissues	brain only?	many tissues
Chromosome location (human)	17	16		19	?	?	?

[a] More detailed explanations and references of each PKC subspecies are given elsewhere.[2]
[b] PS, phosphatidylserine; DG, diacylglycerol; and AA, arachidonic acid.

Differential Expression

PKC with the γ-sequence is expressed only in the brain and spinal cord, but not in other tissues and cell types so far examined. This PKC subspecies develops only post-natally and reaches maximum activity in the rat about three weeks after birth. The highest enzyme activity of the γ-subspecies is found in the hippocampus, cerebral cortex, amygdaloid complex, and cerebellar cortex. This subspecies is present in the nerve endings of the Purkinje cell axon which terminate in the deep cerebellar nuclear cells.[4] Immunoelectron microscopic analysis indicates that the γ-subspecies is associated with most membranous structures present throughout the cell. The nuclei and mitochondria generally lack or poorly express this PKC subspecies.

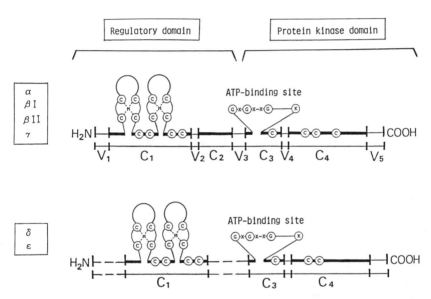

FIGURE 1. Common structure of PKC subspecies. C, G, K, X, and M represent cysteine, glycine, lysine, any amino acid, and metal, respectively. The numbers indicate amino acid residues composed of each subspecies. The ζ-subspecies contains only one set of the cysteine-rich sequence. Other details are described in the text.

On the other hand, PKC with the βI- and βII-sequence are expressed in several neuronal cell types in different ratios. Cytochemical analysis with the antibodies specific to each subspecies indicates their clear distinct cellular expression. For instance, in the rat cerebellar cortex, the βI-subspecies is found primarily in the granular layer, whereas the βII-subspecies is found predominantly in the molecular layer, apparently in the presynaptic nerve endings terminating on the dendrites of and cell body of the Purkinje cell.[5] The activity of the βI-subspecies in the cerebrum and hippocampus is very low.

In contrast, the α-subspecies appears to express commonly in many cell types. Although the structure and genetic identity of the group of the δ-, ε-, ζ-subspecies has not yet been determined, preliminary studies suggest that several subspecies of PKC are co-expressed in a single cell.

Individual Characteristics

PKC is more abundant in the brain than in any other tissues. To date, three distinct fractions, types I, II, and III, have been separated upon hydroxyapatite column chromatography. The structure and genetic identity of each fraction have been determined by comparison with the enzymes that were separately expressed in COS 7 cells after transfection with the respective cDNA-containing plasmids.[6] Type I corresponds to the subspecies encoded by the γ-sequence, type II is an unequal mixture of the βI- and βII-subspecies, and type III has the structure having the α-sequence. Thus far, the βI- and βII-subspecies can be distinguished from each other only by immunochemical methods, and show nearly identical kinetic properties. Type I, II, and III PKCs exhibit different modes of activation and kinetic properties, and slightly different patterns of autophosphorylation, binding of phorbol ester, Ca^{2+} sensitivity, and immunochemical reactivity. Type I PKC is less sensitive to diacylglycerol but significantly activated by relatively low concentrations of free arachidonic acid. Type II PKC exhibits substantial activity without elevated Ca^{2+} levels, and responds well to diacylglycerol. Type III PKC is most sensitive to diacylglycerol that is derived from inositol phospholipids. It is possible that different subspecies of PKC are activated by the series of phospholipid metabolites, such as diacylglycerol and arachidonic acid, that are produced in successive phases of the responses of the cell subsequent to stimulation of a cell-surface receptor.

PROSPECTIVES

A synergistic interaction between Ca^{2+} and PKC proposed earlier[1] is now well recognized to underlie a variety of cellular responses to external stimuli. Several important physiological functions have been assigned for PKC, including involvement in secretion and release reactions, modulation of ion conductance, down-regulation of receptors, cross-talk of various receptors and gene expression and cell proliferation.[7] PKC, however, appears to show a dual action, and exerts feedback control on the Ca^{2+}-signaling pathway. For instance, as illustrated in FIGURE 2, PKC may inhibit the receptor-mediated breakdown of inositol phospholipids. A number of reports have appeared to suggest that PKC also activates the Ca^{2+}-transport ATPase and the Na^{2+}/Ca^{2+} exchanger, thereby removing Ca^{2+} from the cytosol.

A negative feedback role of PKC may be extended to cell proliferation. The receptor for EGF has repeatedly been shown to be phosphorylated by PKC, resulting in rapid decrease in high-affinity binding of EGF as well as inhibition of the ligand-induced tyrosine phosphorylation (for a review, see REF. 7). It is plausible that the treatment of cells with 12-O-tetradecanoylphorbol-13-acetate (TPA) causes disappearance of the PKC molecule itself, and thus relieves the cell from down-regulation of the growth factor receptor, so that uncontrolled cell proliferation might occur. Indeed, a rapid, sometimes sustained, disappearance of PKC by treatment with TPA has been recognized for many cell types. Presumably, under physiological conditions, the activation of PKC is transient; otherwise it would be degraded by proteolysis. Several subspecies co-expressed within a single cell disappear at different rates upon treatment with TPA. In our earlier studies,[8] it was shown that calpain I, which is activated at the micromolar range of Ca^{2+}, cleaves PKC in the presence of phosphatidylserine plus diacylglycerol or TPA, implying that the activated from of PKC is a target of calpain action. It is probable that this calpain-dependent proteolysis initiates the degradation of PKC molecules. Within the cell, TPA is more effective than diacylglycerol in initiating this enzyme degradation because of its stable properties. Thus, the tumor-promoter again provides a dual effect, furnishing a

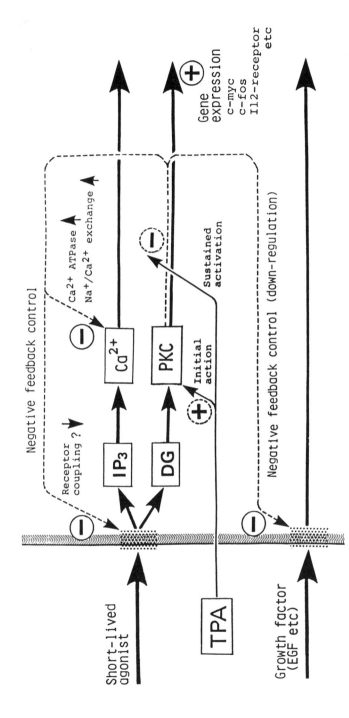

FIGURE 2. Dual action of PKC and of TPA. Unlike short-lived diacylglycerol, TPA appears to show a dual effect with a positive action to initially activate PKC, but with a negative action to degrade the enzyme during sustained activation. The detailed explanation is as given in the text. It does not exclude, however, a possible role for the proteolytically produced catalytically active fragment of PKC. DG represents diacylglycerol; IP$_3$ indicates inositol-1,4,5-trisphosphate.

positive short-term activation of PKC, and then a negative action to initiate the degradation of the enzyme over a long time course (FIG. 2).

Obviously, the negative feedback role of PKC emphasized above does not exclude the existence of a positive forward action of the enzyme. The diverse functions of PKC so far proposed may well involve activation of different subspecies. Although at present, there is little evidence of a specific example, each member of the family probably plays a defined function in processing and modulating various membrane events which occur during cell-to-cell communication.

REFERENCES

1. NISHIZUKA, Y. 1984. The role of protein kinase C in cell surface signal transduction and tumour promotion. Nature **308:** 693–697.
2. NISHIZUKA, Y. 1988. The molecular heterogeneity of protein kinase C and its implications for cellular regulation. Nature **334:** 661–665.
3. ONO, Y., T. FUJII, K. OGITA, U. KIKKAWA, K. IGARASHI & Y. NISHIZUKA. 1988. The structure, expression, and properties of additional members of protein kinase C family. J. Biol. Chem. **263:** 6927–6932.
4. SAITO, N., U. KIKKAWA, Y. NISHIZUKA & C. TANAKA. 1988. Distribution of protein kinase C-like immunoreactive neurons in rat brain. J. Neurosci. **8:** 369–382.
5. ASE, K., N. SAITO, M. S. SHEARMAN, U. KIKKAWA, Y. ONO, K. IGARASHI, C. TANAKA & Y. NISHIZUKA. 1988. Distinct cellular expression of βI- and βII- subspecies of protein kinase C in rat cerebellum. J. Neurosci. **8:** 3850–3856.
6. KIKKAWA, U., Y. ONO, K. OGITA, T. FUJII, Y. ASAOKA, K. SEKIGUCHI, Y. KOSAKA, K. IGARASHI & Y. NISHIZUKA. 1987. Identification of the structures of multiple subspecies of protein kinase C expressed in rat brain. FEBS Lett. **217:** 227–231.
7. NISHIZUKA, Y. 1986. Studies and perspectives of protein kinase C. Science **233:** 305–312.
8. KISHIMOTO, A., K. MIKAWA, K. HASHIMOTO, I. YASUDA, S. TANAKA, M. TOMINAGA, T. KURODA & Y. NISHIZUKA. 1989. Limited proteolysis of protein kinase C subspecies by calcium-dependent neutral protease (calpain). J. Biol. Chem. **264:** 4088–4092.

Calcium- versus G-Protein-Activated Phosphoinositide Hydrolysis in Synaptoneurosomes from Young and Old Rats[a]

L. JUDSON CHANDLER AND FULTON T. CREWS

Department of Pharmacology and Therapeutics
University of Florida College of Medicine
Box J-267, J. Hillis Miller Health Center
Gainesville, Florida 32610

The turnover of membrane inositol phospholipids is thought to represent an important signal transduction pathway in both neuronal and nonneuronal tissues. A large number of neurotransmitter receptors have been shown to activate a phosphoinositide-specific phospholipase C (PLC) that in turn hydrolyzes phosphoinositides. Of particular importance is the hydrolysis of phosphatidlyinositol 4,5-bisphosphate (PIP$_2$), resulting in the formation of two putative intracellular second messengers, diacylglycerol (DG) and inositol (1,4,5-trisphosphate {Ins(1,4,5)P$_3$}. Ins(1,4,5)P$_3$ has been shown to release calcium from ATP-dependent, nonmitochrondrial intracellular calcium stores, whereas DG activates protein kinase C.

There is good evidence to suggest that an as yet unidentified guanine nucleotide binding protein (G-protein) is involved in coupling the agonist-receptor complex to PLC activation (reviewed in REFS. 1,2). However, not all PLC activation appears to be G-protein mediated. Depolarizing concentrations of KCl and calcium ionophores have been shown to stimulate phosphoinositide hydrolysis, leading to the suggestion that calcium influx is also a stimulus for PLC activation.[3,4] An increase in intracellular calcium subsequent to membrane depolarization may directly stimulate phosphoinositide hydrolysis. This scenario is supported by studies with dihydropyridine calcium channel agonists and antagonists that suggest that dihydropyridine-sensitive (L-type) calcium channels can modulate depolarization-dependent phosphoinositide hydrolysis in brain.[4,5] A direct role for calcium in stimulating phosphoinositide hydrolysis is further supported by findings that guanine nucleotides and calcium can stimulate phosphoinositide hydrolysis in cerebral cortical membranes in an additive manner.[6] Phosphoinositide-linked receptors may activate phosphoinositide hydrolysis directly through a receptor-G-protein-PLC interaction and indirectly via stimulation of calcium influx. Recent studies have described two pharmacologically distinct α_1-receptor subtypes (α_{1a} and α_{1b}) that activate phosphoinositide hydrolysis through different mechanisms.[7–9] Alpha$_{1a}$-receptor-stimulated phosphoinositide hydrolysis is dependent upon the presence of extracellular calcium and is inhibited by the dihydropyridine calcium channel antagonist nifedipine. Alpha$_{1b}$-receptor-stimulated phosphoinositide hydrolysis is not dependent upon the presence of extracellular calcium. It may be that α_{1b}-receptors stimulate phosphoinositide hydrolysis through a receptor-G-protein-PLC interaction, whereas α_{1a}-receptor-stimulated phosphoinositide hydrolysis occurs secondary to enhancing calcium influx through a dihydropy-

[a]This work was supported by National Institute on Aging Grant AG06660.

ridine-sensitive calcium channel. We have recently observed in cerebral cortical membranes that guanine nucleotide and calcium-stimulated phosphoinositide hydrolysis have different pH optima and different sensitivities to inhibitors of PLC,[10] possibly indicating two separate membrane-bound PLCs in brain, one linked to activation by a G-protein and the other to activation by calcium.

Regulation of intracellular calcium is of critical importance in modulating cell function, and there is evidence to suggest that changes in intracellular calcium metabolism represents a neuropathological aspect of aging. The activity of a number of calcium regulatory processes in brain, including sodium-calcium exchange,[11] sequestration of intracellular calcium by mitochondria and voltage-sensitive calcium uptake,[12,13] have been reported to decline during senescence. Changes in calcium-mediated signals have also been observed to occur during aging. Meyer et al.[14] found in rat cerebral cortical synaptosomes, an apparent decrease in the sensitivity of the acetylcholine release process to the stimulatory effects of calcium. Ishikawa et al.[15] recently reported that the ability of exogenously added IP$_3$ to stimulate calcium efflux from permeabilized rat parotid cells declines with aging. Thus, many age-related changes in physiological processes may be related to decrements in calcium-regulated signals.

Studies with synaptosomes[16-18] and synaptoneurosomes[19] have shown that intracellular calcium increases when sodium is isosmotically replaced with choline chloride or N-methyl-D-glucamine, and we have recently observed that resuspension of synaptoneurosomes in sodium-free buffer causes a large increase in basal phosphoinositide hydrolysis.[5] It is likely that the increase in phosphoinositide hydrolysis in sodium-free medium is a direct result of increased intracellular calcium. To investigate possible changes in G-protein and calcium-activated phosphoinositide hydrolysis that may occur during aging, we performed experiments with cerebral cortical synaptoneurosomes isolated from young (5-month) and old (26-month) Fisher 344 rats using isotonic buffers with and without sodium. Resuspension of synaptoneurosomes from 5-month-old rats in sodium-free medium (1 mM calcium) caused a large increase in the basal accumulation of [^{3}H]inositol phosphates (FIG. 1). This increase was greater than that produced by 100 μM norepinephrine in sodium-containing medium (compare FIG. 1 with FIG. 2). In sodium-free medium, basal phosphoinositide hydrolysis decreased with decreasing amounts of added calcium and by the addition of 50 μM EGTA and no added calcium. The calcium-dependency of basal phosphoinositide hydrolysis under these ionic conditions provides evidence in support of a stimulatory role of calcium in phosphoinositide hydrolysis. No differences in either the magnitude or calcium dependency of basal phosphoinositide hydrolysis in sodium-free medium were observed in synaptoneurosomes from old versus young rats.

We previously found no age-related decline in either norepinephrine or carbachol-stimulated phosphoinositide hydrolysis in cerebral cortical slices from 3-,12-, and 24-month- old Fisher 344 rats.[21] In agreement with this finding, there does not appear to be an age-related decline in norepinephrine or carbachol-stimulated phosphoinositide hydrolysis in cerebral cortical synaptoneurosomes from 5- and 26-month-old rats (FIG. 2). These data indicate that neither the receptor agonist-activated nor calcium-activated component of phosphoinositide hydrolysis is altered during senescence.

Activation of protein kinase C has been reported to inhibit receptor agonist-stimulated phosphoinositide hydrolysis, and it is hypothesized that protein kinase C is involved in down-regulation of phosphoinositide hydrolysis. Addition of the protein kinase C activator TPA, inhibited both norepinephrine and carbachol-stimulated phosphoinositide hydrolysis in cerebral cortical synaptoneurosomes (FIG. 2). In previous studies, we found that norepinephrine-stimulated phosphoinositide hydrolysis was approximately 10-fold more sensitive to inhibition by phorbol esters than carbachol.[22] To determine if the inhibition by phorbol esters was altered during aging, we treated synaptoneurosomes from

5- and 26-month-old rats with TPA. The inhibition of norepinephrine and carbachol-stimulated phosphoinositide hydrolysis was identical in preparations from young and old rats, suggesting that protein kinase C-mediated inhibition of phosphoinositide hydrolysis was not altered during aging. Phorbol esters have been shown to inhibit guanine nucleotide-stimulated phosphoinositide hydrolysis in digitonin-permeabilized fibroblast[23] and membrane fragments from NG108-15 cells[24] and rat cortical membranes,[25] but to have no effect on IP_3-stimulated calcium release from permeabilized fibroblast.[23] Taken together, these results suggest that protein kinase C inhibition of phosphoinositide hydrolysis occurs distal to receptor occupancy by acting at the level of the G-protein or PLC. In contrast to

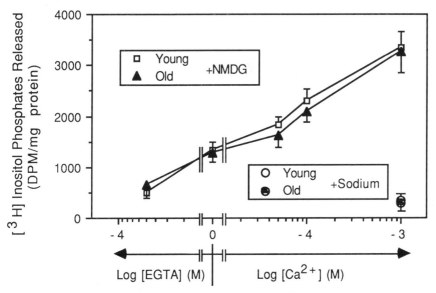

FIGURE 1. Extracellular calcium dependency in sodium-free medium of phosphoinositide hydrolysis in cerebral cortical synaptoneurosomes from 5-month and 26-month-old Fisher 344 rats prepared as previously described.[20] Synaptoneurosomes were prelabelled with [^{3}H]inositol and resuspended in ice-cold incubation medium in which sodium had been isosmotically replaced by N-methyl-D-glucamine (NMDG). Phosphoinositide hydrolysis was assayed by measuring the accumulation of [^{3}H]inositol phosphates over a 60-min time period in the presence of 10 mM lithium and the appropriate calcium or EGTA concentration. The accumulation of [^{3}H]inositol phosphates was determined by Dowex-1 chromatography as previously described.[6] Values are the means ± SEM of 3 separate experiments each performed in triplicate and are expressed per mg of synaptoneurosomal protein.

inhibition of receptor agonist-stimulated phosphoinositide hydrolysis, TPA had no effect on the basal accumulation of [^{3}H]inositol phosphates in either sodium-containing or sodium-free medium. The differences in [^{3}H]inositol phosphate release between sodium-containing and sodium-free medium is likely due to calcium-activated phosphoinositide hydrolysis. Our finding that TPA did not alter the increase in phosphoinositide hydrolysis in sodium-free medium suggests that the calcium-activated component of phosphoinositide hydrolysis is not sensitive to protein kinase C inhibition. Thus, phorbol ester inhibition of phosphoinositide hydrolysis may distinguish G-protein from calcium-activated phosphoinositide signals.

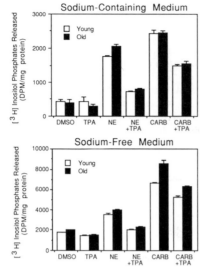

FIGURE 2. Effect of TPA (1.0 μM) upon basal and receptor agonist-stimulated phosphoinositide hydrolysis in sodium-containing and sodium-free medium in 5- and 26-month-old rats. Shown is the release of [³H]inositol phosphates over a 60-min time period in the presence of 1 mM calcium. Values are the means ± SEM of triplicate determinations of a single experiment and are expressed per mg of synaptoneurosomal protein. Note differences in the scale of the Y-axes.

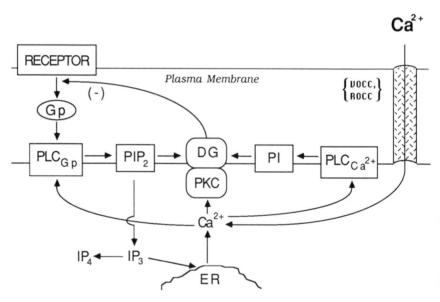

FIGURE 3. Schematic model depicting two separate pathways for stimulation of phosphoinositide hydrolysis. The pathway on the *left* utilizes a G-protein that couples the agonist-receptor complex to a phosphoinositide-specific phospholipase C (PLC_{Gp}). The pathway on the *right* involves an increase in intracellular calcium through either voltage- or receptor-operated calcium channels (VOCC or ROCC). There is evidence of a separate and distinct phospholipase C (PLC_{Ca}^{2+}) that can be directly activated by a rise in intracellular calcium. Although speculative, PLC_{Gp} and PLC_{Ca}^{2+} may show a preferential affinity for hydrolysis of phosphatidylinositol (PI) and phosphatidylinositol 4,5-bisphosphate (PIP_2), respectively. Selective hydrolysis of PI could relate to selective activation of PKC. In addition, small increases in intracellular calcium may stimulate additional calcium mobilization by potentiating PLC_{Gp} hydrolysis of PIP_2.

FIGURE 3 provides a working model of the events involved in stimulation of phosphoinositide hydrolysis in brain. Two separate pathways for stimulating phosphoinositide hydrolysis are proposed. One pathway utilizes a G-protein that couples the agonist-receptor complex to activation of a specific phospholipase C (PLC_{Gp}). In addition, there is evidence of a separate and distinct phospholipase C (PLC_{Ca}^{2+}) that can be activated by an increase in intracellular calcium. Some receptors may act primarily through the G-protein pathway, whereas others may also stimulate phosphoinositide hydrolysis secondary to enhancing calcium influx. Homma et al.[26] recently purified two PLCs from brain that had different calcium sensitivities and different affinities for hydrolysis of PI and PIP_2. Calcium and G-protein-linked PLCs may show a differential affinity for phosphatidylinositol (PI) and phosphatidylinositol 4,5-bisphosphate (PIP_2). A preferential hydrolysis of PIP_2 could lead both to the activation of protein kinase C via diacylglycerol formation, and to mobilization of intracellular calcium via $Ins(1,4,5)P_3$ formation. On the other hand, preferential hydrolysis of PI could lead to a selective activation of protein kinase C. This differential hydrolysis could relate to mediation of separate responses by the two branches of the bifurcating phosphoinositide pathway and also to act to limit $Ins(1,4,5)P_3$ formation (as well as other higher inositol phosphates) by stimulating down-regulation of phosphoinositide hydrolysis via PKC activation. Neither the G-protein-mediated nor calcium-activated component of phosphoinositide hydrolysis appears to be altered in the rat cerebral cortex during aging. A decline in the activity of either of these pathways for stimulating phosphoinositide hydrolysis in brain does not appear to contribute to decrements in synaptic function that are known to occur during aging.

REFERENCES

1. LITOSCH, I. & J. N. FAIN. 1986. Regulation of phosphoinositide breakdown by guanine nucleotides. Life Sci. **39:** 187–194.

2. CHAUNG, D.-M. 1989. Neutrotransmitter receptors and phosphoinositide turnover. Ann. Rev. Pharmacol. Toxicol. **29:** In press.

3. KENDALL, D. A. & S. R. NAHORSKI. 1984. Inositol phospholipid hydrolysis in rat cerebral cortical slices: II. Calcium requirement. J. Neurochem. **42:** 1388–1394.

4. KENDALL, D. A. & S. R. NAHORSKI. 1985. Dihydropyridine calcium channel activators and antagonist influence depolarization-evoked inositol phospholipid hydrolysis in brain. Eur. J. Pharmacol. **115:** 31–36.

5. GONZALES, R. A., L. J. CHANDLER & F. T. CREWS. 1989. Ethanol interactions with phosphoinositide linked signal transduction. In Focus on Biochemistry and Physiology of Substance Abuse. Vol. 1. CRC Press. In press.

6. GONZALES, R. A. & F. T. CREWS. 1985. Guanine nucleotides stimulate production of inositol trisphosphate in rat cortical membranes. Biochem. J. **232:** 799–804.

7. MORROW, A. L. & I. CREESE. 1986. Characterization of α_1-adrenergic receptor subtypes in rat brain: a reevaluation of [3H]WB 4101 and [3H]prazosin binding. Mol. Pharmacol. **29:** 321–330.

8. HAN, C., P. W. ABEL & K. MINNEMAN. 1987. α_1-Adrenergic subtypes linked to different mechanisms for increasing intracellular Ca^{2+} in smooth muscle. Nature **329:** 333–335.

9. MINNEMAN, K. P., C. HAN & P. W. ABEL. 1988. Comparison of α_1-adrenergic receptor subtypes distinguished by chlorethylclonidine and WB 4101. Mol. Pharmacol. **33:** 509–514.

10. GONZALES, R. A. & F. T. CREWS. 1988. Differential regulation of phosphoinositide phosphodiesterase activity in brain membranes by guanine nucleotides and calcium. J. Neurochem. **50:** 1522–1528.

11. MICHAELIS, M. L., K. JOKE & T. E. KITOS. 1984. Age-dependent alterations in synaptic membrane systems for Ca^{2+} regulation. Mech. Ageing Dev. **25:** 215–225.

12. LESLIE, S. W., L. J. CHANDLER, E. M. BARR & R. P. FARRAR. 1985. Reduced calcium uptake by rat brain mitochondria and synaptosomes in response to aging. Brain Res. **343:** 137–145.

13. VITORICA, J. & J. Satrustegui. 1986. Involvement of mitochondria in the age-dependent decrease in calcium uptake of rat brain synaptosomes. Brain Res. **378:** 36–48.

14. MEYER, E. M., F. T. CREWS, D. H. OTERO & K. LARSEN. 1986. Aging decreases the sensitivity of rat cortical synaptosomes to calcium ionophore-induced acetylcholine release. J. Neurochem. **47:** 1244–1246.

15. ISHIKAWA, Y., M. V. GEE, I. S. AMBUDKAR, L. BODNER, B. J. BAUM & G. S. ROTH. 1988. Age-related impairment in rat parotid cell α_1-adrenergic actions at the level of inositol trisphosphate responsiveness. Biochim. Biophys. Acta **968:** 203–210.

16. NACHSHEN, D. A. 1985. Regulation of cytosolic calcium concentration in presynaptic nerve endings isolated from rat brain. J. Physiol. **363:** 87–101.

17. DANIELL, L. C., E. P. BRASS & R. A. HARRIS. 1987. Effects of ethanol on intracellular ionized calcium concentrations in synaptosomes and hepatocytes. Mol. Pharmacol. **32:** 831–837.

18. CHANDLER, L. J. & S. W. LESLIE. 1989. Protein kinase C activation enhances K^+-stimulated endogenous dopamine release in the absence of an increase in cytosolic calcium. J. Neurochem. **52:** 1905–1912.

19. BENAVIDES, J., Y. CLAUSTRE & B. SCATTON. 1988. L-Glutamate increases internal free calcium levels in synaptoneurosomes from immature rat brain via quisqualate receptors. J. Neurosci. **8:** 3607–3615.

20. HOLLINGWORTH, E. B., E. T. MCNEAL, J. L. BURTON, R. J. WILLIAMS, J. W. DALY & C. R. CREVELING. 1985. Biochemical characterization of a filtered synaptoneurosome preparation from guinea pig cerebral cortex: cyclic adenosine $3':5''$-monophosphate generating systems, receptors and enzymes. J. Neurosci. **5:** 2240–2253.

21. CREWS, F. T., R. A. GONZALES, R. PALOVICIK, M. I. PHILLIPS, C. THEISS & M. RAIZADA. 1986. Changes in receptor stimulated phosphoinositide hydrolysis in brain during ethanol administration, aging, and other pathological conditions. Psychopharmacol. Bull. **22:** 775–780.

22. GONZALES, R. A., P. H. GREGER, S. P. BAKER, N. I. GANZ, C. BOLDEN, M. K. RAIZADA & F. T. CREWS. 1987. Phorbol esters inhibit agonist-stimulated phosphoinositide hydrolysis in neuronal primary membranes. Dev. Brain Res. **37:** 59–66.

23. MULDOON, L. L., G. A. JAMIESON & M. L. VILLEREAL. 1987. Calcium mobilization in permeabilized fibroblast: effects of inositol trisphosphate, orthovanadate, mitogens, phorbol ester and guanosine triphosphate. J. Cell. Physiol. **130:** 29–36.

24. OSUGI, T., T. IMAIZUMI, A. MISHUSHIMA, S. UCHIDA & H. YOSHIDA. 1987. Phorbol ester inhibits bradykinin-stimulated inositol trisphosphate formation and calcium mobilization in neuroblastoma x glioma hybrid NG108-15 cells. J. Pharmacol. Exp. Ther. **240:** 617–622.

25. JOPE, R. S., T. L. CASEBOLT & G. V. W. JOHNSON. 1987. Modulation of carbachol-stimulated inositol phospholipid hydrolysis in rat cerebral cortex. Neurochem. Res. **12:** 693–700.

26. HOMMA, Y., J. IMAKI, N. OSAMU & Y. TAKENAWA. 1988. Isolation and characterization of two different forms of inositol phospholipid-specific phospholipase C from rat brain. J. Biol. Chem. **263:** 6592–6598.

Do Activity-Dependent Changes in Expression of Regulatory Proteins Play a Role in the Progression of Central Nervous System Neural Degeneration?

MARY B. KENNEDY

Division of Biology 216-76
California Institute of Technology
Pasadena, California 91125

Cellular responses to the environment are initiated through membrane receptors that are activated by hormones or other extracellular agents. Activated receptors communicate to molecular machinery within the cell either by changing the flux of ions across the membrane or by altering the synthesis or degradation of second messengers such as cAMP, calcium ion, or diacylglycerol. One major mechanism through which these messengers control cell function is the activation of protein kinases.[1] It seems increasingly clear that coordinated control over many cell processes at once is often exerted by networks of protein kinases, perhaps tailored specifically for each cell type. We have only begun to glimpse how these networks might be organized. Most current research is still aimed at the important goal of identifying and thoroughly characterizing the key regulatory proteins (often protein kinases) that mediate regulation by each second messenger in each specialized cell.[2]

Type II Ca^{2+}/Calmodulin-Dependent Protein Kinase

Two major families of protein kinases respond to an increased concentration of calcium ion. One is the well-known family of C-kinases that are activated in a synergistic way by diacylglycerol and/or calcium ion.[3] The other is a more diverse family of Ca^{2+}/calmodulin-dependent protein kinases.[4] This position paper points out a potential role for one of the latter protein kinases, type II CaM kinase, in the progression of central nervous system neural degeneration in general, and in Alzheimer's disease in particular. I shall first summarize the rather complicated molecular structure of this protein kinase, and then describe several characteristics that make it of interest to those studying mechanisms of signalling in the nervous system. Finally I shall outline a hypothetical scenario by which changes in expression of this kinase could contribute to the progression of neural degeneration.

Molecular Structure

Type II CaM kinase was originally discovered and characterized in the brain,[5,6] where it comprises about 1% of total protein.[7] It is a large oligomer composed of varying proportions of at least two structurally similar classes of subunits, α (50 kDal) and β (60 kDal). Both subunits bind calmodulin and are catalytic. It is not known whether there are important functional differences between them. The two subunits are expressed in different proportions in different brain regions, apparently because genes encoding them are controlled differently and independently in different parts of the brain. The average subunit composition in a brain region reflects the levels of expression of messages encoding each subunit.[8] In the forebrain, the α subunit predominates (approx. 9α: 3β) whereas in the cerebellum, the α subunit predominates (approx. 2α: 8β).[9,10] When activated *in vitro*, the CaM kinase phosphorylates a relatively wide range of brain proteins including synapsin I, MAP_2, tyrosine hydroxylase, smooth muscle myosin light chain, and glycogen synthase. It is not clear which, if any, of these proteins are important substrates for the kinase in intact tissue.

Expression and Subcellular Distribution in Brain

Three characteristics of brain type II CaM kinase have attracted the interest of neuroscientists. First; its expression and distribution is not uniform across brain regions. It is much more concentrated in neurons of the hippocampus and parts of the cerebral cortex than it is in those of lower brain regions or the cerebellum.[7] Within the highly organized cerebral cortex of the macaque monkey, it is further concentrated in a subset of neurons arrayed in a layered pattern within the cortex that is not characteristic of any other known functional subset of visual neurons.[11] This precise pattern of expression in sets of neurons suggests that the kinase has a specialized function in those neurons.

Second; within the forebrain, type II CaM kinase has an interesting subcellular distribution. About half of it is soluble and apparently distributed rather evenly throughout the neuronal cytosol. The other half sediments with the particulate fraction. A substantial, but as yet undetermined, fraction of the particulate kinase is apparently associated with postsynaptic densities.[12] The kinase is the single most abundant protein in "purified" postsynaptic densities constituting 20 to 35% of their total protein.[10] Estimates of the effective concentration of the kinase within postsynaptic densities, based upon immunocytochemical staining of postsynaptic densities on synaptosomes, suggest that it is 5 to 10 times more concentrated in densities than in the cytosol (M. B. Kennedy, unpublished observations). It is thus concentrated at a site where it would "detect" increases in calcium concentration produced locally by activation of synaptic receptors.

Regulation by Autophosphorylation.

Third; type II CaM kinase is regulated by its own autophosphorylation in an interesting way.[12] Upon activation by calcium, it begins to phosphorylate exogenous substrates and also phosphorylates itself at specific residues, including threonine residues immediately adjacent to the calmodulin-binding domain ($Thr_{\alpha\text{-}286,\beta\text{-}287}$).[13,14] When these residues are autophosphorylated, the kinase switches to a new state in which it has substantial catalytic activity even in the absence of calcium.[13] The switch to the partially calcium-independent state is highly cooperative; phosphorylation of two to three of the subunits in the dodecameric holoenzyme is sufficient to confer maximum calcium-independent activity on the entire holoenzyme.[12] One consequence of this mechanism is that the action of protein phosphatases is necessary to terminate the action of the CaM kinase once sufficient autophosphorylation has taken place. This means that the CaM kinase may be able to retain information about a prior activating calcium signal until long after the concentration of calcium has fallen back to resting levels. Since many forms of synaptic plasticity require increases in cytosolic calcium concentration, there has been much speculation that the action of the CaM kinase might be necessary to produce some forms of relatively long-lasting changes in synaptic efficacy. One of the most interesting of these is long-term potentiation (LTP) in the hippocampus.[15] Although the type II CaM kinase has often been postulated to play a role in LTP, there is as yet no compelling experimental evidence that it does.

The Level of CaM Kinase Gene Expression May Be Controlled by Electrical Activity

In the course of studies on the distribution of type II CaM kinase in the primate visual system, Hendry and Kennedy[11] made the serendipidous observation that neurons receiving less afferent electrical excitation stained more darkly with antibodies against the α subunit of the kinase than neurons with normal input. The primary visual area (Area 17) of the cortex of the macaque monkey is arranged in 1-mm columns of neurons. Each column is driven primarily by one or the other of the two eyes. These are called ocular dominance columns. They can be visualized by selective staining for cytochrome oxidase activity in sections of brain taken from monkeys deprived of vision in one eye for several weeks. The ocular dominance columns driven by the active eye show denser staining for cytochrome oxidase than those driven by the deprived eye. Hendry[11] made the additional observation that ocular dominance columns previously driven by a deprived eye showed denser staining with antibody against CaM kinase than columns driven by an active eye. Thus, neuronal activity appears to suppress staining for CaM kinase. This observation has since been replicated with animals that have had one eye silenced by tetrodotoxin injection rather than by lid suture or enucleation (Hendry, personal communication). Thus, it seems clear that the crucial variable that influences density of staining for the CaM kinase is loss of afferent electrical activity. Complete denervation is not necessary.

There are several possible mechanistic explanations for this observation and we are presently designing experiments to differentiate among them. The most likely reason for the increase in staining in "deprived-eye columns" is an increase in the amount of kinase protein in the neurons and/or terminals. This may come about through increased expression of one or more genes for the kinase subunits. If this mechanism were correct, we would predict that neuronal activity would downregulate kinase gene expression. Another possibility is that the increase in staining may be caused by a decrease in the rate of proteolysis of kinase molecules leading to a gradual build-up of newly synthesized kinase. It is also formally possible that the increased staining results from a change in disposition of the kinase that makes the epitope more accessible to antibody molecules. This seems the least likely explanation. (The antibody used for staining does not distinguish between phosphorylated and nonphosphorylated kinase molecules.[7])

Changes in Gene Expression of Regulatory Molecules May Contribute to Progressive Neuronal Degeneration

An increase in kinase protein of sufficient magnitude to be observed by immunocytochemistry may have a significant effect on the regulatory "balance" within the neuron. A higher cytosolic concentration of the kinase will cause it to compete more effectively for limiting amounts of Ca^{2+}/calmodulin with other calmodulin-binding proteins. This will bias the response to increases in calcium ion in favor of activation of the CaM kinase, and away from activation of other calmodulin-activated pathways. It is possible that at a certain critical point this will begin to produce pathological responses leading to neuronal dysfunction or death.

Type II CaM kinase phosphorylates cytoskeletal proteins in vitro, including (in decreasing order of catalytic efficiency) MAP_2,[5] tubulin,[6] and tau (Kosik and Kennedy, unpublished observations). Some of these proteins are contained in Alzheimer's neurofibrillary tangles (see REF. 16). The functional significance of phosphorylation of cytoskeletal proteins by CaM kinase *in vivo* is not known. However, it is conceivable that an accumulation of inappropriately phosphorylated sites on these molecules, caused by the gradually increasing concentration of CaM kinase, could lead to their inappropriate association and precipitation as neurofibrillary tangles. FIGURE 1 illustrates this hypothetical scenario.

This speculative hypothesis can be seen as a specific case of a more general hypothetical model. If expression of one regulatory enzyme, CaM kinase, is altered in neurons by long-lasting changes in electrical activity of their afferent inputs, the expression of others (kinases, phosphatases, G-proteins) may also be altered. Thus, subtle but persistent changes in the overall activity in a brain pathway may set in motion slowly progressing, but eventually catastrophic changes in the regulatory machinery of target neurons of the

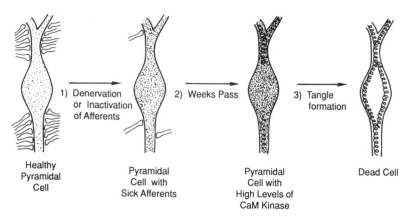

| Healthy | 1) Denervation or Inactivation of Afferents | Pyramidal | 2) Weeks Pass | Pyramidal | 3) Tangle formation | Dead Cell |

Healthy
Pyramidal
Cell

Pyramidal
Cell with
Sick Afferents

Pyramidal
Cell with
High Levels of
CaM Kinase

Dead Cell

FIGURE 1. Hypothetical scheme by which activity-dependent changes in expression of type II CaM kinase could contribute to progression of neuronal degeneration. A loss of afferent input to central nervous system pyramidal neuron may cause slow accumulation of unusually high levels of type II CaM kinase within the neuron.[11] This, in turn, may lead to inappropriate phosphorylation of cytoskeletal proteins, ultimately contributing to formation of neurofibrillary tangles. There is as yet no direct evidence for the occurrence of the second step *in vivo*. This hypothetical scheme may be considered a specific example of a more general model in which long-lasting changes in activity in a brain pathway cause altered expression of regulatory proteins. Shifting levels of such proteins gradually cause neuronal cytoplasmic regulatory networks to be thrown out of balance resulting in pathological neuronal dysfunction or death.

pathway. Such changes could be propagated gradually throughout large areas of the brain, leading to functional deficits like those observed in Alzheimer's disease and other forms of dementia.

REFERENCES

1. HUNTER, T. 1988. A thousand and one protein kinases. Cell **50:** 823–829.
2. EDELMAN, A. M., D. K. BLUMENTHAL & E. G. KREBS. 1987. Protein serine/threonine kinases. Annu. Rev. Biochem. **56:** 567–613.
3. NISHIZUKA, Y. 1984. Turnover of inositol phospholipids and signal transduction. Science **225:** 1365–1373.
4. KENNEDY, M. B., M. K. BENNETT, N. E. ERONDU & S. G. MILLER. 1987. Calcium/cal-

modulin-dependent protein kinases. *In* Calcium and Cell Function. 7th edit. W. Y. Cheung, Ed.: 61–107.

5. BENNETT, M. K., N. E. ERONDU & M. B. KENNEDY. 1983. Purification and characterization of a calmodulin-dependent protein kinase that is highly concentrated in brain. J. Biol. Chem. **258:** 12735–12744.

6. GOLDENRING, J. R., B. GONZALEZ, J. S. MCGUIRE, JR. & R. J. DELORENZO. 1983. Purification and characterization of a calmodulin-dependent kinase from rat brain cytosol able to phosphorylate tubulin and microtubule-associated proteins. J. Biol. Chem. **258:** 12632–12640.

7. ERONDU, N. E. & M. B. KENNEDY. 1985. Regional distribution of Type II Ca^{2+}/calmodulin-dependent protein kinase in rat brain. J. Neurosci. **5:** 3270–3277.

8. BULLEIT, R. B., M. K. BENNETT, S. S. MOLLOY, J. B. HURLEY & M. B. KENNEDY. 1988. Conserved and variable regions in the subunits of brain Type II Ca^{2+}/calmodulin-dependent protein kinase. Neuron **1:** 63–72.

9. MCGUINNESS, T. L., Y. LAI & P. GREENGARD. 1985. Ca^{2+}/calmodulin-dependent protein kinase II—Isozymic forms from rat forebrain and cerebellum. J. Biol. Chem. **260:** 1696–1704.

10. MILLER, S. G. & M. B. KENNEDY. 1985. Distinct forebrain and cerebellar isozymes of Type II Ca^{2+}/calmodulin-dependent protein kinase associate differently with the postsynaptic density fraction. J. Biol. Chem. **260:** 9039–9046.

11. HENDRY, S. H. C. & M. B. KENNEDY. 1986. Immunoreactivity for a calmodulin-dependent protein kinase is selectively increased in macaque striate cortex after monocular deprivation. Proc. Natl. Acad. Sci. USA **83:** 1536–1540.

12. MILLER, S. G. & M. B. KENNEDY. 1986. Regulation of brain Type II Ca^{2+}/calmodulin-dependent protein kinase by autophosphorylation: A Ca^{2+}-triggered molecular switch. Cell **44:** 861–870.

13. MILLER, S. G., B. L. PATTON & M. B. KENNEDY. 1988. Sequences of autophosphorylation sites in neuronal Type II CaM kinase that control Ca^{2+}-independent activity. Neuron **1:** 593–604.

14. SCHWORER, C. M., R. J. COLBRAN, J. R. KEEFER & T. R. SODERLING. 1988. Ca^{2+}/calmodulin-dependent protein kinase II—Identification of a regulatory autophosphorylation site adjacent to the inhibitory and calmodulin-binding domains. J. Biol. Chem. **263:** 13486–13489.

15. NICOLL, R. A., J. A. KAUER & R. C. MALENKA. 1988. The current excitement in long-term potentiation. Neuron **1:** 97–103.

16. ALDERTON, B. H. 1988. Paired helical filaments and the cytoskeleton. Nature **335:** 497–498.

Calcium-Activated Neutral Proteinases as Regulators of Cellular Function

Implications for Alzheimer's Disease Pathogenesis[a]

RALPH A. NIXON

[b]Ralph Lowell Laboratories
Mclean Hospital
115 Mill Street
Belmont, Massachusetts 02178
and
Department of Psychiatry
Harvard Medical School
Boston, Massachusetts 02115

INTRODUCTION

It has been known for many decades that proteases participate in the normal turnover of proteins and provide a level of quality control for the cellular protein pool by eliminating abnormal and potentially toxic proteins.[1-3] Only recently, however, has it become recognized that certain proteases may also be regulatory enzymes that can serve creative as well as catabolic roles.[2,4,5] One such example is the family of calcium-activated neutral proteinases (CANP, calpains). Investigated initially in relation to intracellular protein turnover, CANPs are now also being implicated as regulatory influences on various metabolic cascades[6-11] and in modulating the activities of other regulatory enzymes.[12-16] In this paper, I shall develop further the notion of CANPs as regulatory agents, which act at the crossroads of transmembrane and intracellular signaling pathways and control important dynamic interactions between the cell membrane and the cytoskeleton. The possible influences of CANPs at this level of cellular organization provide an interesting perspective from which to consider the pathogenesis of Alzheimer's disease, a condition in which proteins of the cytoskeleton, membrane cytoskeleton, and membrane are abnormally processed and accumulate in the brain.[17-19] The hypothesis is advanced here that down-regulation (inhibition) of neuronal calcium-mediated proteolysis in Alzheimer's disease is one critical and early step in the development of neurofibrillary degeneration and altered membrane cytoskeleton dynamics, which leads to membrane injury, accumulation of abnormal proteins, and synapse dysfunction.

Given the still meager understanding of the CANP system in neurons, the notion that CANPs are importantly involved in Alzheimer's disease is speculative and derives experimental support largely from studies of nonneural tissues. The principal objectives in presenting such a working hypothesis, therefore, are to draw attention to possibly early molecular events in Alzheimer's disease pathogenesis that might precede the hallmark histopathological abnormalities of the disease and, secondly, to provide one example of a metabolic process which, if impaired, could constitute a unifying basis for the slow,

[a]Studies from this laboratory were supported by Public Health Service Grants AG05604, AG08278, and AG05134.
[b]Address for correspondence.

progressive development of diverse structural and functional abnormalities within neurons. In this discussion, the assumption of a single primary defect in AD pathogenesis is not stressed; instead, down-regulation of the CANP system is viewed here as a necessary early step in pathogenesis and a common outcome of varied metabolic challenges resulting from the events of normal aging, environmental insults, genetic vulnerabilities, etc. The notion that the emergence of Alzheimer's disease depends on a confluence of etiologic factors, each of which by itself may be insufficient to induce pathology, was proposed earlier.[20]

CANP Structure-Function Relationships

CANPs are intracellular cysteine proteinases, each of which is composed of an 80-kDa catalytic subunit and a smaller 30-kDa subunit.[11,21] In addition to having a typical cysteine proteinase domain, the catalytic subunit contains within its structure four potential calcium-binding sites.[22,23] The small subunit has a similar calcium-binding domain as well as a glycine-rich hydrophobic region which may relate to the ability of CANPs to bind to membranes.[22,23] Since the small subunit of CANP appears to be identical in different members of the CANP family,[24,25] differences among the several types of CANPs are attributed to variations in the amino acid sequences of certain domains of the large subunit.[26] Two major classes of CANPs encoded by independent genes correspond to two types of purified CANPs displaying different calcium requirements for activation. One type, referred to as μCANP or CANP I, is optimally active at calcium concentrations in the micromolar range (5–100 μM); the other, designated mCANP or CANP II, requires in its purified state nearly millimolar calcium levels for full activity.[7,24,27,28] The latter CANP type probably represents a latent form of the protease that, under certain conditions, is converted to a "derived" μCANP active at micromolar or submicromolar calcium levels by intramolecular autolysis[29,30] which may be facilitated under various conditions (see below).

The physiological role of mCANPs was earlier questioned because of the high calcium levels required to activate the isolated enzyme; however, the sensitivities of both μCANPs and mCANPs to calcium are now known to be increased by autolysis,[11,29,31] interaction with membrane lipids,[32–34] and other factors[35,36] including properties of the protein substrates.[37–41] These many influences on the calcium requirement and catalytic properties of the enzyme emphasize the importance of studying CANPs within their normal cellular milieu to appreciate fully their complex behavior and functions in vivo.[42,43]

CANPs as Regulators of Cytoskeleton and Membrane Dynamics

An intimate relationship between CANPs and the cytoskeleton has been suspected from the particular susceptibility of intermediate filament proteins,[42,44,45] microtubule-associated proteins,[46–48] fodrin (spectrin)[39,40,49,50] and other high-molecular-weight cytoskeleton-associated proteins[38] to purified CANPs in vitro. More recently, the processing of these proteins by CANP has been observed in vivo in relation to morphologic alterations of the cytoskeleton and plasma membrane, which are part of the cell's response to calcium and second messenger stimulation. For instance, normal tissues contain proteolytic fragments of neurofilaments, fodrin, and MAPs[50–52] similar to those generated by CANP in vitro, suggesting ongoing physiological turnover. Stimulating hippocampal neurons in vivo by administering excitatory amino acids systemically or intraventricularly increases the levels of proteolytic fragments of fodrin and MAP 2, which resemble cleavage products generated by purified CANPs.[52] These events are believed to be linked, in turn, to changes in cytoskeleton organization within synapses, which are essential to the induction of long-term potentiation (LTP)[53] (Lynch et al., these proceedings). By comparison, the administration of leupeptin, a peptide inhibitor of thiol proteases including CANP, raises the levels of neurofilaments[54] (Shea and Nixon, unpublished data) and fodrin[50] in some neural systems.

The influence of CANPs on microtubule dynamics is especially evident from studies of mitosis.[55] Cell division, an event associated with calcium sequestering phenomena and calcium transients, requires the disassembly of microtubules composing the mitotic spindle and a major reorganization of the associated cytoskeletal proteins.[56–58] As cells progress from interphase through different stages of mitosis, CANP II relocates from the plasma membrane to specific intracellular sites with a characteristic timing.[55] The association of CANP activity with specific mitotic stages is further suggested by the ability of CANP II, when microinjected intracellularly, to promote metaphase in cells at interphase and to induce the precocious onset of anaphase in cells in late metaphase.[55]

The possible influences of CANPs extend to the skeletal network under the plasmalemma which regulate many interactions between the membrane and the intracellular milieu.[59,60] The membrane-associated proteins α-actinin, P235,[61] spectrin,[39,40,49,50] band 3, band 4.1,[42] as well as certain transmembrane protein components of platelets and erythrocytes[63,64] are excellent CANP substrates in vitro. The physiological significance of interactions between CANPs and certain of these proteins may be reflected in the fusion of membranes during myogenesis[65,66] or erythrocyte differentiation.[67] As myoblasts begin to fuse in response to a calcium influx,[68] CANP relocates from a dispersed distribution to the membrane concomitantly with a presumed hydrolysis of certain membrane-associated proteins.[65,66] Similar events are proposed for the fusion of the nuclear membrane with the plasma membrane in differentiating erythrocytes.[67]

CANPs as Regulators of Protein Kinases

In addition to influencing cytoskeletal protein behavior directly, CANPs may amplify their effects by activating other regulatory enzymes.[12–16] Most notably, purified CANPs display the curious ability of cleaving certain second-messenger-regulated protein kinases and phosphatases specifically between the catalytic and regulatory domains of the polypeptide, thereby producing constitutively active forms of these enzymes. The number of enzymes that are so activated, including protein kinase C (PKC),[13] calcium- and calmodulin-dependent protein kinase (CaM kinase) (Kwiatkowski and King, unpublished work, cited in REF. 107), and calcium- and calmodulin-dependent phosphatase (calcineurin),[13] suggests that this pattern of limited proteolysis is more than coincidental and may represent an important means of regulating calcium-activated protein kinases and phosphatases.

Some possible functional correlates of this modulating effect of CANPs on PKC have been recently suggested (FIG. 1). These studies are particularly interesting from the perspective of Alzheimer pathogenesis because they suggest how the effects of abnormally lowered CANP activity might be amplified to cause widespread alterations of the cytoskeleton and the plasma membrane. In human platelets or neutrophils, certain functional states can be induced by treating the intact cells with the tumor-promoting phorbol ester, phorbol 12-myristate 13-acetate (PMA). By stabilizing the association of PKC with membranes, PMA promotes the cleavage of PKC by membrane-bound CANP, thereby releasing the catalytic domain of PKC into the cytoplasm.[9,69,73,118,119] This soluble, modified 50-kDa protein kinase, designated PKM, no longer requires calcium or phospholipid for activity[70] and appears to have a different specificity than PKC toward some, but not all, substrates.[71] Although the effect of PKM may be short-lived under normal conditions,[15] its ability to diffuse into the cytosol may allow the modified kinase to act transiently on a new group of substrates. PKM, for example, phosphorylates a group of cytosolic proteins involved in triggering mitosis[72] and could mediate the effects of CANP on mitosis discussed above. Thus, by converting PKC to PKM, CANP could possibly trigger shifts in phosphorylation events from the membrane to locations within the cytoplasmic cytoskeletons, which are essential for the morphologic response to calcium stimulation.

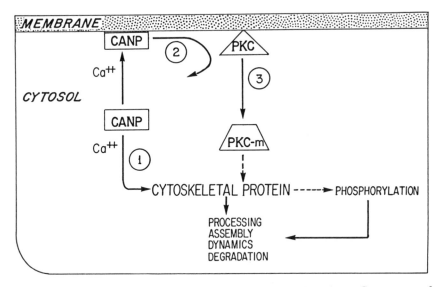

FIGURE 1. Actions of CANP at the level of the cytoskeleton and membrane: Consequences of down-regulation (see text for references). In response to calcium, CANP increasingly binds to the plasma membrane which facilitates autolytic activation of the protease. Membrane-bound CANP is able to cleave the regulatory domains of PKC and release a constitutively active, but likely unstable, form of the kinase, PKM (PKC-m in Figure). The direct action of CANPs on cytoskeletal proteins and indirect actions through the activation of PKM and possibly other protein kinases and phosphatases (see text) regulate the posttranslational fates of proteins comprising the cytoplasmic and membrane-associated skeletons. Down-regulation of CANP at *site 1* is expected to promote accumulation of specific cytoskeletal proteins and promote the disorganization of the cytoskeleton. Altering the activities of cytoplasmic and cytoskeleton-associated protein kinases (*step 3*) promotes similar effects. Reduced proteolysis of protein kinase C (*step 2*) favors increased association of PKC with the membrane and hyperphosphorylation of membrane and membrane-associated proteins, resulting in membrane injury. Further details are discussed in the text.

Further indication that interactions between CANP and PKC are physiological and that defective interactions may result in pathological consequences is provided by studies on CANP inhibition in intact human neutrophils. Phorbol esters elicit a characteristic sequence of responses in neutrophils, which includes the enhanced production of superoxide anion and hydrogen peroxide attributed to PKC-mediated phosphorylation of membrane proteins,[73,74] the release of a membrane-associated serine proteinase,[75] and the exocytosis of granule contents.[76,77] The latter process requires the phosphorylation of certain cytoskeletal proteins apparently by PKM.[71] When CANP activity is inhibited by inducing the cells to take up a monoclonal anti-CANP antibody, the conversion of PKC to PKM decreases, superoxide anion production increases due to a prolonged association of PKC with the membrane, and exocytosis is inhibited due to diminished cytoskeletal protein phosphorylation by PKM.[9] The limited proteolysis of specific cytoskeletal proteins by CANP, which normally accompanies exocytosis, is also blocked.[71]

CANP Down-Regulation and Alzheimer's Disease

The consequences of CANP down-regulation described above, if played out in neurons, could be relevant to Alzheimer's disease pathogenesis (FIG. 1). Brain tissue contains relatively high levels of mCANP,[78] and this relative tissue localization is even more significant when it is considered that CANPs are preferentially concentrated in certain neurons including neocortical and hippocampal pyramidal neurons[79] (Cataldo and Nixon, unpublished data). Within neurons, mCANP and μCANP are predominantly localized immunocytochemically to neuronal perikarya and proximal dendrites.[79] This localization is interesting in view of the fact that these are sites of intense and complex processing and segregation of cytoskeletal polypeptides destined for dendrites and axons[80,81] and the perikaryon is the principal site of neurofibrillary tangle (NFT) accumulation in many neurodegenerative diseases.

Neurofibrillary Tangles

In Alzheimer's disease and other neurofibrillary disorders, the proteins that become disarrayed and accumulate in neuronal perikarya and proximal dendrites are cytoskeletal elements that are likely substrates of CANP *in vivo* (see above). Where studied, the synthesis of some of the proteins that accumulate (*e.g.*, neurofilaments) has been shown to remain unchanged or even to decrease in the diseased neurons,[82,83] indicating that proteolysis is not keeping pace with the generation of these normal or abnormal cytoskeletal proteins within the perikaryon. Abnormally down-regulated CANP activity by one of many possible routes (see below) might be expected to promote the accumulation of cytoskeletal proteins, particularly if the proteins are modified in a way that renders them relatively more resistant to the protease. For example, aluminum, a toxin that induces an accumulation of neurofilaments in neuronal perikarya[83] inhibits the digestion of neurofilament proteins by purified CANP.[84,85] This effect is due in part to direct actions of aluminum on the enzyme[86] but, more importantly, the binding of aluminum to neurofilament proteins causes them to form high molecular weight complexes, which are more resistant to CANPs.

The polypeptides that accumulate as neurofibrillary tangles and paired helical filaments in Alzheimer's disease are also abnormally phosphorylated.[17,87] Paired helical filaments (PHF) are composed largely of a modified form of the tau protein that resembles a hyperphosphorylated fetal variant.[88,89] The appearance of PHFs is associated with or preceded by disorganization of the cytoskeleton, which may represent a more widespread abnormality involving microtubules[90] and membrane-associated proteins.[91] Since tau, MAP 2, neurofilament proteins, as well as many other membrane cytoskeleton proteins, are normally regulated by multiple protein kinases, including protein kinase C,[89,92] down-regulation of CANP would be expected to give rise to various abnormalities of phosphorylation reflecting in part the abnormal persistent association of PKC with the membrane and the reduced phosphorylation of putative PKM substrates. If CANPs are important in regulating the activity of other kinases and phosphatases either by proteolytic activation or degradation,[93] then other phosphorylation events involving the cytoskeleton could also be enhanced or diminished by down-regulating CANPs. Given the increasing evidence that many events of cytoskeleton assembly and dynamics are regulated by phosphorylation,[94-96] it is possible that certain of the abnormalities of protein phosphorylation in Alzheimer's disease are causally related to the disorganization. Others are likely to occur as a result of the disruption. In either case, changes in phosphorylation state can, in addition, alter the susceptibility of these proteins to proteolysis.[37,97] For example, once formed, PHF are known to be highly resistant to purified CANPs (Nixon, unpublished results).

Amyloid

Additional prominent structural alterations in Alzheimer's disease stem from the focal accumulation of amyloid, a 40-residue polypeptide (A4 protein or β-amyloid protein), which is derived from the larger 70-kDa protein, the A4 precursor protein (APP).[99-101] APP is believed to be an integral membrane protein, and the observation that A4 comprises a portion of the putative transmembrane region of the polypeptide has suggested that proteolytic cleavages to generate A4 occur in a part of the molecule normally embedded within the membrane.[102,103] Although the cellular origin of APP-derived polypeptides in senile plaques is controversial, the protein is expressed in neurons.[104,105] If its ability to serve as a substrate for protein kinase C *in vitro*[106] is physiologically relevant, CANP down-regulation in Alzheimer's disease would be expected to promote hyperphosphorylation of APP and other membrane proteins due to the persistent association of PKC with the membrane and thereby alter membrane protein organization and processing.

Long-Term Potentiation

Impaired synaptic plasticity has been considered a possible molecular basis for the striking memory impairment that characterizes patients with Alzheimer's disease. Among the mechanisms critical to this synaptic function is long-term potentiation (LTP), which involves persisting functional or structural modifications of the presynaptic terminal, postsynaptic terminal, or both. Proposed postsynaptic events in LTP include a receptor-mediated activation of CANPs.[107] Subsequent cleavage of cytoskeletal proteins, including fodrin and MAP 2, gives rise to persisting structural alterations of the postsynaptic membrane that facilitate synaptic transmission (Lynch, these proceedings). Presynaptic models of LTP emphasize PKC activation by calcium and/or diacylglycerol or activation of calcium/calmodulin-dependent protein kinase (CaM kinase) followed by an as-yet-unknown sequence of phosphorylation steps.[108] In both cases, persistent protein kinase activity underlies the maintenance of LTP.[107] If, as believed, a constitutively activated protein kinase is required for the presynaptic events of LTP,[108] CANPs, which can effect such modulation *in vitro* and are present in synaptic terminals,[109] could be critical to the process. Given the possibility that CANPs may be involved either directly or indirectly in LTP, CANP down-regulation should be detrimental to this manifestation of synaptic plasticity as suggested by recent experimental data.[110]

CANP Regulation—A Vulnerable Target

The multitude and complexity of mechanisms controlling CANP activity reinforce the notion of the CANP system as a regulatory influence.[10,11,71] The factors modulating CANP activity (TABLE 1), in turn, are affected by processes involved in cellular aging and by other suspected influences on the development of Alzheimer's disease. Many of these favor the down-regulation of CANP-mediated proteolysis. For example, a substantial case has been made for diminished calcium availability within cells as a function of age[111] and for abnormalities of calcium homeostasis in Alzheimer's disease.[112] In this regard, cultured skin fibroblasts from Alzheimer's patients display abnormalities of calcium homeostasis, including decreased free cytosolic calcium levels[113] (Shelanski, these proceedings) which would favor lowered levels of CANP activity. An associated impairment of cytoskeletal dynamics is suggested by the reduced ability of these cells to spread in culture, an abnormality which is abrogated by treating the cells with calcium ionophore.[113] CANP activity is also regulated by the balance between protease levels and the level of an endogenous protein inhibitor, calpastatin.[10,11,114,115] The recent finding of multiple unrelated polypeptides with specific CANP inhibitory activity in brain[116] suggests that this mode of CANP regulation may be more complex than previously recognized and potentially subject to interference from more than one direction. Changes in the expression of multiple genes for CANP, calpastatin, and CANP activators[35,36] may also be effected by various routes.[114,115] In this regard, levels of activatible mCANP and μCANP fall sharply during early postnatal development in specific brain regions while calpastatin levels remain relatively unaffected.[50,117] These are findings consistent with a development- or maturation-related

TABLE 1. Potential Mechanisms of CANP Regulation[a]

Calcium flux, cytosolic free calcium[7,24,27,28]
Autolytic activation[11,29,31]
Endogenous protein inhibitors, calpastatins[10,11]
Activating factors:
  Proteins[35,36,40]
  Phospholipids[32,33,118,119]
Posttranslational modification:
  Protein substrates[37,38,97]
  CANP phosphorylation[120]
  Calmodulin interaction[39,40,41]
Gene expression (CANP, calpastatins, CANP activators)

[a]See text for further discussion.

down-regulation of the CANP system. As a final example, aluminum, a neurotoxin implicated in experimental neurofibrillary degeneration and as a possible etiologic factor in the Parkinson-dementia complex of Guam,[83] inhibits the degradation of neurofilament proteins by CANPs in vitro.[85] The fact that aluminum accumulates in neurofibrillary tangle-bearing neurons in Alzheimer's disease and other neurofibrillary degenerations,[83] even as an epiphenomenon of the primary disease process, favors the down-regulation of CANP-mediated proteolysis. Cumulative inhibitory influences on CANP from multiple sources, such as the ones mentioned above, might be necessary to down-regulate CANP activity enough to cause pathological changes. Furthermore, certain of these same influences (e.g., age-related calcium alterations, toxins, etc.) are likely to alter processes (e.g., PKC and other calcium-dependent kinases) that are modulated in part by CANPs. Interference at various additional metabolic sites downstream from the action of CANPs would be expected to accentuate or mitigate the pathological consequences of CANP down-regulation.

SUMMARY

Evidence is emerging that calcium-activated neutral proteinases (CANPs) not only participate in intracellular protein turnover but help to regulate the functional reorganization of cytoskeletal proteins in response to calcium and second-messenger stimulation. The high concentration of CANPs in certain neurons has suggested prominent roles for this proteolytic system in neuronal and synaptic function. In addition to acting directly on specific constituents of the cytoplasmic and membrane-associated cytoskeletal networks, CANP may amplify its effects by modulating the activities of protein kinase C and possibly other kinases and phosphatases by limited proteolysis. Given its suspected involvement at the cytoskeleton-membrane interface, calcium-mediated proteolysis is an example of a metabolic process which, if impaired, could provide a unifying basis for the slow progressive development of diverse structural and functional abnormalities within neurons. The multiplicity of mechanisms regulating its activity makes the CANP system a vulnerable target for disruption from various sources. A working hypothesis is advanced that down-regulation (inhibition) of neuronal calcium-mediated proteolysis in Alzheimer's disease is one critical and early step in the development of neurofibrillary degeneration and altered membrane cytoskeleton dynamics, which leads to membrane injury, accumulation of abnormal proteins, and synaptic dysfunction.

ACKNOWLEDGMENTS

I wish to thank Drs. Alfred Pope and Ram Sihag for helpful discussions and Mrs. Johanne Khan for assistance in preparing and typing the manuscript.

REFERENCES

1. POPE, A. & R. A. NIXON. 1984. Neurochem. Res. **7:** 291–323.
2. BOND, J. S. & P. E. BUTLER. 1987. Annu. Rev. Biochem. **56:** 333–364.
3. GOLDBERG, A. L. & A. C. ST. JOHN. 1976. Annu. Rev. Biochem. **45:** 747–801.
4. HOLZER, H. & P. C. HEINRICH. 1980. Annu. Rev. Biochem. **49:** 63–91.
5. LOH, Y. P., M. J. BROWNSTEIN & H. GAINER. 1984. Annu. Rev. Neurosci. **7:** 189–222.
6. ZIMMERMAN, U.-J. P. & W. W. SCHLAEPFER. 1984. Prog. Neurobiol. **23:** 63–78.
7. MURACHI, T. 1983. Trends Biochem. Sci. **8:** 167–169.
8. PONTREMOLI, S. & E. MELLONI. 1986. Annu. Rev. Biochem. **55:** 455–481.
9. PONTREMOLI, S., E. MELLONI, G. DAMIANI, F. SALAMINO, B. SPARATORE, M. MICHETTI & B. L. HORECKER. 1988. **263:** 1915–1919.
10. MELLGREN, R. L. 1987. FASEB J. **1:** 110–115.
11. SUZUKI, K., S. IMAJOH, Y. EMORI, H. KAWASAKI, Y. MINAMI & S. OHNO. 1987. FEBS Lett. **220:** 271–727.
12. TOGARI, A., S. ICHIKAWA & T. NAGATUS. 1986. Biochem. Biophys. Res. Commun. **134:** 749–754.
13. INOUE, M., A. KISHIMOTO, Y. TAKAI & Y. NISHIZUKA. 1977. J. Biol. Chem. **252:** 7610–7616.
14. TALLANT, E. A., L. M. BRUMLEY & R. W. WALLACE. 1988. Biochemistry **27:** 2205–2211.
15. MURRAY, A. W., A. FOURNIER & S. J. HARDY. 1987. Trends Biochem. Sci. **12:** 53–54.
16. LYNCH, C. J., G. E. SOBO & J. H. EXTON. 1986. Biochim. Biophys. Acta **855:** 110–120.
17. ANDERSON, B. H. 1988. Nature **335:** 497–498.
18. MILLER, C., M. HAUGH, J. KAHN & B. ANDERTON. 1986. Trends Neurosci. **9:** 76–81.

19. VINTERS, H. V., B. L. MILLER & W. W. PARDRIDGE. 1988. Ann. Int. Med. **109:** 41–54.
20. KHACHATURIAN, Z. S. 1987. Neurobiol. Aging **8:** 345–346.
21. MELLGREN, R. L., R. D. LANE & S. S. KAKER. 1987. Biochem. Biophys. Res. Commun. **142:** 1025–1031.
22. SAKIHAMA, T., H. KAKIDANI, K. ZENITA, N. YUMOTO, T. KIKUCHI, T. SASAKI, R. KANNAGI, S. NAKANISHI, M. OHMORI, K. TAKIO, K. TITANI, T. MURACHI. 1985. Proc. Natl. Acad. Sci. USA **82:** 6075–6079.
23. OHNO, S., Y. EMORI, S. IMAJOH, H. KAWASAKI, M. KISARAGI & K. SUZUKI. 1984. Nature **312:** 566–570.
24. WHEELOCK, M. J. 1982. J. Biol. Chem. **257:** 12471–12474.
25. KITAHARA, A., T. SASAKI, T. KIKUCHI, N. YUMOTO, N. HOSHIMURA, M. HATANAKA & T. MURACHI. 1984. J. Biochem. (Tokyo) **95:** 1759–1766.
26. KAWASAKI, H., S. IMAJOH, S. KAWASHIMA, H. HAYASHI & K. SUZUKI 1986. J. Biochem. **99:** 1525–1532.
27. MELLGREN, R. L. 1980. FEBS Lett. **109:** 129–133.
28. KISHIMOTO, A., N. KAJIKAWA, H. TABUCHI, M. SHIOTA & Y. NISHIZUKA. 1981. J. Biochem. (Tokyo) **90:** 889–892.
29. DAYTON, W. R. 1982. Biochim. Biophys. Acta **709:** 166–172.
30. MALIK, M. N., M. D. FENKO, A. M. SHEIKH, R. J. KASCSAK, M. S. TONNA-DEMASI & H. M. WISNIEWSKI. 1987. Biochim. Biophys. Acta **916:** 135–144.
31. DEMARTINO, G. N., C. A. HUFF & D. E. CROALL. 1986. J. Biol. Chem. **261:** 12047–12052.
32. COOLICAN, S. A. & D. R. HATHAWAY. 1984. J. Biol. Chem. **259:** 11627–11630.
33. IMAJOH, S., H. KAWASAKI & K. SUZUKI. 1986. J. Biochem. **99:** 1281–1284.
34. MELLONI, E., S. PONTREMOLI, M. MICHETTI, O. SACCO, B. SPARATORE & F. SALAMINO. 1985. Proc. Natl. Acad. Sci. USA **82:** 6435–6439.
35. TAKEYAMA Y., H. NAKANISHI, Y. URATSUJI, A. KISHIMOTO & Y. NISHIZUKA. 1986. FEBS Lett. **194:** 110–114.
36. PONTREMOLI, S., E. MELLONI, M. MICHETTI, F. SALAMINO, B. SPARATORE & B. L. HORECKER. 1988. Proc. Natl. Acad. Sci. USA **85:** 1740–1743.
37. GOLDSTEIN, M. E., N. H. STERNBERGER & L. A. STERNBERGER. 1987. J. Neuroimmunol. **14:** 149–169.
38. NIXON, R. A., R. QUACKENBUSH & A. VITTO. 1986. J. Neurosci. **6:** 1252–1263.
39. HARRIS, A. S. & J. S. MORROW. 1988. J. Neurosci. **8:** 2640–2651.
40. HARRIS, A. S. & J. S. MORROW. 1988. J. Cell Biol. **107:** 25a.
41. SEUBERT, P., M. BAUDRY, S. DUDEK & G. LYNCH. 1987. Synapse **1:** 20–24.
42. NIXON, R. A. 1983. *In* Neurofilaments. C. A. Marotta, Ed.: 117–154. University of Minnesota Press. Minneapolis, MN.
43. NIXON, R. A., S. E. LEWIS & C. A. MAROTTA. 1987. J. Neurosci. **7:** 1145–1158.
44. SCHLAEPFER, W. W. 1983. *In* Neurofilaments. C. A. Marotta, Ed.: 57–85. University of Minnesota Press. Minneapolis, MN.
45. VORGIAS, C. E. & P. TRAUB. 1986. Biosci. Rep. **6:** 57–64.
46. SANDOVAL, I. V. & K. WEBER. 1978. Eur. J. Biochem. **92:** 463–470.
47. BILLGER, M., M. WALLIN & J.-O. KARLSSON. 1988. Cell Calcium **9:** 33–44.
48. YOSHIMURA, N., I. TSUKAHARA & T. MURACHI. 1984. Biochem. J. **223:** 47–51.
49. SIMAN, R., M. BAUDRY & G. LYNCH. 1984. Proc. Natl. Acad. Sci. USA **81:** 3572–3576.
50. SIMAN, R., M. AHDOOT & G. LYNCH. 1987. J. Neurosci. **7:** 55–64.
51. SCHLAEPFER, W. W., C. LEE, V. M-Y. LEE & U.-J. P. ZIMMERMAN. 1985. J. Neurochem. **44:** 502–509.
52. SIMAN, R. & J. C. NOSZEK. 1988. Neuron **1:** 279–287.
53. SEUBERT, P., G. IVY, J. LARSON, J. LEE, K. SHAHI, M. BAUDRY & G. LYNCH. 1988. Brain Res. **459:** 226–232.
54. ROOTS, B. I. 1983. Science **221:** 971–972.
55. SCHOLLMEYER, J. E. 1988. Science **240:** 911–913.
56. KEITH, C. H., F. R. MAXFIELD & M. L. SHELANSKI. 1985. Proc. Natl. Acad. Sci. USA **82:** 800–804.
57. POENIE, M., J. ALDERTON, R. STEINHARDT & R. TSIEN. 1986. Science **233:** 886–889.
58. KIEHART, D. P. 1981. J. Cell Biol. **88:** 604–617.

59. ISHIKAWA, H. 1988. Arch. Histol. Cytol. **51:** 127–145.
60. BENNETT, V. 1985. Annu. Rev. Biochem. **54:** 273–304.
61. BECKERLE, M. C., T. O'HALLORAN & K. BURRIDGE. 1986. J. Cell Biochem. **30:** 259–270.
62. PANT, H. C., M. VIRMANI & P. E. GALLANT. 1983. Biochem. Biophys. Res. Commun. **117:** 372–377.
63. MCGOWAN, E. B., K. T. YEO & T. C. DETWILER. 1983. Arch. Biochem. Biophys. **227:** 287–301.
64. YAMAMOTO, K., G. KOSAKI, K. SUZUKI, K. TANOUE & H. YAMAZAKI. 1986. Thromb. Res. **43:** 41–55.
65. SCHOLLMEYER, J. E. 1986. Exp. Cell Res. **162:** 411–422.
66. SCHOLLMEYER, J. E. 1986. Exp. Cell Res. **163:** 413–422.
67. THOMAS, P., A. R. LIMBRICK & D. ALLAN. 1983. Biochim. Biophys. Acta **730:** 351–358.
68. PAUW, P. G. & J. D. DAVID. 1979. Dev. Biol. **70:** 27–38.
69. TAPLEY, P. M. & A. W. MURRAY. 1984. Biochem. Biophys. Res. Commun. **118:** 835–841.
70. NISHIZUKA, Y. 1986. Science **233:** 305–312.
71. PONTREMOLI, S., E. MELLONI, M. MICHETTI, B. SPARATORE, F. SALAMINO, O. SACCO & B. L. HORECKER. 1987. Proc. Natl. Acad. Sci. USA **84:** 3604–3608.
72. GUY, G. R., J. GORDON, L. WALKER, R. H. MICHELL & G. BROWN. 1986. Biochem. Biophys. Res. Commun. **135:** 146–153.
73. PONTREMOLI, S., E. MELLONI, M. MICHETTI, O. SACCO, F. SALAMINO, B. SPARATORE & B. L. HORECKER. 1986. J. Biol. Chem. **261:** 8309–8313.
74. BADWEY, J. A. & M. L. KARNOVSKY. 1986. *In* Current Topics in Cellular Regulation, Vol. 28: 123–208. Academic Press. New York, NY.
75. PONTREMOLI, S., E. MELLONI, M. MICHETTI, O. SACCO, B. SPARATORE, F. SALAMINO, G. DAMIANI & B. L. HORECKER. 1986. Proc. Natl. Acad. Sci. USA **83:** 1685–1689.
76. POZZAN, T., D. P. LEW, C. B. WOLLHEIM & R. Y. TSIEN. 1983. Science **221:** 1413–1415.
77. WHITE, J. R., C.-K. HUANG, J. M. HILL, JR., P. H. NACCACHE, E. L. BECKER & R. I. SHAAFI. 1984. J. Biol. Chem. **259:** 8605–8611.
78. KAWASHIMA, S., M. HAYASHI, Y. SAITO, Y. KASAI & K. IMAHORI. 1988. Biochim. Biophys. Acta **965:** 130–135.
79. HAMAKUBO, T., R. KANNAGI, T. MURACHI & A. MATUS. 1986. J. Neurosci. **6:** 3103–3111.
80. BLACK, M. M. & V.-M. LEE. 1988. J. Neurosci. **8:** 3296–3305.
81. MATUS, A., G. HUBER & R. BERNHARDT. 1983. Cold Spring Harbor Symp. Quant. Biol. **48:** 775–782.
82. MUMA, N. A., J. C. TRONCOSO, P. N. HOFFMAN, E. H. KOO & D. L. PRICE. 1988. Mol. Brain Res. **3:** 115–122.
83. CRAPPER-MCLACHLAN, D. R. 1986. Neurobiol. Aging **7:** 525–532.
84. NIXON, R. A., K. B. LOGVINENKO & J. CLARKE. 1986. Soc. Neurosci. Abstr. **12:** 99.
85. NIXON, R. A., J. F. CLARKE, K. B. LOGVINENKO & M. K. H. TAN. Submitted.
86. VITTO, A. & R. A. NIXON. 1986. J. Neurochem. **47:** 1039–1051.
87. GRUNDKE-IQBAL, I., K. IQBAL, Y.-C. TUNG, W. QUINLAN, H. M. WISNIEWSKI & L. I. BINDER. 1986. Proc. Natl. Acad. Sci. USA **83:** 4913–4917.
88. KOSIK, K. S., C. L. JOACHIM & D. J. SELKOE. 1986. Proc. Natl. Acad. Sci. USA **83:** 4044–4048.
89. BAUDIER, J. & R. D. COLE. 1987. J. Biol. Chem. **262:** 17577–17583.
90. PAULA-BARBOSA, M., M. A. TAVARES & A. CADETO-LEITE. 1987. Brain Res. **417:** 139–142.
91. GRAY, E. G., M. PAULA-BARBOSA & A. ROHER. 1987. Neuropathol. Appl. Neurobiol. **13:** 91–110.
92. SIHAG, R. K., A. Y. JENG & R. A. NIXON. 1988. FEBS Lett. **233:** 181–185.
93. WASTERLAIN, C. G., R. NIXON, D. B. FARBER & J. BRONSTEIN. 1988. Trans. Soc. Neurochem. **19:** 146.
94. NIXON, R. A. & S. E. LEWIS. 1987. *In* Molecular Mechanisms of Neuronal Responsiveness. Y. H. Ehrlich, R. H. Lenox, E. Kornecki & W. O. Berry, Eds.: 167–186. Plenum. New York, NY.
95. LEWIS, S. E. & R. A. NIXON. 1988. J. Cell Biol. **107:** 2689–2702.

96. NESTLER, E. J. & P. GREENGARD. 1984. Protein Phosphorylation in the Nervous System.
     John Wiley & Sons. New York, NY.
97. PANT, H. 1988. Biochem. J. 256: 665–668.
98. MASTERS, C. L., G. SIMMS, N. A. WEINMAN, G. MULTHAUP, B. L. MCDONALD & K.
     BEYREUTHER. 1985. Proc. Natl. Acad. Sci. USA 82: 4245–4249.
99. SELKOE, D. J., C. R. ABRAHAM, M. B. PODLISNY & L. K. DUFFY. 1986. J. Neurochem.
     46: 1820–1834.
100. GLENNER, G. G. & C. W. WONG. 1984. Biochem. Biophys. Res. Commun. 120: 885–890.
101. KANG, J., H. G. LEMAIRE, A. UNTERBECK, et al. 1987. Nature 325: 733–736.
102. DYRKS, T., A. WEIDEMANN, G. MULTHAUP, J. M. SALBAUM, H. G. LEMAIRE, J. KANG, B.
     MÜLLER-HILL, C. L. MASTERS & K. BEYREUTHER. 1988. EMBO J. 7: 949–957.
103. HIGGINS, C. A., D. A. LEWIS, S. BAHMANYAR, D. GOLDGABER. 1988. Proc. Natl. Acad. Sci.
     USA 85: 1297–1301.
104. COHEN, M. L., T. E. GOLDE, M. F. USIAK, L. H. YOUNKIN & S. G. YOUNKIN. 1988. Proc.
     Natl. Acad. Sci. USA 85: 1227–1231.
105. GANDY, S., A. J. CZERNIK & P. GREENGARD. 1988. Proc. Natl. Acad. Sci. USA 85:
     6218–6221.
106. LYNCH, G. & M. BAUDRY. 1984. Science 224: 1057–1063.
107. KENNEDY, M. B. 1988. Nature 335: 770–772.
108. MALINOW, R., D. V. MADISON & R. W. TSIEN. 1988. Nature 335: 830–824.
109. GUSTAVSSON, S. & J.-O. KARLSSON. 1986. Neurosci. Lett. 63: 221–224.
110. STAUBLI, U., J. LARSON, O. THIBAULT, M. BAUDRY & G. LYNCH. 1988. Brain Res. 444:
     153–158.
111. GIBSON, G. E. & C. PETERSON. 1987. Neurobiol. Aging 8: 329–343.
112. KHACHATURIAN, Z. S. 1984. In Handbook of Studies on Psychiatry and Old Age. D. W. Kay
     & G. D. Burrows, Eds.: 7–30. Elsevier. New York, NY.
113. PETERSON, C., R. R. RATAN, M. L. SHELANSKI & J. E. GOLDMAN. 1986. Proc. Natl. Acad.
     Sci. USA 83: 7999–8001.
114. PONTREMOLI, S., E. MELLONI, F. SALAMINO, B. SPARATORE, M. MICHETTI, O. SACCO & G.
     BIANCHI. 1986. Biochem. Biophys. Res. Commun. 139: 341–347.
115. PONTREMOLI, S., E. MELLONI, F. SALAMINO, B. SPARATORE, P. VIOTTI, M. MICHETTI, L.
     DUZZI & G. BIANCHI. 1986. Biochem. Biophys. Res. Commun. 138: 1370–1375.
116. MINKOVITZ, J. B., N. FLEMING & R. A. NIXON. J. Neurochem. 48(Suppl.): S149C.
117. SIMONSON, L., M. BAUDRY, R. SIMAN & G. LYNCH. 1985. Brain Res. 327: 153–159.
118. MELLONI, E., S. PONTREMOLI, F. SALAMINO, B. SPARATORE, M. MICHETTI, O. SACCO & B.
     L. HORECKER. 1986. J. Biol. Chem. 261: 11437–11439.
119. MELLONI, E., S. PONTREMOLI, M. MICHETTI, O. SACCO, B. SPARATORE & B. L. HORECKER.
     1986. J. Biol. Chem. 261: 4101–4105.
120. ZIMMERMAN, U.-J. P. & W. W. SCHLAEPFER. 1985. Biochem. Biophys. Res. Commun.
     129: 804–811.

On the Mechanism whereby Phosphorylation Modulates Protein Folding

Relevance to Protein Tangles and Plaques of Alzheimer's Disease[a]

DAN W. URRY,[b] D. K. CHANG, AND K. U. PRASAD

Laboratory of Molecular Biophysics
University of Alabama at Birmingham
P.O. Box 300
UAB Station
Birmingham, Alabama 35294

INTRODUCTION

In the macromolecular tangles of Alzheimer's disease there are abnormally phosphorylated proteins. These phosphorylated proteins are components of the intraneuronal paired helical filaments and extraneuronal amyloid deposits characteristic of Alzheimer's disease,[1,2] and they include the microtubule-associated tau proteins[3,4] and the amyloid precursor proteins.[5] In the present report, the sensitivity of protein folding and assembly to phosphorylation is demonstrated in an elastic protein element model. The mechanism whereby phosphorylation modulates protein structure and function is not the often discussed electrostatic interaction, but rather is what appears to be a more efficient mechanism for the chemical modulation of protein structure. This can be demonstrated in synthetic model contractile proteins which are capable of chemomechanical transduction[6] in which the amount of chemical energy to perform a given amount of mechanical work can be compared with that of mechanochemical systems based on the electrostatic repulsion mechanism.[7]

The Model Protein System and Inverse Temperature Transitions

In the single protein chain that comprises the elastic protein, elastin, the longest sequence between cross-links in porcine[8] and bovine[9] elastins is a 57-residue sequence that is comprised entirely of a repeating pentamer. This may be written poly(VPGVG) where V = Val, P = Pro, and G = Gly or (L-Val^1-L-Pro^2-Gly^{3}L-Val^4-Gly^5)$_n$. This sequential polypeptide is soluble in water in all proportions below 25°C, but on raising the temperature, it has been shown to self-assemble into fibers comprised of fibrils which are, in turn, comprised of parallel aligned twisted filaments.[10,11] The individual components of the twisted filaments are dynamic helical structures containing repeating β-turns which are secondary structural features involving Val^1 C-O $\cdots$ H - N Val^4 hydrogen bonds. This

[a]Supported in part by National Institutes of Health Grants HL 29578 and GM 26898.
[b]To whom correspondence should be addressed.

dynamic helical structure of recurring β-turns is called a β-spiral. While the hydrogen bond is present below 25°C, what clearly happens on raising the temperature is the wrapping up of largely extended relatively disordered polypeptide chains into regular helical structures which then associate to form filaments, fibrils, and fibers. This ordering process on raising the temperature is called an inverse temperature transition.

An inverse temperature transition exhibited by a polypeptide in an aqueous system is necessarily understood in terms of both the structure of the polypeptide and of the structure of the interfacial water. In particular, hydration of a hydrophobic group such as a Val residue is exothermic (*i.e.*, a negative enthalpy, ΔH) resulting in the formation of a special net-like water structure called clathrate water.[12,13] The clathrate water is in the form of structures such as pentagonal dodecahedra,[14] the formation of which results in a substantial negative entropy change (ΔS). Thus, the Gibb's free energy of hydration, ΔG = ΔH - T ΔS, is comprised of two oppositely signed terms, and the low solubility of hydrophobic residues in water indicates that the two terms are of similar magnitude. The folding of a protein wherein the hydrophobic side chains are turned inward away from the solvent is the result of changes in the relative magnitudes of these two terms. Protein folding occurs when ΔG for the process is negative. As dissolution of hydrophobic groups is exothermic (has a negative ΔH), the folding of a protein with the turning inward of hydrophobic groups contributes a positive ΔH component to the Gibb's free energy. Thus,

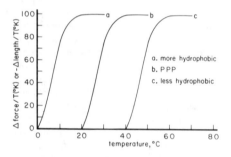

FIGURE 1. Schematic representation of the temperature-normalized decrease in length (contraction) or increase in force as a function of temperature for model elastomeric proteins of different mean hydrophobicities (polarites). (From #.[16a] Reprinted by permission from the *International Journal of Quantum Chemistry Symposium.*)

a. more hydrophobic
b. PPP
c. less hydrophobic

on raising the temperature, the negative TΔS term for folding increases in magnitude such that at a transition temperature the destructuring of the clathrate water (a relatively large positive ΔS) can allow for the smaller negative ΔS due to the increase in order of the protein part of the system as the result of folding. In a delicately balanced system in which the transition temperature is near physiological temperatures, small changes in the relative magnitude of the two terms can result in the folding or unfolding or in the association or dissociation of protein.

Chemical Modulation of the Inverse Temperature Transition

The temperature at which the inverse temperature transition occurs for poly(VPGVG) is near 30°C.[11] If the Val residue in position one is replaced by an Ile residue which represents the addition of a CH_2 moiety, to give poly(IPGVG), there is no difference in the conformations either below the transition or above the transition; however, the transition for poly(IPGVG) occurs near 10°C.[15] The more hydrophobic the polypeptide is, the lower the temperature is at which the transition occurs. Also, it has been shown that the less hydrophobic polypeptide, poly(VPGG), exhibits its transition near 50°C[16] (see FIG. 1).

Now if, instead of synthesizing a new polymer to achieve a change in hydrophobicity, an amino acid residue is included that can exist in both a more hydrophobic (less polar) and a less hydrophobic (more polar) state, then by simply chemically changing from one state to the other, folding and unfolding should be achieved at a constant intermediate temperature.

A model protein system capable of this is poly[4(VPGVG),(VPGEG)], where the E = Glu residue can exist in the more hydrophobic (less polar) COOH state and in the less hydrophobic (more polar) COO$^-$ state.[6] In phosphate buffered saline (0.15 N NaCl, 0.01 M phosphate) at pH 2.1 where all Glu side chains are as COOH, the inverse temperature transition occurs near 20°C, whereas at pH 7 where all of the side chains are the carboxylate anion, COO$^-$, the inverse temperature transition occurs near 70°C (see FIG. 2). In fact, when the temperature is maintained at 37°C, simply changing the pH from 4.3 to 3.3 will cause the folding transition to occur. Differential scanning calorimetric studies have shown that the endothermal heat of the transition, interpreted to be primarily the heat required to destructure the clathrate water, is reduced to one-fourth on raising the pH from 2.5 to 4.0.[17] This pH change converts one COOH to a COO$^-$ per 100 residues. Therefore, it appears that the chemical work required to remove a proton from one carboxyl moiety per 100 residues results, from the thermodynamic point of view, in the destructuring of three-fourths of the clathrate water. What had been achieved thermally (the melting of clathrate water to achieve the folding of an inverse temperature transition) was achieved chemically by formation of a more polar species, that is, the COO$^-$ resulted in the destructuring of clathrate-like water which at a temperature above that of the inverse temperature transition had been the thermodynamic driving force for folding. This is phenomenologically called chemical modulation of an inverse temperature transition. Somewhat mechanistically, it can be called chemical modulation of the hydrophobic effect. At a more mechanistic level the chemical effect is considered to arise from hydration mediated apolar–polar repulsion free energies. This interaction free energy is apparent in the effective COO$^-$ destructuring of clathrate water, or, when hydrophobicity of the sequential polypeptide is increased, in the raising of the pKa of a carboxyl moiety.[18] The latter is seen on comparing the pKa value for poly[4(VPGVG),(VPGEG)] which is 4.4 with that of the more hydrophobic poly[4(IPGVG),(IPGEG)] which is 5.4.

Relative Efficacy of Aqueous Electrostatic and Apolar–Polar Interaction Mechanisms

Comparison of the relative efficiencies of the electrostatic repulsion mechanism in water and of the hydration-mediated apolar–polar interaction free energy mechanism for driving protein structural changes can be made with mechanochemical coupling systems. The model system for the electrostatic repulsion mechanism can be the polyelectrolyte, polymethacrylic acid [-(CH$_3$)C(COOH)-CH$_2$-]$_n$.[7,19] This polyelectrolyte can be formed into an elastomeric matrix by cross-linking occasional COOH side chains, *e.g.*, one in every 100 repeats. A polymethacrylic acid chain becomes essentially fully extended when 50% ionized, *i.e.*, 50% COO$^-$. Collapsing of the extended chain occurs on lowering the pH until say a state of 10% COO$^-$ is reached with an associated decrease in length to about one-half the extended length. The cross-linked matrix in the process can pick up weights that are several thousand times the dry weight of the matrix. Once γ-irradiation cross-linked, poly[4(VPGVG),(VPGEG)], on going through its inverse temperature transition, contracts to less than one-half its extended length and can also pick up weights that are several thousand times its dry weight.[6] However, in the process only four COO$^-$ moieties are converted to COOH, whereas, for the polymethacrylic acid system it appears to be necessary to convert ten times as many COO$^-$ moieties to COOH to perform a

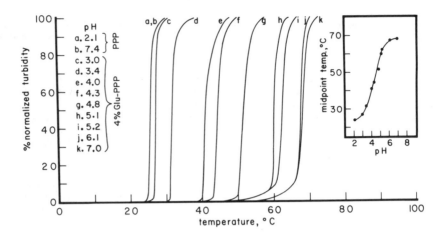

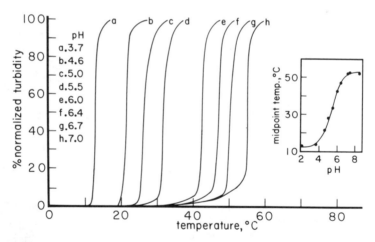

FIGURE 2. Temperature dependence of the inverse temperature transitions, followed as the development of turbidity (aggregation), as a function of pH (*above*) for the less hydrophobic 4% Glu-polypentapeptide, poly[4(VPGVG),(VPGEG)], and (*below*) for the more hydrophobic poly-[4(IPGVG),(IPGEG)]. (From #.[16a] Reprinted by permission from the *International Journal of Quantum Chemistry Symposium.*)

similar amount of work. Therefore, the mechanism of chemical modulation of the hydrophobic effect, *i.e.*, the hydration-mediated apolar–polar repulsion free energy mechanism, appears to be an order of magnitude more efficient in converting chemical energy into mechanical work than the electrostatic repulsion mechanism. If this is the case, then it is to be expected that living organisms would have evolved to utilize the most efficient mechanism possible for its chemical composition. Accordingly, it is not unreasonable to expect that modulation of protein structure by phosphorylation/dephosphorylation would involve in large measure the mechanism of chemical modulation of the hydrophobic effect.

Modulation of the Inverse Temperature Transition by Phosphorylation

Phosphorylation of proteins most commonly occurs on Ser, Thr, or Tyr residues. As a model system to determine the effect of phosphorylation on the temperature of the inverse temperature transition, poly[9(VPGVG),(VPGTG)], where T = Thr, was synthesized. This is also called 2% Thr-PPP. It has two threonine residues per 100 residues of polypentapeptide. Efforts at chemically preparing the phosphothreonine Thr(OP) = T(OP), have been carried out in order to prepare poly[9(VPGVG),(VPGT(OP)G)], also referred to as 2% Thr(OP)-PPP. In the procedure for attempted synthesis of the VPGT(OP)G-containing polypentapeptide, the phosphoryl moiety was protected by two phenyl moieties; the pentamer was prepared, mixed in the ratio of 9(VPGVG) to one pentamer containing the protected phosphothreonine, and polymerized. This product is indicated as 2% Thr(OPϕ_2)-PPP. After this polypentapeptide was prepared, efforts were then made to deprotect the phosphate group to have the phosphorylated threonine in the sequential polypeptide. From proton magnetic resonance data (see FIG. 3), it appears that the removal of the blocking phenyl moieties was essentially complete. In the deblocking process, however, partial replacement of the phenyl moieties by a NCH$_2$CH$_2$CH$_2$CH$_3$ moiety appeared to occur. The latter proton resonances are indicated in FIGURE 3C where

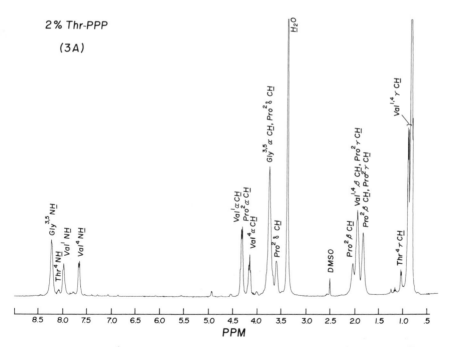

FIGURE 3. 400 MHz ^{1}H NMR spectra for (A) 2% Thr-PPP, (B) 2% Thr(OPϕ_2)-PPP, and (C) 2% Thr(OPX)-PPP in DMSO-d$_6$ solutions at room temperature. The symbol ϕ stands for the phenyl moiety and X stands for the NHCH$_2$CH$_2$CH$_2$CH$_3$ grouping.

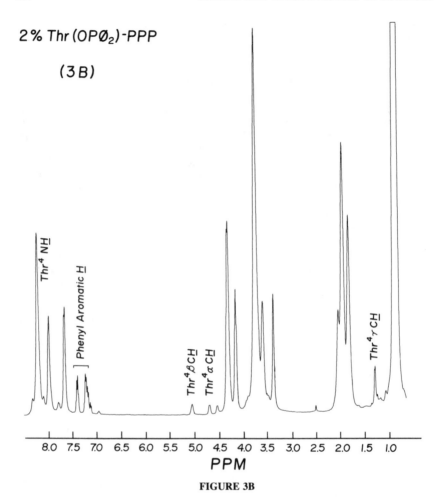

FIGURE 3B

there appear to be more than one such substitution per phosphate group. The phosphorus NMR spectra are given in FIGURE 4 for 2% Thr(OP$_{\phi 2}$)-PPP and for the product of the deblocking effort 2% Thr(OPX)-PPP. As may be seen in FIGURE 4B, the product from the deprotection effort gives three P-31 NMR resonances. If the lowest field resonance were the totally deblocked phosphate, this would mean that about 0.8% Thr(OP)-PPP has been prepared, that is, less than one phosphothreonine per 100 residues.

The temperature profiles for aggregation, which is a means of determining the temperature of the inverse temperature transition, are given in FIGURE 5 for 2% Thr-PPP, curve a, 2% Thr(OPϕ_2)-PPP, curve b, and 2% Thr(OPX)-PPP, curve c. Considering the NCH$_2$CH$_2$CH$_2$CH$_3$ substitutions as replacements of the very hydrophobic phenyl moieties, the effect of this decrease in hydrophobicity is to raise the temperature of the transition from 32°C to 52°C. Obviously, if the very polar, highly nonhydrophobic phosphate moiety had been produced, the transition temperature would have been raised even higher.

From these results, it is clear that phosphorylation of the folded aggregated polypentapeptide fibrillar state would result in disassembly and an unfolding. By analogy to the COOH to COO⁻ conversion, the phosphorylation would function to destructure the clathrate water that forms around exposed hydrophobic groups. The thermodynamic driving force which keeps the elastic protein element folded is eliminated and the polypeptide unfolds.

Relevance to Alzheimer's Disease

It has been demonstrated that the proteins of the macromolecular tangles of Alzheimer's disease are hyperphosphorylated.[3,4] This has been significantly noted by Pettegrew and co-workers who have reported most relevant phosphorus-31 nuclear magnetic reso-

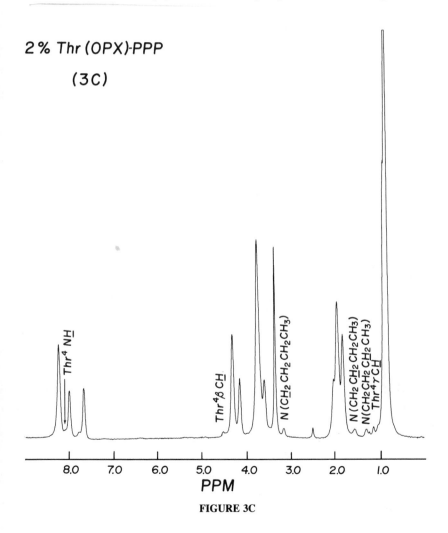

FIGURE 3C

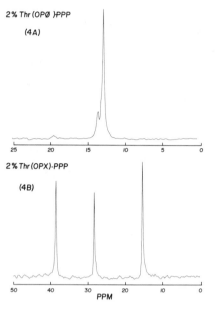

FIGURE 4. 40 MHz ^{31}P NMR spectra for **(A)** 2% Thr (OPϕ_2)-PPP and **(B)** 2% Thr(DPX)-PPP in DMSO-d$_6$ solutions at room temperature, where ϕ stands for the phenyl moiety and X stands for the NHCH$_2$CH$_2$CH$_2$CH$_3$ grouping.

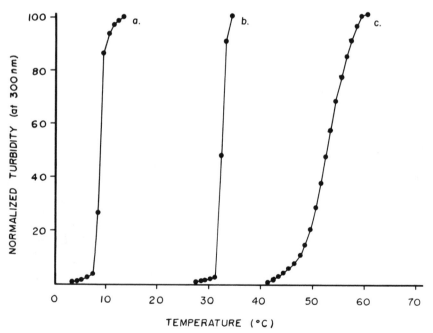

FIGURE 5. Temperature profiles for aggregation. (a) 2% Thr-PPP; (b) 2% Thr(OPϕ_2)-PPP, and (c) 2% Thr(OPX)-PPP.

nance studies elsewhere and in this volume.[20-23] From the above analyses and data, it can be expected that once inappropriately phosphorylated, the proteins of Alzheimer's disease would no longer be able to insert into a membranous site of function or to assemble into microtubules to perform their axonal transport function. Futhermore, once hyperphosphorlyated, the proteins could be expected to present binding sites for multivalent cations which would induce their aggregation into the protein tangles of Alzheimer's disease.

ACKNOWLEDGMENTS

The author wishes to thank Cynthia McIntyre and Huey-Ru Liao for technical assistance in obtaining curves for FIGURE 5.

REFERENCES

1. ANDERTON, B. H. 1987. Tangled genes and proteins. Nature **329:** 106–107.
2. GLENNER, G. G. 1987. Alzheimer's disease: Its proteins and genes. Cell **52:** 307.
3. GRUNDKE-IQBAL, I., K. IQBAL, Y.-C. TUNG, M. QUINLAN, H. M. WISNIEWSKI & L. I. BINDER. 1986. Abnormal phosphorylation of the microtubule-associated protein τ (tau) in Alzheimer cytoskeletal pathology. Proc. Natl. Acad. Sci. USA **83:** 4913–4917.
4. IQBAL, K., I. IQBAL, T. ZAIDI, P. A. MERZ, G. Y. WEN, S. S. SHAIKH, H. M. WISNIEWSKI, I. ALAFUZOFF & B. WINBLAD. 1986. Defective brain microtubule assembly in Alzheimer's disease. Lancet Aug. **23:** 421.
5. GANDY, S., A. J. CZERNIK & P. GREENGARD. 1988. Phosphorylation of Alzheimer disease amyloid precursor peptide by protein kinase C and Ca^{2+}/calmodulin-dependent protein kinase II. Proc. Natl. Acad. Sci. USA **85:** 6218.
6. URRY, D. W., B. HAYNES, H. ZHANG, R. D. HARRIS & K. U. PRASAD. 1988. Mechanochemical coupling in synthetic polypeptides by modulation of an inverse temperature transition. Proc. Natl. Acad. Sci. USA **85:** 3407.
7. KATCHALSKY, A., S. LIFSON, I. MICHAELI & M. ZWICK. 1960. *In* Size and Shape of Contractile Polymers: Conversion of Chemical and Mechanical Energy. A. Wasserman, Ed.: 1. Pergamon. New York, NY.
8. SANDERBERG, L. B., J. G. LESLIE, C. T. LEACH, V. L. TORRES, A. R. SMITH & D. W. SMITH. Elastin co-valent structure as determined by solid phase amino acid sequencing. 1985. Pathol. Biol. **33:** 266.
9. YEH, H., N. ORNSTEIN-GOLDSTEIN, Z. INDIK, P. SHEPPARD, N. ANDERSON, J. C. ROSENBLOOM, G. CICILA, K. YOON & J. ROSENBLOOM. 1987. Alternative splicing of human elastin mRNA indicated by sequence analysis of cloned genomic and complementary DNA. Collagen Relat. Res. **7:** 235.
10. URRY, D. W. 1982. Characterization of soluble peptides of elastin by physical techiques. *In* Methods in Enzymology. L. W. CUNNINGHAM & D. W. FREDERIKSEN, Eds. **82:** 673–716. Academic Press. New York, NY.
11. URRY, D. W. 1988. Entropic elastic processes in protein mechanisms. I. Elastic structure due to an inverse temperature transition and elasticity due to internal chain dynamics. J. Protein Chem. **7:** 1–34.
12. FRANK, H. S. & M. W. EVANS. 1945. Entropy of vaporization in liquids and the pictorial theory of the liquid states. J. Chem. Phys. **13:** 493–507.
13. KAUZMANN, W. 1957. Some factors in the interpretation of protein denaturation. Adv. Protein Chem. **14:** 1–63.
14. SWAMINATHAN, S., S. W. HARRISON & D. BEVERIDGE. 1978. Monte Carlo studies on the structure of a dilute aqueous solution of methane. J. Am. Chem. Soc. **100:** 5705–5711.
15. URRY, D. W., M. M. LONG, R. D. HARRIS & K. U. PRASAD. 1986. Temperature correlated force and structure development in elastomeric polypeptides: The IIe^1 analog of the polypentapeptide of elastin. Biopolymer **25:** 1939–1953.

16. URRY, D. W., R. D. HARRIS, M. M. LONG & K. U. PRASAD. 1986. Polytetrapeptide of elastin: Temperature correlated elastometric force and structure development. Int. J. Pept. Protein Res. **28:** 649–660.

16a. 1988. Int. J. Quantum Chem. Symp. **15:** 235–245.

17. URRY, D. W., C. H. LUAN, R. D. HARRIS & K. U. PRASAD. Chemical modulation of the hydrophobic effect, a driving force for protein folding and assembly. In preparation.

18. URRY, D. W., D. K. CHANG, H. ZHANG & K. U. PRASAD. 1988. pK shift of functional group in mechanochemical coupling due to hydrophobic effect. Biochem. Biophys. Res. Commun. **153:** 832–839.

19. KUHN, W., B. HARGITAY, A. KATCHALSKY & H. EISENBERG. 1950. Reversible dilation and contraction by changing the state of ionization of high-polymer acid networks. Nature (London) **165:** 514–516.

20. PETTEGREW, J. W., K. PANCHALINGAM, J. MOSSY, J. MARTINEZ, G. RAO & F. BOLLER. 1988. Correlation of phosphorus-31 magnetic resonance spectroscopy and morphologic findings in Alzheimer's disease. Arch. Neurol. **45:** 1093–1096.

21. PETTEGREW, J. W., J. MOOSSY, G. WITHERS, D. MCKEAG & K. PANCHALINGAM. 1987. [31]P nuclear magnetic resonance study of the brain in Alzheimer's disease. J. Neuropathol. Exp. Neurol. **47:** 235–248.

22. PETTEGREW, J. W., G. WITHERS, K. PANCHALINGAM & J. F. M. POST. 1987. [31]P disease. J. Neural Transm. **24:** 261–268.

23. PETTEGREW, J. W. This volume.

Glutamate Neurotoxicity, Calcium, and Zinc[a]

DENNIS W. CHOI, JOHN H. WEISS, JAE-YOUNG KOH, CHADWICK W. CHRISTINE, AND MATTHIAS C. KURTH

Department of Neurology
Stanford University Medical Center
Stanford, California 94305

A specific link between brain calcium and diseases of aging may be the neurotoxicity of the endogenous dicarboxylic amino acid, glutamate, and related compounds. Glutamate is present in high concentrations throughout the brain, and likely mediates most normal excitatory synaptic transmission. However excess exposure to glutamate can destroy neurons,[1] and may contribute to the neuronal cell loss associated with several age-related diseases of the central nervous system.[2] Glutamate neurotoxicity is most strongly implicated in acute brain injury, such as that produced by stroke, but also might be a factor in certain chronic neurodegenerative diseases, such as Huntington's disease, motor neuron disease, or Alzheimer's disease.

Here, we shall briefly review some evidence that the neuronal injury induced by glutamate is mediated by a toxic calcium influx. In the spirit of speculation encouraged by this conference, we shall end by considering another endogenous divalent cation— zinc—that may importantly modify the glutamate system and influence glutamate neurotoxicity.

Mediation of Glutamate Neurotoxicity by Calcium Influx

The mechanisms by which excess exposure to glutamate can produce neuronal cell injury are not fully understood, but recent studies have indicated an important role for extracellular calcium.[3]

Nearly 20 years ago, Olney and colleagues noted a general correspondence between the neuroexcitatory and neurotoxic potential of many compounds, and developed the ''excitotoxicity'' hypothesis:[4] that injury was a direct consequence of excessive neuroexcitation. Subsequent *in vitro* experiments supported the existence of such an excitotoxic mechanism. Cultured hippocampal neurons[5] and neurons in the isolated chick retina[6] exhibited acute toxic swelling upon 30 min exposure to excitatory amino acids. Replacement of extracellular sodium with an impermeant cation, which substantially attenuates the neuroexcitatory action of excitatory amino acids, blocked this acute toxic swelling; similar protection was also achieved by replacing extracellular chloride with an impermeant anion, a maneuver expected to secondarily decrease net sodium (and water) influx by limiting availability of an anion partner for this influx. Furthermore, the acute neuronal swelling produced by excitatory amino acid exposure could be mimicked by other depolarizing agents such as high potassium or veratridine.[5]

However the potent neuroexcitant kainate is a surprisingly weak neurotoxin on most cortical neurons[7] (with brief exposure) dependent on endogenous glutamate projections

[a]Supported by National Institutes of Health Grants NS12151 and NS26907.

for *in vivo* neurotoxicity.[8] In addition, while replacement of extracellular sodium abolished the acute neuronal swelling seen in cortical cell cultures following brief (5 min) exposure to glutamate, most neurons still developed delayed degeneration over the following day.[9] On the other hand, if extracellular calcium was removed from the exposure solution, acute neuronal swelling was actually enhanced, but neuronal loss the next day was markedly reduced. If both sodium and calcium were removed, cortical neurons exposed to glutamate showed neither acute swelling nor delayed degeneration. Furthermore, the neurotoxicity of low concentrations of glutamate could be enhanced by augmenting extracellular calcium.

These observations suggested a separation of glutamate-induced neuronal injury into two components: an acute, sodium-dependent component marked by immediate cell swelling, and a delayed, calcium-dependent component marked by delayed cell degeneration. Both components, at least *in vitro* where cell swelling is unlimited, are capable of acting in isolation to produce irreversible neuronal injury. However, the calcium-dependent component likely predominates at lower toxic exposures.

A similar dependence on extracellular calcium characterizes the toxicity of homocysteate[10] and quinolinate[11] on cortical neurons, as well as glutamate- or N-methyl-D-aspartate (NMDA)-induced toxicity in hippocampal cultures,[12] cerebellar slices,[13] and neuronal cell line cultures.[14] The basis for this dependence on extracellular calcium remains to be delineated, although mediation (or triggering) of excitatory amino acid-induced neurotoxic injury by a calcium influx seems most likely. Delayed cortical neuronal disintegration can be mimicked by the calcium ionophore A23187.[9] Several types of studies (see REF. 3 for review) have established that calcium entry accompanies excitatory amino acid action on central neurons. Furthermore, quantitative examination of glutamate-induced [45]calcium accumulation by cortical neurons shows that this index, measured at different levels of toxic exposure, exhibits a close correlation with resultant neuronal degeneration[15] (also, M. Kurth and D. Choi, unpublished observations).

The influx of extracellular calcium, together with any calcium release triggered from intracellular stores, would elevate cytosolic free calcium. Such an elevation if sustained would likely be cytotoxic for many reasons, reflecting derangements in the many intracellular processes regulated by calcium availability;[16,17] in particular the uncontrolled activation of catabolic enzymes may be most directly responsible for cell destruction. A calcium-activated neutral protease (calpain) is involved in injury induced by cholinergic agonists on muscle,[18] and calpain I activation has been specifically linked to glutamate receptors in rat hippocampus.[19] Elevated cytosolic calcium would also activate phospholipases capable of breaking down the cell membrane and liberating arachidonic acid; metabolism of arachidonic acid by oxidases leads to the production of oxygen free radicals that can trigger peroxidative degradation of lipid membranes and other destructive events.[20] Superoxide radicals may also be generated by the enzymatic action of xanthine oxidase, generated from xanthine dehydrogenase by the action of a calcium-activated protease.[21] Another important effect of elevated cytosolic calcium might be to act in concert with diacylglycerol to activate protein kinase C, leading to alterations in membrane channels that could substantially enhance cell excitability and augment calcium influx.[22,23]

In addition, an influx of calcium might act to propagate glutamate neurotoxicity in a positive-feedback fashion, by further stimulating the release of transmitter glutamate. Some reduction of neurotoxicity can be achieved when glutamate antagonists are added "late" after conclusion of glutamate exposure,[12,24] possibly due to interruption of this positive-feedback loop.

The calcium influx produced by glutamate could occur both directly through the NMDA receptor-gate ion channel,[25] and indirectly through one of several routes linked to sodium influx: 1) voltage-gated channels activated by membrane depolarization; 2) nonspecific membrane leak conductance associated with cell swelling; or 3) reverse operation of sodium-calcium membrane transport exchanger (review, REF. 3). Pharmacological studies suggest that NMDA receptor activation is essential to most glutamate neurotoxicity,[24] a conclusion consistent with the observation that NMDA receptor activation is critical to the increase in intracellular free calcium produced by excitatory amino acids on single striatal neurons,[26] and the observation that glutamate neurotoxicity can be blocked by magnesium and potentiated by glycine.[27] Further study will be needed to see if the calcium influx mediated by the NMDA receptor-activated channel accounts completely for resultant elevations in intracellular free calcium, or if it is simply a required triggering step for other important events. Other sources of intracellular free calcium—for example, release from intracellular stores—could also prove essential to glutamate neurotoxicity.

A host of different specific abnormalities, affecting excitatory synapses either presynaptically or postsynaptically, could account for gradual neuronal destruction in a given neurodegenerative condition. Presynaptically, there might be excessive neuronal activity, or excessive release of glutamate (or related agonists), or reduced glutamate uptake into nerve terminals or glia. Postsynaptically, there could be abnormalities in the number or behavior of glutamate receptors, reduced average resting potential (leading to decreased magnesium block of NMDA receptors), or increased vulnerability to calcium-mediated damage (e.g., due to reduced buffering capability). In addition, NMDA receptor-mediated toxicity could also be produced by abnormalities in modulatory factors for that receptor: for example, reduced synaptic zinc (see below), excess synaptic glycine (or a reduction in an endogenous glycine antagonist), or a reduction in an endogenous ligand for the phencyclidine (PCP) site.

Two approaches probably will be useful in the future in establishing excitotoxic involvement in a given neurodegenerative condition: 1) attempting to reproduce certain features of the disease in animals by means of excitotoxic exposure, either by exogenous administration of an excitotoxic agonist, or by manipulation of endogenous excitatory amino acid systems to produce excitotoxicity; and 2) scrutiny of the disease biology to pinpoint alterations capable of producing the appropriate excitotoxic condition. This scrutiny will likely require investigation at several different levels, including human neuropathology, cell culture, and molecular biology. If a disease gene can be isolated, functional studies of expression in cell and transgenic animal models may be needed to delineate the possibly complex alterations in synaptic function that could result in excitotoxic exposure.

Modification of Glutamate Action by Zinc

Carbonic anhydrase and many other vital metalloenzymes require zinc for normal function; thus the central nervous system, like other tissues, has a broad-based metabolic dependence on zinc. However, growing evidence now points to an additional special involvement of zinc as a mediator of central neural signaling at excitatory synapses. Large amounts of chelatable zinc are present throughout the brain, especially in the neuropil of the hippocampus and neocortex.[28-31] Ultrastructural studies have suggested that much of this free zinc is localized to a subpopulation of synaptic vesicles within excitatory boutons.[32] In the hippocampus, zinc is concentrated in the terminals of the excitatory mossy fiber projections which originate from dentate granule cells and synapse with CA3 pyramidal cells.[29] Several lines of investigation suggest that presynaptic neuronal firing produces a calcium-dependent release of this endogenous zinc; with intense activity, zinc concentrations of several hundred micromolar may be achieved in synaptic clefts.[33,34]

Following release, zinc is probably actively reaccumulated back into presynaptic terminals.[34]

By these features of presence, release, and uptake removal, free zinc has the potential to serve as a widespread signaling substance at central excitatory synapses, modulating glutamate neurotransmission. Recently, investigations of zinc action on cultured cortical neurons,[35] as well as on cultured hippocampal neurons,[36] have demonstrated a possible specific basis for this modulation: zinc selectively blocks excitation mediated by the NMDA subclass of glutamate receptors, while modestly potentiating excitation mediated by the quisqualate subclass of glutamate receptors. The co-release of zinc with glutamate would therefore be expected to reduce the proportion of NMDA channels relative to non-NMDA channels activated by glutamate. Since NMDA channels have special properties, including high calcium permeability and voltage-dependent blockade by magnesium, this reduction could serve to modify the character of the postsynaptic response, and thus dynamically regulate the nature of excitatory neurotransmission. For example, a buildup of zinc in the synaptic cleft with repetitive excitatory synaptic activity could serve as a negative feedback "brake," countering the magnesium-induced tendency of NMDA channels to amplify synaptic excitation. In addition, since zinc may interact with the open NMDA receptor-activated channel (C. Christine and D. Choi, unpublished observations) and thereby enter neurons,[37] it is possible that zinc may serve as a "transsynaptic messenger." The entry of synaptically released zinc into postsynaptic neurons could importantly modify intracellular metabolism, for example activating protein kinase C,[38] and thus lead to lasting changes in synaptic behavior such as long-term potentiation.

The effect of zinc on glutamate receptor-mediated effects is not limited to neuroexcitation, but also extends to neurotoxicity: zinc attenuates NMDA receptor-mediated neurotoxicity, while potentiating non-NMDA receptor-mediated toxicity.[39] The amount of zinc available at postsynaptic glutamate receptors may therefore be a key determinant of neurotoxic injury under conditions of excess exposure to glutamate. A paucity of zinc might contribute to selective overactivation of NMDA receptors, such as has been postulated to occur in Huntington's disease.[40,41] On the other hand, an excess of zinc might contribute to selective overactivation of non-NMDA receptors and the preferential loss of neurons vulnerable to such overactivation, for example, neurons containing somatostatin and NADPH-diaphorase.[39] Early damage to these somatostatin/NADPH-diaphorase-containing neurons has been described in Alzheimer's disease,[42-45] and has been found in cortical cultures exposed to low concentrations of the cycad toxin, beta-N-methylamino-L-alanine,[46] a toxin linked to the pathogenesis of selective motor neuronal degeneration in humans.[47]

REFERENCES

1. OLNEY, J. W. & L. G. SHARPE. 1969. Brain lesions in an infant rhesus monkey treated with monosodium glutamate. Science 166: 386–388.
2. CHOI, D. W. 1988. Glutamate neurotoxicity and diseases of the nervous system. Neuron 1: 623–634.
3. CHOI, D. W. 1988. Calcium-mediated neurotoxicity: Relationship to specific channel types and role in ischemic damage. Trends Neurosci. 11: 465–469.
4. OLNEY, J. W., R. C. COLLINS & R. S. SLOVITER. 1986. Excitotoxic mechanisms of epileptic brain damage. Adv. Neurol. 44: 857–877.
5. ROTHMAN, S. M. 1985. The neurotoxicity of excitatory amino acids is produced by passive chloride influx. J. Neurosci. 5: 1483–1489.
6. OLNEY, J. W., M. T. PRICE, L. SAMSON & J. LABRUYERE. 1986. The role of specific ions in glutamate neurotoxicity. Neurosci. Lett. 65: 65–71.
7. CHOI, D. W., M. A. MAULUCCI-GEDDE & A. R. KRIEGSTEIN. 1987. Glutamate neurotoxicity in cortical cell culture. J. Neurosci. 7: 357–368.

8. BIZIERE, K. & J. T. COYLE. 1978. Influence of cortico-striatal afferents on striatal kainic acid neurotoxicity. Neurosci. Lett. **8:** 303–310.

9. CHOI, D. W. 1987. Ionic dependence of glutamate neurotoxicity in cortical cell culture. J. Neurosci. **7:** 369–379.

10. KIM, J. P., J. Y. KOH & D. W. CHOI. 1987. L-homocysteate is a potent neurotoxin on cultured cortical neurons. Brain Res. **437:** 103–110.

11. KIM, J. P. & D. W. CHOI. 1987. Quinolinate neurotoxicity in cortical cell culture. Neuroscience **23:** 423–432.

12. ROTHMAN, S. M., J. H. THURSTON & R. E. HAUHART. 1987. Delayed neurotoxicity of excitatory amino acids in vitro. Neuroscience **22:** 471–480.

13. GARTHWAITE, G. & J. GARTHWAITE. 1986. Neurotoxicity of excitatory amino acid receptor agonists in rat cerebellar slices: Dependence on calcium concentration. Neurosci. Lett. **66:** 193–198.

14. MURPHY, T. H., A. T. MALOUF, A. SASTRE, R. L. SCHNAAR & J. T. COYLE. 1988. Calcium-dependent glutamate cytotoxicity in a neuronal cell line. Brain Res. **444:** 325–332.

15. MARCOUX, F. W., J. E. GOODRICH, A. W. PROBERT & M. A. DOMINICK. 1986. Ketamine prevents glutamate-induced calcium influx and ischemia nerve cell injury. *In* Sigma and Phencyclidine-Like Compounds as Molecular Probes in Biology. E. F. Domino & J. Kamenka, Eds.: 735–746. NPP Books. Ann Arbor, MI.

16. CHEUNG, J. Y., J. V. BONVENTRE, C. D. MALIS & A. LEAF. 1986. Calcium and ischemic injury. New Engl. J. Med. **314:** 1670–1676.

17. SIESJO, B. K. 1988. Historical overview. Calcium, ischemia, and death of brain cells. Ann. N. Y. Acad. Sci. **522:** 638–661.

18. LEONARD, J. P. & M. M. SALPETER. 1982. Calcium-mediated myopathy at neuromuscular junctions of normal and dystrophic muscle. Exp. Neurol. **76:** 121–138.

19. SIMAN, R. & J. C. NOSZEK. 1988. Excitatory amino acids activate calpain I and induce structural protein breakdown in vivo. Neuron **1:** 279–287.

20. CHAN, P. H. & R. A. FISHMAN. 1978. Transient formation of superoxide radicals in polyunsaturated fatty acid induced brain swelling. J. Neurochem. **35:** 1004–1007.

21. DYKENS, J. A., A. STERN & E. TRENKNER. 1987. Mechanism of kainate toxicity to cerebellar neurons in vitro is analogous to reperfusion tissue injury. J. Neurochem. **49:** 1222–1228.

22. CONNOR, J. A., W. J. WADMAN, P. E. HOCKBERGER & R. K. WONG. 1988. Sustained dendritic gradients of Ca2 + induced by excitatory amino acids in CA1 hippocampal neurons. Science **240:** 649–653.

23. KACZMAREK, L. K. 1987. The role of protein kinase C in the regulation of ion channels and neurotransmitter release. Trends Neurosci. **10:** 30–34.

24. CHOI, D. W., J. KOH & S. PETERS. 1988. Pharmacology of glutamate neurotoxicity in cortical cell culture: Attenuation by NMDA antagonists. J. Neurosci. **8:** 185–196.

25. MACDERMOTT, A. B., M. L. MAYER, G. L. WESTBROOK, S. J. SMITH & J. L. BARKER. 1986. NMDA-receptor activation increases cytoplasmic calcium concentration in cultured spinal cord neurones. Nature **321:** 519–522.

26. MURPHY, S. N., S. A. THAYER & R. J. MILLER. 1987. The effects of excitatory amino acids on intracellular calcium in single mouse striatal neurons in vitro. J. Neurosci. **7:** 4145–4158.

27. FINKBEINER, S. & C. F. STEVENS. 1988. Applications of quantitative measurements for assessing glutamate neurotoxicity. Proc. Natl. Acad. Sci. USA **85:** 4071–4074.

28. WONG, P. Y. & K. FRITZE. 1969. Determination by neutron activation of copper, manganese, and zinc in the pineal body and other areas of brain tissue. J. Neurochem. **16:** 1231–1234.

29. HAUG, F. M. S. 1973. Heavy metals in the brain: A light microscope study of the rat with Timm's sulphide silver method. Methodological considerations and cytological and regional staining patterns. Adv. Anat. Embryol. Cell Biol. **47:** 1–71.

30. DONALDSON, J., T. ST. PIERRE, J. L. MINNICH & A. BARBEAU. 1973. Determination of Na +, K +, Mg2 +, Cu2 +, Zn2 +, and Mn2 + in rat brain regions. Can. J. Biochem. **51:** 87–92.

31. DANSCHER, G., G. HOWELL, J. PEREZ-CLAUSELL & N. HERTEL. 1985. The dithizone, Timm's sulphide silver and the selenium methods demonstrate a chelatable pool of zinc in CNS. A proton activation (PIXE) analysis of carbon tetrachloride extracts from rat brains and spinal cords intravitally treated with dithizone. Histochemistry **83:** 419–422.

32. PEREZ-CLAUSELL, J. & G. DANSCHER. 1985. Intravesicular localization of zinc in rat telencephalic boutons. A histochemical study. Brain Res. **337:** 91–98.
33. ASSAF, S. Y. & S. CHUNG. 1984. Release of endogenous Zn^{2+} from brain tissue during activity. Nature **308:** 734–736.
34. HOWELL, G. A., M. G. WELCH & C. J. FREDERICKSON. 1984. Stimulation-induced uptake and release of zinc in hippocampal slices. Nature **308:** 736–738.
35. PETERS, S., J. KOH & D. W. CHOI. 1987. Zinc selectively blocks the action of N-methyl-D-aspartate on cortical neurons. Science **236:** 589–593.
36. WESTBROOK, G. L. & M. L. MAYER. 1987. Micromolar concentrations of Zn^{2+} antagonize NMDA and GABA responses of hippocampal neurons. Nature **328:** 640–643.
37. CHOI, D. W. & J. KOH. 1988. Zinc central neurotoxicity may require open NMDA channels. Soc. Neurosci. Abstr. **14:** 417.
38. CSERMELY, P., M. SZAMEL, K. RESCH & J. SOMOGYI. 1988. Zinc can increase the activity of protein kinase C and contributes to its binding to plasma membranes in T lymphocytes. J. Biol. Chem. **263:** 6487–6490.
39. KOH, J. & D. W. CHOI. 1988. Zinc alters excitatory amino acid neurotoxicity on cortical neurons. J. Neurosci. **8:** 2164–2171.
40. KOH, J., S. PETERS & D. W. CHOI. 1986. Neurons containing NADPH-diaphorase are selectively resistant to quinolinate toxicity. Science **234:** 73–76.
41. YOUNG, A. B., J. T. GREENAMYRE, Z. HOLLINGSWORTH, R. ALBIN, C. D'AMATO, I. SHOULSON & J. B. PENNEY. 1988. NMDA receptor losses in putamen from patients with Huntington's Disease. Science **241:** 981–983.
42. DAVIES, P., R. KATZMAN & R. D. TERRY. 1980. Reduced somatostatin-like immunoreactivity in cerebral cortex from cases of Alzheimer disease and Alzheimer senile dementa. Nature **288:** 279–280.
43. ROSSOR, M. N., P. C. EMSON, C. Q. MOUNTJOY, M. ROTH & L. L. IVERSEN. 1980. Reduced amounts of immunoreactive somatostatin in the temporal cortex in senile dementia of Alzheimer type. Neurosci. Lett. **20:** 373–377.
44. MORRISON, J. H., J. ROGERS, S. SCHERR, R. BENOIT & F. E. BLOOM. 1985. Somatostatin immunoreactivity in neuritic plaques of Alzheimer's patients. Nature **314:** 90–92.
45. ROBERTS, G. W., T. J. CROW & J. M. POLAK. 1985. Location of neuronal tangles in somatostatin neurones in Alzheimer's disease. Nature **314:** 92–94.
46. WEISS, J. H., J. KOH & D. W. CHOI. 1988. Beta-N-methylamino-L-alanine (BMAA) neurotoxicity on murine cortical neurons. Soc. Neurosci. Abstr. **14:** 417.
47. SPENCER, P. S., P. B. NUNN, J. HUGON, A. C. LUDOLPH, S. M. ROSS, D. N. ROY & R. C. ROBERTSON. 1987. Guam amyotrophic lateral sclerosis-Parkinsonism-dementia linked to a plant excitant neurotoxin. Science **237:** 517–522.

Neurotoxicity at the N-Methyl-D-Aspartate Receptor in Energy-Compromised Neurons

An Hypothesis for Cell Death in Aging and Disease

R. C. HENNEBERRY, A. NOVELLI, J. A. COX,
AND P. G. LYSKO

Molecular Neurobiology Section
Laboratory of Molecular Biology
National Institute of Neurological and Communicative
Disorders & Stroke
Building 36, Room 3-D-02
National Institutes of Health
Bethesda, Maryland 20892

INTRODUCTION

The mechanisms responsible for the progressive loss of neurons during normal aging may also be involved in the accelerated loss of neurons characteristic of neurodegenerative diseases such as Alzheimer's and Huntington's. Although these mechanisms are not well understood, evidence is accumulating that the excitatory amino acids (EAAs), in addition to their function as neurotransmitters in the healthy brain, can play a deleterious role under certain undefined, adverse conditions. As neurotoxins, the EAAs may be of primary importance in a variety of neuropathologic conditions.

There is an emerging consensus that L-glutamate is the major excitatory neurotransmitter, and that it is probably the endogenous agonist at the N-methyl-D-aspartate (NMDA) subtype of glutamate receptor.[1] It is also generally agreed that the NMDA receptor is not involved in fast synaptic neurotransmission, but a major role has been suggested for it in the establishment of certain types of long-term potentiation (LTP) during memory formation in the healthy brain.[2-4]

Recognition of the fact that glutamate can function as a potent neurotoxin can be traced to the pioneering studies of Lucas and Newhouse more than 30 years ago when they demonstrated that systemically injected glutamate caused retinal degeneration.[5] In 1969, Olney showed that exogenous glutamate can cause central nervous system lesions,[6] and subsequently published his "excitotoxic" hypothesis of glutamate's action.[7] Within the last 5 years a considerable body of evidence has appeared to support the excitotoxic hypothesis, as summarized in several recent reviews of the subject.[8-10] Numerous investigators have demonstrated toxicity of EAAs in neurons *in vivo* and *in vitro;* the many explanations offered for glutamate's transition from neurotransmitter to neurotoxin have focused on the availability of unusually high concentrations of the EAA at the receptor due to excess release, abnormal leakage, or impaired uptake (*e.g.* REFS. 11–15). Based on our studies with rat cerebellar granule cells in primary culture, we have recently described an alternate mechanism which emphasizes an increased response of a particular subtype of glutamate receptor to low concentrations of agonist (REF. 16; Lysko *et al.*, submitted; Cox *et al.*, submitted).

Our results indicate that relief of the Mg^{++} block of the NMDA receptor channel plays a pivotal role in potentiating EAA neurotoxicity in cerebellar granule cells. Electrophysiological studies have shown that the ion channel associated with the NMDA receptor is blocked by Mg^{++} when the neuron is at a normal resting potential.[17] The Mg^{++} block has been characterized as voltage-dependent; *i.e.*, the block is relieved when the neuron partially depolarizes.[18,19] We have found that glucose starvation or other metabolic perturbations lead to sufficient depolarization to relieve the Mg^{++} block, permitting NMDA receptor agonists to cause a rapid influx of Na^+ and Ca^{++}. In this energy-deprived condition the ion pumps are unable to maintain homeostasis and cell death follows. Based on these studies, we propose a general hypothesis for glutamate neurotoxicity at the NMDA receptor. According to this hypothesis, any series of events which reduces the energy level or otherwise causes excessive depolarization in any subset of neurons expressing the NMDA receptor would render those neurons susceptible to the toxic effects of low concentrations of endogenous glutamate.

MATERIALS AND METHODS

Primary cultures of rat cerebellar neurons were prepared from 8-day-old Sprague-Dawley rat pups by a previously described procedure,[16] grown in basal Eagle's medium (BME; Quality Biologicals, Gaithersburg, MD), supplemented with 10% fetal calf serum (FCS, heat-inactivated for 30 min at 56°C), 100 µg/ml gentamycin, 2 mM glutamine, 25 mM KCl, and seeded (3×10^6 cells) in 35-mm dishes (Nunc, Roskilde, Denmark) coated with poly-L-lysine (5 µg/ml H_2O, incubated at least 15 min, then removed). Cultures were incubated at 37°C and 5% CO_2 at 100% humidity, and cytosine arabinoside (10 µM) was added 20–24 hr later to prevent growth of nonneuronal cells. Cultures generated by this method have been characterized and shown to contain more than 90% granule cells. Cells were used for experiments 8–9 days after plating.

For toxicity experiments, cells were washed at 37°C with 1 ml of buffer (154 mM NaCl, 5.6 mM KCl, 1 mM $MgCl_2$, 2.3 mM $CaCl_2$, 8.6 mM HEPES, 5.6 mM glucose, pH 7.4); we refer to this as complete incubation buffer. Glucose or Mg^{2+} were omitted as indicated. The cells were incubated in buffer for 10 min, a fresh 1 ml of buffer was added, and incubation continued for an additional 30 min. Glutamate (10^{-4} M unless otherwise noted) was added, and the cells were incubated for a further 30 min. When the antagonists MK-801 or AP5 were used, they were added 20 min prior to the addition of glutamate. After the 30-min incubation with glutamate, the incubation buffer was removed, and 1 ml of complete incubation buffer was added. The ^{51}Cr uptake assay was adapted from Neville.[24] Neurons were incubated 24 hr at 37°C and 5% CO_2 in incubation buffer with the 30-min treatment with agonist described above; then 1 µCi of sodium [^{51}Cr]chromate was added to each dish and incubation at 37°C continued for exactly 2 hr. Each dish was washed twice with 1 ml Dulbecco's PBS (Gibco, Grand Island, NY), and 1 ml NaOH (0.2 N) was added and the solution counted for radioactivity by scintillation spectrophotometry. When viewed by phase microscopy prior to addition of ^{51}Cr, neurons not exposed to glutamate presented the typical, healthy appearance of cerebellar granule cells in culture, whereas neurons exposed to sufficient glutamate to be scored as dead by ^{51}Cr uptake invariably were totally disintegrated.

The percent cytotoxicity was calculated as:

$$\% \text{ cytotoxicity} = 1 - (E/T) \times 100$$

where E is equal to cpm of ^{51}Cr uptake by experimentally treated cells and T is cpm of ^{51}Cr uptake by control cells, and the protective effects of the indicated additions was calculated as:

$$\% \ protection \ = \ 100 \ - \ \% \ cytotoxicity.$$

For fluorescein diacetate (FDA) staining, the incubation buffer was aspirated after glutamate treatment and FDA (5 μg/ml) was added in complete buffer; culture dishes were incubated 5 min at 37°C; the staining solution was aspirated and replaced with 1 ml complete incubation buffer; and the cells were examined immediately by fluorescence microscopy. Viability is expressed in semiquantitative terms as an estimate of the percent of the total number of cells retaining fluorescein in randomly selected microscope fields.

The following abbreviations are used: AP5 = (±)-2-amino-5-phosphonopentanoic acid; AP7 = (±)-2-amino-7-phosphonoheptanoic acid; CPP = (±)-3-(2-carboxypiper-azin-4-yl)-propyl-1-phosphonic acid; KAM = ketamine; DMO = dextromethorphan; PCP = phencyclidine; and MK-801 = (+)-10,11-dihydro-5-methyl-5H-dibenzo[a,d]-cyclohepten-5,10-imine hydrogen maleate.

TABLE 1. Glutamate Is Neurotoxic in the Absence of Glucose[a]

Agonist	Mg^{++}	Glucose	Toxicity
None	+	+	5%
None	−	+	5%
None	+	−	5%
None	−	−	5%
L-Glutamate, 1 mM	−	+	95%
L-Glutamate, 20 μM	+	−	95%
L-Glutamate, 5 μM	−	−	95%
L-Aspartate, 200 μM	+	−	95%
NMDA, 500 μM	+	−	95%
Ibotenate, 500 μM	+	−	95%

[a]Neurons were cultured, treated, and scored for viability by FDA staining as described in Methods. Where present, glucose was 5.6 mM and Mg^{++} was 1 mM.

RESULTS

Cerebellar neurons showed no signs of toxicity when L-glutamate at concentrations as high as 5 mM was added to incubation buffer containing both Mg^{++} (1 mM) and glucose (5.6 mM). When either Mg^{++} or glucose was omitted from the buffer, neurons remained undamaged in the absence of glutamate. However, addition of 1 mM glutamate in the absence of Mg^{++} led to progressive loss of cellular refractility, inability to exclude trypan blue, and inability to contain fluorescein diacetate (FDA), all of which are considered indicators of cell death. These results are summarized in TABLE 1 and are consistent with a Mg^{++} block of the NMDA receptor channel;[17] we have previously shown this channel to be present on granule cells.[20] Others have demonstrated the ability of Mg^{++} to prevent EAA-mediated neurotoxicity in hippocampal neurons[21,22] and in cerebellar slices.[23]

Surprisingly, addition of as little as 20 μM glutamate led to rapid cell death in the absence of glucose and the presence of 1 mM Mg^{++}. Omission of both glucose and Mg^{++} from the incubation buffer potentiated the toxicity of glutamate at concentrations as low as 5 μM. These results are summarized in TABLE 1 in which neuronal viability was scored by staining with FDA[23] and in FIGURE 1 in which viability was scored by ^{51}Cr uptake.[24] These two methods of assessing neurotoxicity were in very close agreement. In all cases, neurons exposed to conditions which led to cell death as scored by FDA staining

TABLE 2. Protective Effects of Competitive and Noncompetitive NMDA Antagonists on Glutamate-Induced Neurotoxicity[a]

Glutamate	Antagonist	Toxicity
100 μM	None	95%
100 μM	AP5, 500 μM	5%
100 μM	AP7, 500 μM	5%
100 μM	CPP, 100 μM	5%
100 μM	KAM, 50 μM	5%
100 μM	DMO, 10 μM	5%
100 μM	PCP, 1 μM	5%
100 μM	MK-801, 20 nM	5%

[a]Neurons were cultured, treated, and scored for toxicity by FDA staining as described in Methods. Abbreviations for antagonists are defined in Methods.

or ^{51}Cr uptake continued to deteriorate with time and by 12 hr had completely disintegrated when examined with phase optics.

Other NMDA receptor agonists were also neurotoxic for granule cells when glucose was absent but Mg^{++} present, although glutamate was at least 10-fold more potent than any other agonist tested (TABLE 1). As shown in TABLE 2, several competitive and noncompetitive antagonists selective for NMDA receptors were tested for their ability to block glutamate neurotoxicity in the absence of glucose and found to be quite potent. CPP is thought to be the most potent competitive antagonist of the NMDA receptor;[25] it completely prevented glutamate toxicity in granule cells at 100 μM. MK-801 is reported to be the most potent noncompetitive NMDA antagonist;[14] it completely prevented glutamate toxicity at 20 nM (TABLE 2). Using the ^{51}Cr assay for neurotoxicity, dose-response curves were generated for the antagonists vs the toxicity of 100 μM glutamate. 50% protective doses (PD_{50}) were calculated for MK-801 and AP5 at 6 nM and 100 μM, respectively.

Assuming that glucose protected against glutamate neurotoxicity by providing a source of energy in the form of highly phosphorylated nucleotides (ATP) to maintain the

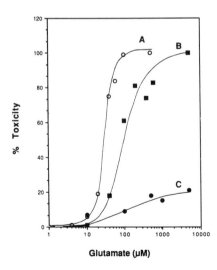

FIGURE 1. Effects of glucose and Mg^{++} on the dose-response curve for glutamate neurotoxicity. Effect of glutamate concentration on toxicity for cerebellar granule cells in: (A) the absence of glucose with 1 mM Mg^{++} present; (B) the absence of Mg^{++} with 5.6 mM glucose present; and (C) the presence of both 5.6 mM glucose and 1 mM Mg^{++}. Granule cells were cultured, tested, and scored for toxicity by the uptake of ^{51}Cr as described in Methods.

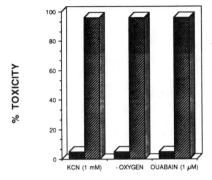

FIGURE 2. Effect of disruption of energy metabolism on glutamate neurotoxicity. Neurons were cultured, tested, and scored for viability by FDA staining as described in Methods. Mg^{++} was present at 1 mM and glucose at 5.6 mM throughout. *Hatched bars* represent cultures treated as indicated and then exposed to: KCN + 100 μM glutamate; O_2-deprivation + 100 μM glutamate; and Ouabain + 10 μM glutamate. *Solid bars* indicate the same treatment but without glutamate. Oxygen deprivation was achieved by adding degassed incubation buffer and incubating culture dishes in an atmosphere of 100% argon.

membrane potential and thereby preserve the Mg^{++} block of the NMDA receptor channel, we examined the effects of disruption of energy metabolism on toxicity. As shown in FIGURE 2, addition of KCN (1 mM) alone was not toxic for granule cells, but enabled 100 μM glutamate to express its toxicity in the presence of both glucose and Mg^{++}. Similarly, O_2-deprivation was innocuous by itself, but potentiated the toxicity of 100 μM glutamate even with both glucose and Mg^{++} present. Further, assuming that ATP plays a protective role by serving as a substrate for the Na^+/K^+ ATPases involved in maintaining the membrane potential, we examined the effects of the Na^+/K^+ ATPase inhibitor, ouabain. As shown in FIGURE 2, 1 μM ouabain was not toxic alone but potentiated the toxicity of 10 μM glutamate in the presence of both Mg^{++} and glucose.

As described above, our results enabled us to conclude that glutamate is toxic for cerebellar granule cells by its action at the NMDA receptor. Further, we conclude that glutamate becomes toxic in energy-starved neurons due to relief of the Mg^{++} block of the NMDA channel, which is known to be voltage-dependent from electrophysiologic studies.[18,19] To confirm our assumption that glucose-starvation for 40 min significantly reduced the energy level of the neurons, adenine nucleotides were measured by HPLC and expressed as adenylate energy charge (A.E.C.; REF. 26; FIG. 3). In the presence of glucose (5.6 mM), addition of 100 μM glutamate had a slight effect on energy charge,

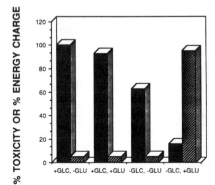

FIGURE 3. Reduced adenylate energy charge correlates with glutamate neurotoxicity. Neurons were cultured, treated, and stained for viability with FDA, and adenine nucleotides were measured by HPLC as described in REF. 16. Adenylate energy charge = (ATP + 1/2 ADP)/(ATP + ADP + AMP) *(solid bars)*. % Toxicity *(hatched bars)*. Mg^{++} (1 mM) was present throughout; glucose (GLC; 5.6 mM) and glutamate (GLU; 100 μM) were present where indicated.

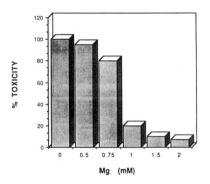

FIGURE 4. Protective effect of Mg^{++} on glutamate toxicity in presence of glucose. Neurons were
cultured, treated, and stained for viability with FDA
as described in Methods. Glucose (5.6 mM), EDTA
(1 mM), and glutamate (100 μM) were present
throughout.

suggesting a slight increase in energy demand due to a small amount of ion influx through
those channels opened by agonist and momentarily unblocked by Mg^{++}. In the absence
of glucose the energy charge was reduced by about 40%; addition of 100 μM glutamate
to these neurons led to massive cell death and disintegration reflected in an energy charge
of 0.13 (FIG. 3). Future plans include correlations of neurotoxicity with more detailed
measurements of ATP, ADP, and creatine phosphate.

As a further test that glutamate neurotoxicity in granule cells was mediated by partial
depolarization sufficient to relieve the Mg^{++} block of the NMDA channel, we demonstrated that high, but nonlethal, concentrations of veratridine potentiated the toxicity of
100 μM glutamate (Cox et al., submitted). This result, together with our finding that
glutamate did not seriously depress the neuron's energy charge when glucose was present,
suggests that the EAA causes insignificant depolarization of cerebellar granule cells when
they are metabolically competent and provided with an oxidizable source of energy.

We assumed that changes in membrane potential affect the Mg^{++} block by altering
the affinity of the NMDA channel binding site for Mg^{++}. This assumption predicts that
toxicity, at a given dose of EAA, will be inversely proportional to the concentration of
Mg^{++}; results in agreement with this prediction are shown in FIGURE 4 for toxicity
(scored by FDA staining) of 1 mM glutamate in the presence of 5.6 mM glucose and 1
mM EDTA. Comparable results are shown in FIGURE 5 with toxicity scored by both FDA
staining and ^{51}Cr uptake; in this case high concentrations of Mg^{++} block the toxicity of
100 μM glutamate in the absence of glucose. These two conditions represent Mg^{++}
binding to its NMDA channel site at "normal" membrane potential and partial depolarization, respectively. We interpret this result as supporting the hypothesis that glutamate
becomes neurotoxic at the NMDA receptor when the neuron depolarizes sufficiently to
relieve the voltage-dependent Mg^{++} block of the NMDA channel. Our results with
cerebellar granule cells in primary culture are consistent with those of Garthwaite and
Garthwaite,[23] who reported a twofold potentiation of glutamate toxicity for granule cells
in cerebellar slices; they attributed this to depolarization-induced relief of the voltage-
dependent Mg^{++} block of the NMDA channel.

DISCUSSION

The refractory nature of cerebellar granule cells to high concentrations of glutamate in
the presence of glucose and Mg^{++} was unexpected, especially since these cells also
contain the kainate subtype of glutamate receptor[20] and are killed by kainate (unpublished
data). Glutamate might be expected to stimulate Na^+ influx via the kainate receptor

channel, thus causing partial depolarization and relieving the voltage-dependent NMDA channel block. Resistance to high concentrations of glutamate is also in sharp contrast to results reported by others; *e.g.*, neocortical neurons in culture are killed by a 5 min exposure to 50–100 μM glutamate with both glucose and Mg^{++} present.[27] A possible explanation for this apparent disparity is that cerebellar granule cells express only one type of receptor responsive to glutamate, *viz.*, the NMDA receptor, and that receptor has a voltage-dependent Mg^{++} block. The low incidence of open, unblocked channels causes insufficient depolarization to open voltage-dependent Ca^{++} channels or to relieve the voltage-dependent block of the NMDA channel. If cerebellar granule cells express typical glutamate receptor responses, implications of this interpretation are that: (a) glutamate is incapable of acting as a neurotoxin at the kainate receptor *in vivo,* and/or (b) glutamate is not the endogenous ligand for the kainate receptor. These results also suggest that glutamate cannot open a significant number of NMDA channels in energy-competent neurons without an independent depolarizing influence.

Glutamate is toxic for granule cells in the presence of glucose when Mg^{++} is absent. Significantly, this toxicity is completely blocked by AP5 and MK-801. The requirement for a high glutamate concentration (1 mM) for toxicity in the absence of Mg^{++} may reflect the ability of energy-competent neurons to cope with substantial ion influx; 1 mM glutamate may sufficiently increase the proportion of open NMDA channels to overwhelm this ability.

Glutamate is toxic for granule cells with Mg^{++} present when the cells are starved for glucose for 40 min prior to the addition of glutamate. In protection experiments in the presence of Mg^{++}, glucose and pyruvate were found to have PD_{50}s of 50 μM and 200 μM, respectively. The results described in this paper, together with other reports from this laboratory, have led us to conclude that glutamate is toxic for cerebellar granule cells via its action at the NMDA receptor. Further, we conclude that glutamate becomes toxic in glucose-starved neurons due to decreases in ATP leading to relief of the Mg^{++} block of the NMDA channel, which is known to be voltage-dependent. This hypothesis is supported by our results with KCN, O_2-deprivation, and ouabain potentiating the toxicity of glutamate in the presence of glucose and Mg^{++}. The relationship between intracellular ATP levels and membrane potential has been described (*e.g.*, REFS. 28, 29), and is presently being studied in detail in our model system.

Due to certain unique characteristics which may relate to the subtypes of glutamate receptors expressed, the cerebellar granule cell in primary culture offers a suitable model for analysis of NMDA receptor action. While results obtained with this model may not necessarily translate directly to all areas of the brain, the information gained should

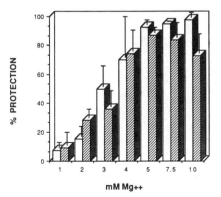

FIGURE 5. Protective effect of Mg^{++} on glutamate toxicity in absence of glucose. Neurons were cultured, treated, and scored for viability with FDA *(solid bars)* or by ^{51}Cr uptake *(hatched bars)* as described in Methods. Glucose was absent and glutamate (100 μM) was present throughout.

increase our basic understanding of the NMDA receptor. At the practical level, this model appears to be useful for the screening of drugs with the potential for antagonizing NMDA receptor activity (Henneberry *et al.*, unpublished data).

SUMMARY

Our results demonstrated that the neurotoxicity of glutamate and closely related agonists was mediated by the NMDA receptor in rat cerebellar granule cells. Evidence was presented to support our hypothesis[16] that the pivotal event in the transition of these EAAs from neurotransmitters to neurotoxins is relief of the voltage-dependent Mg^{++} block of the NMDA channel due to changes in membrane potential which can be caused by depletion of highly phosphorylated nucleotides or by other depolarizing stimuli. Persistent stimulation of NMDA receptors whose channels are unblocked by Mg^{++} can permit excessive influx of Na^+ and Ca^{++} and neuronal death can follow by a mechanism not yet understood. Glutamate is not toxic at kainate receptors although they are present on these cells. These findings underline the potential importance of perturbations in energy metabolism in a variety of neurodegenerative disorders and in the normal process of aging which share the common feature of the loss of neurons.

REFERENCES

1. OLVERMAN, H. J., A. W. JONES & J. C. WATKINS. 1988. [³H] D-2-amino-5-phosphonopentanoate as a ligand for N-methyl-D-aspartate receptors in the mammalian central nervous system. Neuroscience **26**: 1–15.
2. LYNCH, G. & M. BAUDRY. 1984. The biochemistry of memory: A new and specific hypothesis. Science **224**: 1057–1063.
3. COLLINGSRIDGE, G. L. & T. V. P. BLISS. 1987. NMDA receptors—their role in long-term potentiation. Trends Neurosci. **10**: 288–293.
4. COLLINGSRIDGE, G. L. 1987. The role of NMDA receptors in learning and memory. Nature (London) **330**: 604–605.
5. LUCAS, D. R. & J. P. NEWHOUSE. 1957. The toxic effect of sodium L-glutamate on the inner layers of the retina. Arch. Opthalmol. **58**: 193–204.
6. OLNEY, J. W. 1969. Brain lesions, obesity and other disturbances in mice treated with monosodium glutamate. Science **164**: 719–721.
7. OLNEY, J. W., O. L. HO & V. RHEE. 1971. Cytotoxic effects of acidic and sulphur containing amino acids on the infant mouse central nervous system. Exp. Brain Res. **14**: 61–76.
8. MELDRUM, B. 1985. Possible therapeutic applications of antagonists of excitatory amino acid neurotransmitters. Clin. Sci. **68**: 113–122.
9. MAYER, M. L. & G. L. WESTBROOK. 1987. Cellular mechanisms underlying excitotoxicity. Trends Neurosci. **10**: 59–61.
10. ROTHMAN, S. M. & J. W. OLNEY. 1987. Excitotoxicity and the NMDA receptor. Trends Neurosci. **10**: 299–302.
11. MCGEER, E. G. & P. L. MCGEER. 1976. Duplication of biochemical changes of Huntington's chorea by intrastriatal injections of glutamate and kainic acids. Nature (London) **263**: 517–519.
12. COYLE, J. T., S. J. BIRD, R. H. EVANS, R. L. GULLEY, J. V. NADLER, W. J. NICKLAS & J. W. OLNEY. 1981. Excitatory amino acid neurotoxins: Selectivity, specificity, and mechanisms of action. Neurosci. Res. Program Bull. **19**: 331–427.
13. MARAGOS, W., J. T. GREENAMYRE, J. B. PENNY, JR. & A. B. YOUNG. 1987. Glutamate dysfunction in Alzheimer's disease: An hypothesis. Trends Neurosci. **10**: 65–68.
14. KEMP, J. A., A. C. FOSTER, R. GILL & G. N. WOODRUFF. 1987. MK801, NMDA receptors, and ischemia-induced neurodegeneration. Trends Pharmacol. Sci. **8**: 414–415.
15. BARNES, D. 1988. NMDA receptors trigger excitement. Science **239**: 254–256.

16. NOVELLI, A., A. J. REILLY, P. G. LYSKO & R. C. HENNEBERRY. 1988. Glutamate becomes neurotoxic via the N-methyl-D-aspartate receptor when intracellular energy levels are reduced. Brain Res. **451:** 205–212.
17. AULT, B., R. H. EVANS, A. S. FRANCIS, D. J. OAKES & J. C. WATKINS. 1980. Selective depression of excitatory amino acid induced depolarizations by magnesium ions in isolated spinal cord preparations. J. Physiol. **307:** 413–428.
18. MAYER, M. L., G. L. WESTBROOK & P. B. GUTHRIE. 1984. Voltage-dependent block by Mg^{++} of NMDA responses in spinal cord neurons. Nature (London) **309:** 261–263.
19. NOWAK, L., P. BREGESTOVSKI, P. ASHER, A. HERBET & A. PROCHIANTZ. 1984. Magnesium gates glutamate-activated channels in mouse central neurons. Nature (London) **307:** 462–465.
20. NOVELLI, A. & R. C. HENNEBERRY. 1987. cGMP synthesis in cultured cerebellar neurons is stimulated by glutamate via a Ca^{++}-mediated, differentiation-dependent mechanism. Dev. Brain Res. **34:** 307–310.
21. ROTHMAN, S. M. 1983. Synaptic activity mediates death of hypoxic neurons. Science **220:** 536–539.
22. FINKBEINER, S. & C. F. STEVENS. 1988. Applications of quantitative measurements for assessing glutamate neurotoxicity. Proc. Natl. Acad. Sci. USA **85:** 4071–4087.
23. GARTHWAITE, G. & J. GARTHWAITE. 1987. Receptor-linked ionic channels mediate N-methyl-D-aspartate neurotoxicity in rat cerebellar slices. Neurosci. Lett. **83:** 241–246.
24. NEVILLE, M. E. 1987. ^{51}Cr uptake assay: A sensitive and reliable method to quantitate cell viability and cell death. J. Immunol. Methods **99:** 77–82.
25. DAVIES, J., R. H. EVANS, P. L. HERRLING, A. W. JONES, H. J. OLVERMAN, P. POOK & J. C. WATKINS. 1986. CPP, a new potent and selective NMDA antagonist. Depression of central neuron responses, affinity for D-AP5 binding sites on brain membranes and anticonvulsant activity. Brain Res. **382:** 169–173.
26. ATKINSON, D. E. & G. M. WALTON. 1967. Adenosine triphosphate conservation in metabolic regulation. J. Biol. Chem. **242:** 3239–3241.
27. CHOI, D. W., M. MAULUCCI-GEDDE & A. R. KRIEGSTEIN. 1987. Glutamate neurotoxicity in cortical cell culture. J. Neurosci. **7:** 357–368.
28. HANSEN, A. J., J. HOUNSGAARD & J. JAHNSEN. 1982. Anoxia increases potassium conductance in hippocampal nerve cells. Acta Physiol. Scand. **115:** 301–310.
29. FUJIWARA, N., H. HIGASHI, K. SHIMOJI & M. YOSHIMURA. 1987. Effects of hypoxia in rat hippocampal neurones in vitro. J. Physiol. **384:** 131–151.

Calcium, Excitotoxins, and Neuronal Death in the Brain[a]

BO K. SIESJÖ,[b] FINN BENGTSSON,[b,c] WOLFGANG GRAMPP,[d]
AND STEN THEANDER[b,d]

[b]Laboratory for Experimental Brain Research
and
Departments of [c]Clinical Pharmacology and [d]Physiology
University of Lund
Floor EA-5
Lund Hospital
S-221 85 Lund, Sweden

Around 1970, Fleckenstein postulated that cardiac muscle damage following ischemia or excessive stimulation with catecholamines was due to calcium entry and mitochondrial calcium overload (*e.g.*, REF. 1). Somewhat later, a similar hypothesis was advanced to explain cell necrosis in dystrophic skeletal muscle.[2] It soon emerged that lethal calcium influx could occur, not only via voltage-sensitive calcium channels (VSCCs), but also via those operated by agonists (AOCCs).[3] Subsequently, Schanne *et al.*[4] suggested that liver cell damage due to a variety of toxins was mediated by calcium influx (see also REF. 5). These various hypotheses of calcium-related cell necrosis thus have in common the assumption that cell calcium homeostasis is upset by enhanced influx into the cell, the routes of entry varying with the condition.

Brain damage usually occurs in one of two forms: selective neuronal vulnerability, affecting groups of neurons with a characteristic distribution within the brain, and infarction or pan-necrosis, affecting not only neurons but also glial cells and vascular elements. When it was originally thought that selective neuronal vulnerability in ischemia, hypoglycemia, and status epilepticus is a calcium-related phenomenon, it was tacitly assumed that calcium entered through VSCCs in the apical dendrites of vulnerable (burst-firing) neurons.[6] It was soon reported that, following a transient ischemic insult, both net calcium accumulation and neuronal necrosis may be considerably delayed (see REF. 7). Since calcium accumulation seemed to precede light microscopical evidence of cell necrosis, the hypothesis was advanced that increased calcium cycling across damaged membranes causes gradual mitochondrial overload and, after varying delays, cell necrosis.[8] However, although this may be one of the mechanisms involved, the general consensus is that some of the most important consequences of a disordered calcium homeostasis are activation of lipases, proteases, and endonucleases (see REFS. 6,9–12 and, for corresponding data on liver tissue, 13,14).

Although doubts have been raised that calcium is an obligatory mediator of anoxic (or toxic) cell death,[15,16] recent results strongly indicate that necrosis of neurons, exposed to anoxia or excitotoxins, is calcium-mediated.[17–19] In fact, it now seems likely that anoxic neuronal damage, at least in part, represents an excitotoxic lesion caused by enhanced

[a]Studies from the authors' own laboratory were financially supported by grants from the Crafoordska Foundation, Lund, Sweden, the Thelma Zoégas Fund and the Medical Faculty, Lund University, Lund, Sweden, the Swedish Medical Research Council, and the United States Public Health Service through the National Institutes of Health.

release and/or diminished uptake of glutamate or a related excitatory amino acid (EAA), and that the toxic action of EAAs is caused mainly by enhanced calcium influx, notably via a channel gated by EAA receptors.[17-20] Probably, brain damage arising as a result of epilepsy[10,21] and hypoglycemia[22] occurs by similar mechanisms. Although the mechanisms are probably different, it is also conceivable that neuronal dysfunction and death in aging and Alzheimer's disease are related to a disordered calcium homeostasis.[23-25]

In support of these contentions, several studies demonstrate that antagonists which block glutamate receptors, or the channels they gate, ameliorate anoxic and ischemic damage (for literature, see REF. 18). Similar results have been obtained in hypoglycemia.[22,26] As will be discussed below, though, the area is controversial since some studies fail to show that glutamate antagonists ameliorate ischemic brain damage, or block calcium influx into cells during anoxia or hypoglycemia. The controversial results raise two equestions. *First,* under what conditions will antagonists of EAA receptors ameliorate brain damage? *Second,* do these drugs act primarily by hindering calcium influx into vulnerable cells?

It is the objective of this paper to discuss available information in this field and to attempt explaining the discrepant findings in terms of a unifying hypothesis. We recently presented such a hypothesis, based on the assumption that the massive ionic fluxes occurring in anoxia/ischemia or hypoglycemia, as well as in spreading depression (SD), occurs via a nonspecific cation conductance mechanism, which is activated by a rise in $Ca^{2+}{}_i$.[27,28] The present account represents an updating of the previous ones. Specifically, we have explored the possibility that the discrepant results in the literature can be explained by a modified hypothesis, one which does not require the operation of a special, calcium-activated conductance mechanism. It seems justified that we begin by recalling the salient features of cellular calcium homeostasis.

Cell Calcium Homeostasis

Since calcium is a regulator of metabolic pathways, and serves important functions as a second messenger, the free cytosolic calcium concentration $(Ca^{2+}{}_i)$ must be tightly regulated at around 10^{-7} M. The major mechanisms, responsible for influx, efflux, intracellular release, and binding/sequestration, are illustrated in FIG. 1 (see REFS. 29–37). Influx occurs by VSCCs and by AOCCs, while release from the endoplasmic reticulum (ER) is triggered by inositol trisphosphate (IP_3), the latter arising as a result of stimulation of receptors coupled to phospholipase C. Since the ER may be refilled from the outside by a separate "channel,"[38,39] there are principally three routes of calcium entry, each being independently regulated. We have previously postulated that yet another channel may be expressed under certain conditions. This channel, provisionally denoted NSCC (nonspecific cation channel) will be further discussed below.

Efflux of calcium occurs by a high affinity–low capacity ATPase (which is calmodulin-dependent) and by a low affinity–high capacity, electrogenic $3Na^+/Ca^{2+}$ exchanger, the latter being driven by the Na^+ gradient and the membrane potential. Any calcium entering the cell, or released within the cell, is buffered by calcium-binding proteins and a host of other compounds, and some is taken up by the ER in an ATP-dependent reaction. When $Ca^{2+}{}_i$ increases above about 1 μM, the mitochondria become important in calcium sequestration due to their large capacity. It is noteworthy that since calcium sequestration within the cell is dependent on ATP and/or oxygen, the calcium-buffering capacity must fall in anoxia (see REFS. 40,41). It is also to be expected that anoxia, by causing accumulation of H^+, could displace Ca^{2+} from binding places in the cell, thereby contributing to the rise in $Ca^{2+}{}_i$ (see REF. 9).

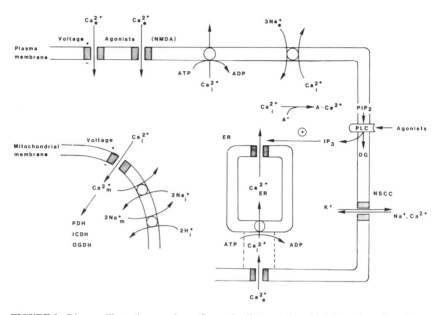

FIGURE 1. Diagram illustrating membrane flux and cellular cycling of calcium. Normally, almost all intracellular calcium (Ca^{2+}_i) is bound or sequestered. Numerous high- and low-molecular weight compounds with negative sites (A^-; *upper right*) is present to bind Ca^{2+}_i. By an electrophoretic mechanism Ca^{2+}_i can be sequestered into the mitochondria (*middle left*), and by an ATP-dependent reaction into the endoplasmatic reticulum (ER; *middle right*). Efflux of calcium from these latter sites occurs through different mechanisms shown in the figure (see below, and in the text). Extrusion of calcium from the cell normally occurs through a calcium-activated ATPase (*top middle*) and through a Na^+/Ca^+ exchange (*top right*). Normally, cellular calcium entry occurs through voltage-sensitive and agonist-operated (N-methyl-D-aspartate; NMDA receptor-gated) calcium channels (*top left*). However, agonist overactivation of membrane-bound phospholipase C (PLC; *right-upper*) may produce inositoltrisphosphate (IP_3)-triggered release of calcium from the ER. This released ER calcium may be replaced by external calcium entry through a separate channel (*bottom*). Two additional routes for calcium entry may appear under pathological conditions. The electrogenic $3Na^+/Ca^{2+}$ antiporter (*top right*) may be reversed after depolarization, and it is speculated that a large nonselective cation (NSCC) conductance can be activated (*right-lower*). This NSCC is opened to allow rapid dissipative cation fluxes when Ca^{2+}_i reaches above a certain threshold level. (Modified after Siesjö.[27])

Voltage-Sensitive and Agonist-Operated Calcium Channels

In order to understand how calcium enters cells under pathological conditions, influx through VSCCs and AOCCs must be specified (see REFS. 42–47). As suggested by Nowycky *et al.*[43]; (however, see also REF. 47) influx may occur through three types of VSCCs, designated L (long-lasting), T (transient), and N (neither L nor T). Based on data discussed in the articles quoted (see also REF. 18) we have assumed that presynaptic calcium influx occurs via an N type of channel, also that the L type is found postsynaptically on dendrites and soma membranes, but not on dendritic spines (FIG. 2). It is commonly assumed that only the L types of channel is blocked by calcium antagonists of the dihydropyridine (DHP) and other types (however, see REF. 47).

An important route of calcium entry is provided by AOCCs, notably those gated by receptors activated by glutamate and related EAAs. Three major subtypes of such receptors have been described, selectively activated by N-methyl-D-aspartate (NMDA), by kainate (K), or by quisqualate (Q) (see REFS. 42, 44, 48–51). We have assumed that these receptors are co-localized on apical dendrites and dendritic spines. Two separate ion channels are gated by these receptors. One, linked to the K/Q type of receptor, opens a channel for monovalent cations. This channel will thus allow K^+ to leave, and Na^+ to enter the cell and, very likely, it also provides a conductance mechanism for H^+. The other channel, gated by the NMDA receptor, allows calcium ions to enter the cell. One unique feature of this NMDA-linked channel is that it is blocked by Mg^{2+} in a voltage-dependent manner.[49,52,53] When the membrane depolarizes, e.g., due to Na^+ influx through the channel linked to the K/Q-receptor, the Mg^{2+} block is relieved. Another interesting feature is that the NMDA response, and thereby the influx of calcium, is potentiated by glycine.[54] We recognize that since glutamate is a mixed agonist, acting on both K/Q and NMDA receptors, it leads to calcium influx by a sequential mechanism which involves Na^+ influx via the K/Q-receptor-linked channel, depolarization, relief of the Mg^{2+} block, and calcium influx through the NMDA-receptor-linked channel.

The glutamate-activated channels are antagonized by compounds, different from those blocking VSCCs. Competitive NMDA antagonists, i.e., those blocking the receptor, encompass phosphonates such as 2-amino-5-phosphono-valeric acid (APV), while non-

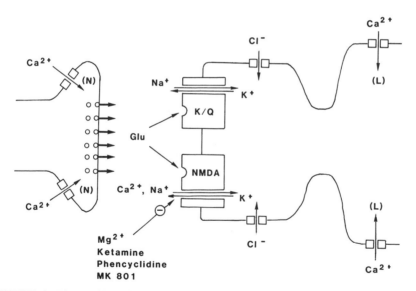

FIGURE 2. Diagram illustrating presynaptic voltage-sensitive calcium channels (VSCCs) and postsynaptic agonist-operated calcium channels (AOCCs). Presynaptically, VSCCs involved in transmitter release are assumed to be of the N type (N), which is not sensitive to calcium antagonists. At the postsynaptic site, kainate/quisqualate-operated channels (K/Q) are assumed to allow Na^+ influx and thereby to cause depolarization, while N-methyl-D-aspartate-operated channels (NMDA) are assumed to allow calcium influx. The ionic shifts at the K/Q and NMDA-gated channels occur when the K/Q and NMDA receptors are activated by glutamate (Glu), respectively. When Na^+ enters the cell postsynaptically, electrostatic forces are assumed to cause Cl^- influx along voltage-sensitive or agonist-operated anion channels. Na^+ and Cl^- entry is accompanied by osmotically obliged water that can result in swelling of apical dendrites. On the central dendrite or cell-soma VSCCs of the L type (L) are assumed to be located to allow Ca^{2+} entry. (Modified after Siesjö.[9])

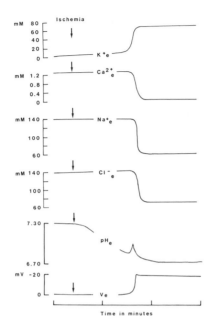

FIGURE 3. Schematic diagram illustrating changes in cerebral cortical concentrations of extracellular K^+, Ca^{2+}, Na^+, Cl^-, and pH, as well as (*bottom*) the development of a negative DC potential shift between brain and extracranial sites following complete ischemia. Although K^+_e shows an early, slow rise in concentration, and pH_e steadily falls after the induction of ischemia, the major ion fluxes occur about 90 sec after interruption of circulation. At the time of these rapid ion fluxes, a transient alkaline shift of pH_e is observed. This indicates the possibility that the cation conductance mechanism allows equilibration of H^+ or an anion conductance mechanism a corresponding equilibration of HCO_3^-. (Data derived from sources quoted by REFS. 55,56,57,58,61).

competitive NMDA antagonists, which block the receptor-gated channel, include phencyclidine, ketamine, and MK-801.

Loss of Ion Homeostasis

Changes in extracellular ion concentrations accompanying anoxia/ischemia have been described in detail, particularly for rat cerebellum and neocortex.[55-58] As FIGURE 3 shows, the first phase is characterized by a slow rise in extracellular K^+ (K^+_e) with little or no changes in Ca^{2+}_e, Na^+_e, or Cl^-_e. However, pH_e falls steadily, and there is some reduction in extracellular fluid (ECF) volume (not shown). After about 90 sec a sudden efflux of K^+ occurs, with an equally sudden cellular uptake of Ca^{2+}, Na^+, Cl^-, and water. This "shock" opening of ionic gates occurs *pari passu* with the development of a negative shift in the DC potential (brain versus an extracranial site). Since these events are accompanied by a transient alkaline shift of pH_e, it seems likely that the cation conductance mechanism allows equilibration of H^+ or an anion conductance mechanism a corresponding equilibration of HCO_3^- (see REFS. 28, 59). Thus, before the membrane potential has collapsed the electrochemical gradients favor influx of H^+, and efflux of HCO_3^-. Obviously, the conductance mechanisms opened seem to permit all cations and anions to move rapidly in the direction of their electrochemical gradients. However, since it appears more likely that calcium influx is, at least *in vitro,* responsible for delayed neuronal necrosis, it seems warranted to specifically examine the situations in which calcium influx occurs *in vivo.*

FIGURE 4 also illustrates calcium influx during ischemia,[55-57] hypoglycemia,[60] SD,[56,61] and focal ischemia due to experimental stroke. In the latter conditions, calcium fluxes were measured in the perifocal, "penumbra" zone.[62] The results demonstrate that dense ischemia leads to a rapid, permanent fall in Ca^{2+}_e, hypoglycemia to a corresponding fall preceded by an abortive and transient recovery, SD by spontaneous recovery each time a SD wave is elicited by strong electrical stimulation (S), and focal ischemia to trains of spontaneously occurring SD-like calcium transients. The last three conditions differ from the first in that efflux of K^+ and influx of Ca^{2+} occur suddenly while, as has already been mentioned, Ca^{2+} influx in ischemia is preceded by a slow rise of K^+_e to values of 10–15 mM.

Routes of Calcium Entry, and Means of Intracellular Release

The calcium fluxes described, and the way they may be influenced pharmacologically, can best be understood if we examine how agonist-receptor interactions and depolarization modulate calcium fluxes across the membranes of normal cells. Two types of experiments have given valuable information on the routes by which calcium enters cells. One is the measurement of Ca^{2+}_i in neurons from primary culture.[63-65] Such results have shown that many neurons contain VSCCs which open in response to depolarization and DHP agonists, and which can be blocked by DHP antagonists. They have also shown that calcium influx occurs via AOCCs, opened in response to receptor activation by glutamate, by NMDA, and by other glutamate analogues. The general picture which emerges from such studies is that it may take lesser degrees of depolarization to relieve the Mg^{2+} block of the NMDA-gated channel, allowing calcium to enter upon stimulation of the receptor, than to open VSCCs. Another impression gained is that only the L type of VSCC has a conductance which could allow appreciable amounts of calcium to enter. Furthermore, VSCCs may inactivate more readily during sustained depolarization/receptor acti-

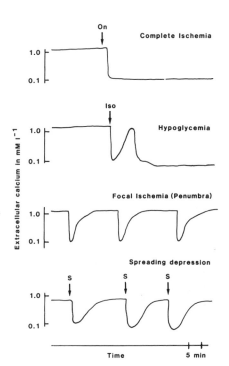

FIGURE 4. Extracellular (interstitial) calcium concentrations measured in the rat parietal cortex with Ca^{2+}-sensitive microelectrodes. From *top* to *bottom* we observe changes during acute, global ischemia induced by cardiac arrest ("On"); insulin-induced hypoglycemia with an isoelectric EEG ("Iso"); spontaneous repetitive, reversible changes in the penumbra zone after focal ischemia, and repetitive, reversible changes at stimulation ("S") in spreading depression. (Modified after Zhang et al.[74])

vation than AOCCs, favoring the latter as the main entry routes. Finally, the observation that MK-801 and phencyclidine completely block the NMDA-induced calcium influx, but not that caused by glutamate, suggests that glutamate activation of K/Q receptors somehow triggers calcium influx. There are three possible explanations: the depolarization caused by Na^+ influx allows calcium entry through VSCCs, reverses the normal direction of the (electrogenic) $3Na^+/Ca^{2+}$ antiporter, or relieves the channel block exerted by noncompetitive NMDA antagonists (see below).

The other type of experiments which has given valuable information on this issue is the measurements of Ca^{2+}_e, Na^+_e, and K^+_e during ionophoretic release of glutamate and related EAAs.[66-68] Such results have shown that release of glutamate and related EAAs cause marked reductions in Ca^{2+}_e in the neocortex, which are maximal in layers 2 and 5. Following release of NMDA, Ca^{2+}_e decreased maximally by about 1.1 mM and Na^+_e by 50–60 mM, while the maximal increase in K^+_e was 13 mM.[68] Glutamate gave similar, albeit smaller changes. The competitive NMDA antagonist APV blocked the NMDA-induced calcium and sodium changes, but not the decrease in Na^+_e caused by Q or K. Furthermore, whereas APV blocked the glutamate-induced reduction in Ca^{2+}_e, it left the reduction in Na^+_e virtually unaffected. These and other results, including those pertaining to the participation of VSCCs, strongly suggest that EAA-induced decreases in Ca^{2+}_e are mainly due to influx of calcium through NMDA-operated channels, while Na^+ influx occurs through both NMDA- and K/Q-operated channels. A further implication of the findings is that VSCCs may have too low a conductance, or inactivate too quickly, to significantly reduce Ca^{2+}_e.

As will be described below, influx of calcium during adverse conditions may be modulated by changes in Ca^{2+}_i, caused by intracellular calcium release. Release of calcium from the ER by IP_3 is known to occur in response to activation of receptors, which are activated by several agonists, e.g., acetylcholine and noradrenaline. Recently, it has been shown that EAAs stimulate breakdown of polyphosphoinositides.[69-72] Such "metabolotropic" receptors seems to be of the Q-preferring type. Interestingly, activation of NMDA receptors appears to be able to blunt or prevent the response, probably by triggering influx of calcium.[51]

Glutamate Release and Ionic Fluxes during Anoxia, Hypoglycemia, and SD

Almost 30 years ago, in a deduction of remarkable foresight, van Harreveld (see, e.g., REF. 73) concluded that release of glutamate during asphyxia and SD activates a postsynaptic Na^+ conductance mechanism, allowing influx of Na^+ and Cl^-, with osmotically obliged water, thereby causing swelling of apical dendrites. Quite recently, this idea was revived and the suggestion made that EAAs, when accumulating at postsynaptic sites, cause osmolytic, dendritic damage.[19]

It is tempting to assume that when the rise in K^+_e during ischemia is sufficient to depolarize presynaptic endings, and cause release of glutamate, both K/Q and NMDA receptors are stimulated. This will lead to activation of a conductance mechanism for Na^+, K^+, and Ca^{2+} and, secondary to the rise in Ca^{2+}_i or the depolarization, also of one for Cl^- and HCO_3^-. In this way, release of glutamate or a related EAA could account for the "shock" opening of ionic gates. Obviously, a similar release of glutamate may then explain the loss of ion homeostasis in hypoglycemia, and in SD, whether elicited electrically in normal tissue or occurring spontaneously in the ischemic penumbra. One can also speculate that SD causes sudden loss of ion homeostasis, without a preceding rise in K^+_e, because the electrical stimulation leads to an instantaneous release of transmitter. This would also explain why electrical stimulation during anoxia (see REF. 74) shortens the delay before massive ionic fluxes occur.

This interpretation is, at least in part, supported by recent results which demonstrate that deafferentation of CA1 cells prevented Ca^{2+}_e from falling during 10 min of ischemia[75] (FIG. 5). However, it is more difficult to explain why the local injection of a competitive NMDA antagonist only partly prevented cellular uptake of calcium by cells in the CA1 sector,[75] and why MK-801 failed to affect calcium uptake in the neocortex.[74] Furthermore, it seems difficult to account for the sudden ionic fluxes in anoxia, hypoglycemia, and SD only in terms of glutamate release, since these fluxes, especially the efflux K^+, are much more pronounced than those observed in response to ionophoretic release of glutamate (see REFS. 67, 68). Finally, no good explanation has been offered for why NMDA antagonists ameliorate ischemic brain damage in some experimental settings, but not in others. Such antagonists have usually been efficacious in experimental stroke (see REF. 76) but particularly ineffective in models of dense (or complete) ischemia, *i.e.*, ischemia of the type seen in cardiac arrest and forebrain ischemia in the rat (A. Buchan and W. A. Pulsinelli, personal communication; T. Wieloch, personal communication).

A Unifying Hypothesis

As proposed elsewhere[27,28] most, if not all, of the discrepant findings reported in the literature can be explained if one assumes that SD, ischemia, and hypoglycemia lead to the activation of nonselective cation and anion conductance mechanisms, and that such activation occurs when Ca^{2+}_i reaches a certain threshold value. When the hypothesis was formulated, we were inspired by some previous findings. For example, calcium-activated nonselective cation channels seem to be present in many cells (see REFS. 77–79), and calcium-activated Cl^- channels may be equally abundant (REFS. 72, 80; see also REF. 79). Since at least some Cl^- channels are permeable to HCO_3^- (see REF. 81), the anion fluxes may also occur through a nonselective channel, activated by calcium. We also wish to recall that Phillips and Nicholson[82] found that SD was accompanied by the opening of a channel, indiscriminately permeable to anions below a certain size.

We shall now discuss the possibility that available data on fluxes of calcium and other ions can be explained by a modified hypothesis, one which is not crucially dependent on the presence of a nonselective, calcium-activated cation channel. Of importance to this modified hypothesis is the behavior of VSCCs, and the possibility that certain conditions can lead to a reversal of the normal exchange catalyzed by the (electrogenic) $3\ Na^+/Ca^{2+}$ antiporter.

There is no doubt that VSCCs exist at both pre- and postsynaptic neuronal sites, probably also in glial cells (see REFS. 46, 47, 64). However, what is not known is how much calcium they will carry from ECF to ICF during sustained depolarization and in what cellular domains this calcium will be deposited. Some evidence exists that VSCCs do not provide an important route of calcium entry. Thus, depolarization of neocortical cells by superfusion of the tissue with K^+-containing solutions did not materially reduce Ca^{2+}_e in normal tissue,[68] and systemic administration of calcium antagonists had no, or only very small, effects on calcium influx during anoxia.[83,84] However, the first observation does not exclude the possibility that VSCCs form important routes of entry during anoxia (or hypoglycemia), especially since phosphorylation of the channel by kinases may alter its activation/inactivation pattern. Furthermore, the tissue concentrations of calcium antagonists reached in the experiments of Höller *et al.*[83] (see also REF. 84) may have been too low to block flux through VSCCs.

Sodium-calcium exchange by an electrogenic $3Na^+/Ca^{2+}$ antiporter is known to occur in a variety of tissues, and to regulate Ca^{2+}_i in axons and nerve endings.[85–88] Since the exchanger is electrogenic, ionic fluxes depend on the Ca^{2+} and Na^+ gradients, and on the membrane potential (V_m, inside *minus* outside). The driving force for Ca^{2+}

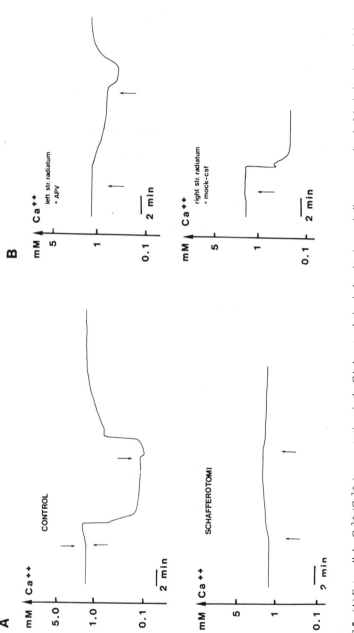

FIGURE 5. (A) Extracellular Ca^{2+} (Ca^{2+}_e) concentrations in the CA-1 sector during ischemia. *Arrows* indicate start and end of the ischemic period in a control rat (*top*) and in a rat with denervated CA-1 region ("Schafferotomi"; *bottom*). Virtually no changes in Ca^{2+}_e are observed in the CA-1 denervated rat. (B) Ca^{2+} concentrations in the CA-1 sector during ischemia. *Arrows* indicate start (*top* and *bottom*) and end (*top*) of the ischemic period in a rat pretreated with APV in the left CA-1 (*top*) and in its control right CA-1 (*bottom*). Only a slow decrease in Ca^{2+}_e, and never more than 45% below the baseline, is observed for the APV-treated left CA-1. (From Benveniste *et al.*[75] Reprinted by permission from *Acta Neurologica Scandinavica*.)

extrusion is given by the difference between the electrochemical potentials for Na^+ (E_{Na}) and Ca^{2+} (E_{Ca}) which are defined as

$$E_{Na} = 3 \left(\frac{RT}{F} \ln \frac{[Na]_e}{[Na]_i} - V_m \right)$$

$$E_{Ca} = \frac{RT}{F} \ln \frac{[Ca]_e}{[Ca]_i} - 2V_m$$

However, this means that a reduction of either the Na^+ gradient or the membrane potential can reverse the direction of the catalyzed transport, and cause calcium influx. Since the direction of the flux critically depends on the $[Ca^{2+}]_i$ / $[Ca^{2+}]_e$ ratio attained it cannot be predicted how the exchanger will handle calcium during, *e.g.*, anoxia or hypoglycemia. Suffice it to say that the exchanger represents a potential cause of cellular calcium entry.

With this as a background we shall describe the cascade of events occurring as a result of depolarization and massive transmitter release. We shall start with complete ischemia and with SD, *i.e.*, the two conditions in which energy production is either completely interrupted or remains unabated.

Complete Ischemia

The results of Benveniste *et al.*[75] suggest that the event triggering the suddenly occurring influx of calcium (and the simultaneous fluxes of K^+ and Na^+) is presynaptic release of transmitters, particularly glutamate or a related EAA. Thus, once glutamate is released and activates the K/Q and the NMDA type of receptors, downhill fluxes of K^+, Ca^{2+}, Na^+, and H^+ will occur via the receptor-linked channels. In this setting, a nonspecific anion conductance mechanism may be activated by either the ensuing depolarization or the rise in Ca^{2+}_i. If glutamate release is the major cause of the massive ion fluxes observed, and if the channels do not rapidly inactivate, the events will be terminated first when the EAAs are taken up by neurons and glial cells in processes which depend on an adequate transmembrane gradient for Na^+. If ATP shortage prevents the Na^+ gradient from providing the necessary driving force, and if ion channels remain in the open state, any ATP formed will be quickly consumed since ions will be pumped against an overwhelmingly large leak. In complete ischemia, of course, no ATP formation will occur. We recognize that since membranes must be depolarized, the leak pathways may be either voltage-sensitive or agonist-operated. The former will include the surface $3 Na^+/Ca^{2+}$ exchanger and, very hypothetically, the latter could encompass a membrane channel connected to the phospholipase C-IP$_3$ system (see above).

If this is the scenario, what triggers release of EAAs and other transmitters? Two alternatives can be envisaged. One is that the slow seepage of K^+ from cells (via Ca^{2+}-activated K^+ channels?) eventually increases K^+_e to values which will depolarize presynaptic endings. The other is that when ATP is reduced to a level, which will open ATP-dependent K^+ channels (see REFS. 89, 90) and the massive K^+ efflux which ensues leads to the required presynaptic depolarization. In this context, we wish to recall that part of the release of EAAs may be calcium-independent and be caused by a reversal of flux via the (electrogenic) $2Na^+$/glutamate symporter at a critical degree of depolarization (see REF. 91).

Obviously, failure of glutamate release would explain why deafferentation prevents Ca^{2+}_e from falling during ischemia.[75] However, why is the calcium uptake only partly blocked by a locally injected, competitive NMDA antagonist, and unaffected by a noncompetitive one? If our hypothesis is correct, we must be able to explain why Ca^{2+}_e falls when NMDA receptors, or the ion channel they gate, are blocked.

It seems easy to explain why deafferentation effectively prevents Ca^{2+}_e from falling since, in the absence of glutamate release, neither the K/Q nor the NMDA receptors will be activated. In the absence of Na^+ influx, and depolarization, we can envisage that VSCCs will not be activated, and the $3Na^+/Ca^{2+}$ exchange will not be reversed. Furthermore, since the receptors coupled to phospholipase C will not be activated, any effect of a rise in Ca^{2+}_i on ion channels will be minimized.

There are two possible reasons why calcium, in the presence of a competitive NMDA antagonist, will continue to enter cells, albeit at a reduced rate. *First,* entry of Na^+ through the K/Q type of receptor must depolarize the postsynaptic membranes, favoring Ca^{2+} influx via VSCCs and by reversal of $3 Na^+/Ca^{2+}$ exchange. *Second,* activation of K/Q and other receptors coupled to phospholipase C by the transmitters released during depolarization could cause calcium-induced alterations of ion channels, maintaining them in the open state (see, *e.g.*, REF. 92). In other words, although blockade of the NMDA receptor may delay or curtail calcium influx, it cannot prevent it altogether.

But why, then, cannot blockade of the NMDA-linked channel by MK-801 prevent the sudden loss of ion homeostasis during anoxia? We have no ready explanation for this but wish to recall two facts. *First,* the results with APV were obtained following local injection of a large dose of the antagonist into the CA1 sector of the hippocampus, while those with MK-801 were obtained on the neocortex, following systemic injection of the drug. *Second,* conceivably, in terms of NMDA-related responses, the densely packed CA1 neurons may behave differently from those in the neocortex. Besides, the active concentration of antagonist may be much lower following systemic injection of MK-801 than following local injection of APV.

In summary, we propose that the event triggering opening of both agonist- and voltage-dependent ion channels is the release of glutamate and the activation of K/Q and NMDA receptors. This would explain why deafferentation virtually prevents calcium influx during ischemia (neither receptor is activated) and why NMDA receptor blockade only curtails it (the K/Q receptor can still be activated). Once depolarization has been triggered, and Na^+ influx has occurred, calcium may enter by VSCCs, and by reversal of the $3Na^+/Ca^{2+}$ exchanger. Possibly, any increase in Ca^{2+}_i, secondary to acidosis or IP_3 release, may contribute to calcium influx by altering directly the phosphorylation state of ion channels. Activation of C-kinase might have similar effects. We recall that C-kinase activation could be the result either of stimulation of receptors coupled to phospholipase C stimulation (see REFS. 29, 104), or of a rise in Ca^{2+}_i (see REF. 93).

Spreading Depression

This condition differs from complete ischemia in that cellular energy production is not compromised. Based on previous *in vitro* and *in vivo* results, we conclude that the major route of calcium entry during SD is the ion channel gated by NMDA receptors. Thus, glutamate release will, by the sequential mechanism involving K/Q and NMDA receptor activation, lead to influx of calcium and, simultaneously, to downhill fluxes of K^+ and Na^+. It remains to be explained, though, why competitive and noncompetitive NMDA antagonists such as AP7,[94] ketamine,[95] and MK-801,[96] by blocking calcium entry, prevent the SD from being elicited (and propagated). Tentatively, one reason is that since blockade of the NMDA-gated channel prevents influx of both Ca^{2+} and Na^+, depolarization is curtailed. This may reduce activation of VSCCs and prevent reversal of the $3Na^+/Ca^{2+}$ exchanger. Reduced influx of Na^+ will also decrease the driving force for such a reversal, and diminished influx of Ca^{2+} may reduce changes in the phosphorylation state of ion channels, perhaps preventing or reducing channel activation. Potentially, blockade of downhill fluxes of Ca^{2+}, Na^+, and K^+ will prevent local consumption of ATP and thus prevent activation of calcium- and ATP-dependent K^+ channels. Furthermore, since the ATP stores are normal, active transport of Na^+, K^+, and Ca^{2+} will continue. As a speculative possibility one may also assume that blockade of presynaptic NMDA receptors prevents calcium influx and transmitter release. If this is so, Na^+ influx via the K/Q-linked channel will be reduced as well. Obviously, all these changes may prevent the massive depolarization and activation of the conductance mechanisms which are required to make the local disturbance a propagated one (*e.g.*, by reducing efflux of K^+ and EAAs from the affected cell).

As already discussed, the hypothesis of calcium-related damage assumes that cell damage is incurred when Ca^{2+}_i reaches critical levels for a certain period of time. An unexplained finding is that SDs, even when repeatedly elicited for a period of 4 hr, do not cause brain damage.[97] Very likely this is because the calcium entering cells is quickly expelled by ATP-driven transport or $3Na^+/Ca^{2+}$ exchange; besides, sequestration into mitochondria and ER may help in dampening the rise in Ca^{2+}_i and in terminating the transient. It follows from this reasoning that should SDs be elicited in an energetically strained tissue, cell damage may well occur. Furthermore, one would predict that NMDA antagonists, by reducing Ca^{2+} flux via a major entry route, could prevent or ameliorate such damage (see below).

Hypoglycemia

The final depolarization during hypoglycemia is usually preceded by one or, occasionally, two SD-like transients.[60,98] FIGURE 4 (see above) illustrates how such a transient leads to partial and transient normalization of Ca^{2+}_e. It seems very likely that the first depolarization activates an agonist- and voltage-dependent conductance mechanism in the same manner as a SD does. However, in the energy-compromised situation prevailing during hypoglycemia, repumping of ions by ATP-driven translocases rapidly consumes the available ATP. In the situation thus arising, loss of energy homeostasis prevails because the ionic leak present causes ATP to be consumed at a rate which exceeds its rate of production. Thus, as long as the ionic leak persists, so does the energy failure. This interpretation is supported by data demonstrating rapid decreases in tissue phosphocreatine and ATP concentrations during the first depolarization.[99] It is also of interest that cerebral oxygen consumption is maintained during hypoglycemic coma, at least during the first 15 min.[100,101] This could be due to uncoupling of oxidative phosphorylation; however, the combination of maintained oxygen consumption and markedly reduced ATP values may well reflect the presence of a large ionic leak.

Obviously, in hypoglycemia, blockade of calcium and sodium influx through the NMDA-gated channel could be what changes the situation to one in which the reduced energy production can cope with the ionic leaks through channels which are still patent. It is understandable, therefore, that competitive and noncompetitive NMDA antagonists may ameliorate hypoglycemic brain damage. It is more difficult to explain why MK-801 failed to affect calcium influx during hypoglycemic coma.[74] However, the difference between these results, obtained in the cerebral cortex, and those showing amelioration of hypoglycemic damage in the caudoputamen may reflect not only differences in brain areas chosen for study, but also the severity of the hypoglycemic insult.

Focal Ischemia

It seems likely that conditions in the central core of an infarct resembles those prevailing in dense global or forebrain ischemia. Predictably, in neither of these conditions would noncompetitive NMDA antagonists materially affect calcium influx, or significantly ameliorate the ischemic damage. This conclusion is supported by results demonstrating the lack of efficacy of such antagonists in global or forebrain ischemia (see above), and by those showing that although MK-801 and other noncompetitive NMDA antagonists reduce infarct size following experimental middle cerebral artery occlusion, they leave the densely ischemic core in the caudoputamen unaffected (*e.g.*, REF. 102).

Probably, conditions are different in the periofocal "penumbra" zone in which flow rates are less reduced. In this zone, depolarization and enhanced transmitter release may be transient events, reflecting the presence of a remaining energy source (see, *e.g.*, REF. 103). In fact, at least in the rat trains of depolarization waves are observed resembling SD.[62] We submit that such irregularly or regularly occurring depolarizing events predispose to progressive neuronal damage, and an extension of the infarct into the penumbra zone. Probably, in this situation noncompetitive NMDA antagonists protect cells by preventing sodium and calcium influx, and K^+ efflux, in much the same way as they protect the tissue during severe hypoglycemia. In summary, NMDA antagonists, while not protecting the ischemic core prevent the infarct from being extended into the penumbra zone, simply because enough ATP is produced to efficiently extrude Ca^{2+}_i and Na^+_i and to accumulate K^+_i when an NMDA antagonist is present to block the major route of entry. A contributory factor may be that the ATP produced may expel enough Na^+, and accumulate enough K^+, to prevent the fall in transmembrane Na^+ gradient and the depolarization which triggers massive presynaptic release of transmitters (see REF. 26).

CONCLUDING REMARKS

Obviously, our hypothesis does not critically depend on the presence of a calcium-activated nonspecific cation conductance mechanism since many of the essential features of the hypothesis would be valid even if other conductance mechanisms are involved. These features encompass, for example, the assumption that the absence or presence of a remaining energy source will determine whether or not NMDA antagonists will be efficacious in a given situation. Thus, if flux through NMDA-gated channels is impeded in the presence of some ATP production, the ionic leaks may be sufficiently reduced to allow the remaining ATP source to extrude whatever calcium is leaking in, to keep cell membranes polarized, and to maintain transmembrane Na^+ gradients, thereby enhancing transmitter uptake and promoting calcium extrusion by $3Na^+/Ca^{2+}$ exchange. However, before we can decide whether the original or the alternative hypothesis best predicts the true membrane events it seems justified to explore the role played by VSCCs and by reversed $3Na^+/Ca^{2+}$ exchange under the pathological conditions discussed. It is also warranted that various antagonists of VSCCs and AOCCs are studied under similar experimental conditions, preferably also combined.

The general hypothesis of calcium-related damage, and the dependence of this damage on the severity of the cellular energy failure, also requires rigorous testing, since the results of properly conducted experiments may have far-reaching implications for clinical therapy of several disorders which are accompanied by loss of cellular calcium homeostasis. Possibly, such disorders encompass age-related diseases, including Alzheimer's disease.

ACKNOWLEDGMENTS

The authors wish to acknowledge the skilled and devoted work of Mrs. Birgit Olsson and Mrs. Erna Björkengren in preparing the manuscript.

REFERENCES

1. FLECKENSTEIN, A., J. JANKE, H. J. DORING & O. LEDER. 1974. Myocardial fiber necrosis due to intracellular Ca overload—A new principle in cardiac hypertrophy. Recent Adv. Stud. Card. Struct. Metab. 4: 563–568.
2. WROGEMANN, K. & S. D. J. PENA. 1976. Mitochondrial calcium overload: A general mechanism for cell necrosis in muscle diseases. Lancet 1: 672–674.
3. LEONARD, J. P. & M. M. SALPETER. 1979. Agonist-induced myopathy at the neuromuscular junction is mediated by calcium. J. Cell Biol. 82: 811–819.
4. SCHANNE, F. A. X., A. B. KANE, E. E. YOUNG & J. L. FARBER. 1979. Calcium dependence of toxic cell death: A final common pathway. Science 206: 700–702.
5. FARBER, J. L. 1981. The role of calcium in cell death. Life Sci. 29: 1289–1295.
6. SIESJÖ, B. K. 1981. Cell damage in the brain: A speculative synthesis. J. Cereb. Blood Flow Metab. 1: 155–185.
7. DIENEL, G. A. 1984. Regional accumulation of calcium in postischemic rat brain. J. Neurochem. 43: 913–925.
8. DESHPANDE, J. K., B. K. SIESJÖ & T. WIELOCK. 1987. Calcium accumulation and neuronal damage in the rat hippocampus following cerebral ischemia. J. Cereb. Blood Flow Metab. 7: 89–95.
9. SIESJÖ, B. K. 1988. Historical overview. Calcium, ischemia, and death of brain cells. Ann. N.Y. Acad. Sci. 522: 638–661.
10. MELDRUM, B. S. 1983. Metabolic factors during prolonged seizures and their relation to nerve cell death. Adv. Neurol. 34: 261–275.
11. RAICHLE, M. 1983. The pathophysiology of brain ischemia. Ann. Neurol. 13: 2–10.
12. SIESJÖ, B. K. & T. WIELOCK. 1985. Cerebral metabolism in ischemia: Neurochemical basis for therapy. Br. J. Anaesth. 57: 47–62.
13. NICOTERA, P., P. HARZELL, G. DAVIS & S. ORRENIUS. 1986. The formation of plasma membrane blebs in hepatocytes exposed to agents that increase cytosolic Ca^{2+} is mediated by the activation of a non-lysosomal proteolytic system. FEBS Lett. 209: 139–144.
14. ORRENIUS, S., D. J. McCONKEY, D. P. JONES & P. NICOTERA. 1988. Ca^{2+}-activated mechanisms in toxicity and programmed cell death. ISI Atlas of Science. Pharmacology 2: 319–324.
15. CHEUNG, J. Y., J. V. BONVENTRE, C. D. MALIS & A. LEAF. 1986. Calcium and ischemic injury. N. Engl. J. Med. 314: 1670–1676.
16. LEMASTERS, J. J., J. DIGUISEPPI, A. L. NIEMINEN & B. HERMAN. 1987. Blebbing, free Ca^{2+} and mitochondrial membrane potential preceding cell death in hepatocytes. Nature (London) 325: 78–81.
17. CHOI, D. W. 1985. Glutamate neurotoxicity in cortical cell culture is calcium dependent. Neurosci. Lett. 58: 293–297.
18. CHOI, D. W. 1988. Glutamate neurotoxicity and diseases of the nervous system. Neuron 1. In press.
19. ROTHMAN, S. M. & J. W. OLNEY. 1986. Glutamate and the patholphysiology of hypoxic-ischemic brain damage. Ann. Neurol. 19: 105–111.
20. MAYER, M. L. & G. L. WESTBROOK. 1987. The physiology of excitatory amino acids in the vertebrate central nervous system. Prog. Neurobiol. 28: 197–276.
21. SLOVITER, R. S. 1983. "Epileptic" brain damage in rats induced by sustained electrical stimulation of the perforant path. I. Acute electrophysiological and light microscopic studies. Brain Res. Bull. 10: 675–697.
22. WIELOCH, T. 1985. Hypoglycemia-induced neuronal damage is prevented by a N-methyl-D-aspartate receptor antagonist. Science 230: 681–683.

23. KHACHATURIAN, Z. S. 1984. Towards theories of brain ageing. *In* Handbook of Studies in Psychiatry and Old Age. D. W. Kay and G. D. Burrows, Eds. 7–30. Elsevier. Amsterdam.
24. GIBSON, G. E. & C. PETERSON. 1987. Calcium and the aging nervous system. Neurobiol. Aging 8: 329–343.
25. MARAGOS, W. F., J. T. GREENAMYRE, J. B. PENNEY, JR. & A. B. YOUNG. 1987. Glutamate dysfunction in Alzheimer's disease: An hypothesis. Trends Neurosci. 10: 65–66.
26. WESTERBERG, E., J. KEHR, U. UNGERSTEDT & T. WIELOCH. 1988. The NMDA-antagonist MK-801 reduces extracellular amino acid levels during hypoglycemia and prevents striatal damage. Neurosci. Res. Commun. 3(3): 151–158.
27. SIESJÖ, B. K. & F. BENGTSSON. 1989. Calcium, calcium antagonists, and ischemic cell death in the brain. I. Krieglstein, Ed.: 23–29. CRC Press. Boca Raton, FL.
28. SIESJÖ, B. K. & F. BENGTSSON. 1989. Calcium, calcium antagonists, and calcium-related pathology in brain ischemia, hypoglycemia and spreading depression: A unifying hypothesis. J. Cereb. Blood Flow Metab. 9. In press.
29. BERRIDGE, M. J. 1984. Inositol triphosphate and diacylglycerol as second messengers. Biochem. J. 221: 345–360.
30. BERRIDGE, M. J. & R. F. IRVINE. 1984. Inositol trisphosphate, a novel second messenger in cellular signal transduction. Nature (London) 312: 315–321.
31. CARAFOLI, E. 1982. The regulation of intracellular calcium. Adv. Exp. Med. Biol. 151: 461–472.
32. CARAFOLI, E. 1987. Intracellular calcium homeostasis. Annu. Rev. Biochem. 56: 395–433.
33. RASMUSSEN, H. & D. M. WAISMAN. 1983. Modulation of cell function in the calcium messenger system. Rev. Physiol. Biochem. Pharmacol. 95: 111–148.
34. BLAUSTEIN, M. P. 1984. The energetics and kinetics of sodium-calcium exchange in barnacle muscles, squid axons, and mammalian heart: The role of ATP. *In* Electrogenic Transport: Fundamental Principles and Physiological Implications. M. P. Blaustein & M. Lieberman, Eds.: 129–147. Raven Press. New York, NY.
35. DENTON, R. M. & J. G. MCCORMACK. 1985. Ca^{2+} transport by mammalian mitochondria and its role in hormone action. Am. J. Physiol. 249: E543–E554.
36. HANSFORD, R. G. 1985. Relation between mitochondrial calcium transport and control of energy metabolism. Rev. Physiol. Biochem. Pharmacol. 102: 1–72.
37. NICHOLLS, D. G. 1986. Intracellular calcium homeostasis. Br. Med. Bull. 42: 353–358.
38. PUTNEY, J. W., JR. 1986. A model for receptor-regulated calcium entry. Cell Calcium 7: 1–12.
39. TAYLOR, C. W. 1987. Receptor regulation of calcium entry. Trends Pharmacol. Sci. 8: 79–80.
40. ABERCROMBIE, R. F. & C. E. HART. 1986. Calcium and proton buffering and diffusion in isolated cytoplasm from Myxicola axons. Am. J. Physiol. 250: C391–C405.
41. BAKER, P. F. & J. A. UMBACH. 1987. Calcium buffering in axons and axoplasm of loligo. J. Physiol. 383: 369–394.
42. COLLINGRIDGE, G. J. 1985. Long-term potentiation in the hippocampus: Mechanisms of initiation and modulation by neurotransmitters. Trends Phamacol. Sci. 6: 407–411.
43. NOWYCKY, M. C., A. P. FOX & R. W. TISEN. 1985. Three types of neuronal calcium channel with different calcium agonist sensitivity. Nature (London) 316: 440–443.
44. FOSTER, A. C. & G. E. FAGG. 1987. Taking apart NMDA receptors. Nature (London) 329: 395–396.
45. MAYER, M. 1987. Two channels reduced to one. Nature (London) 325: 480–481.
46. MILLER, R. J. 1987. Multiple calcium channels and neuronal function. Science 235: 46–52.
47. YAARI, Y., B. HAMON & H. D. LUX. 1987. Development of two types of calcium channels in cultured mammalian hippocampal neurons. Science 235: 680–682.
48. WATKINS, J. C. & R. H. EVANS. 1981. Excitatory amino acid transmitters. Annu. Rev. Pharmacol. Toxicol. 21: 165–204.
49. MAYER, M. L., G. L. WESTBROOK & P. B. GUTHRIE. 1984. Voltage-dependent block by Mg^{2+} of NMDA responses in spinal cord neurones. Nature (London) 309: 261–263.
50. COTMAN, C. W., D. T. MONAGHAN, E. PALMER & J. GEDDES. 1988. Excitatory amino acid neurotransmission: NMDA receptors and hebb-type synaptic plasticity. Annu. Rev. Neurosci. 11. In press.

51. COTMAN, C. W., D. MONAGHAN, E. PALMER & J. GEDDES. 1989. NMDA receptors, Ca^{2+} and Alzheimer's disease. Ann. N.Y. Acad. Sci. In press.
52. MACDERMOTT, A. B., M. L. MAYER, G. L. WESTBROOK, S. J. SMITH & J. L. BAKER. 1986. NMDA-receptor activation increases cytoplasmic calcium concentration in cultured spinal cord neurones. Nature (London) 321: 519–522.
53. NOWAK, L., P. BREGESTOVSKI, P. ASCHER, A. HERBET & A. PROCHIANTZ. 1984. Magnesium gates glutamate-activated channels in mouse central neurones. Nature (London) 307: 462–465.
54. JOHNSON, J. W. & P. ASCHER. 1987. Glycine potentiates the NMDA response in cultured mouse brain neurons. Nature (London) 325: 529–531.
55. NICHOLSON, C., G. T. GRUGGENCATE, R. STEINBERG & H. STÖCKLE. 1977. Calcium modulation in brain extracellular microenvironment demonstrated with ion-selective micropipette. Proc. Natl. Acad. Sci. USA 74: 1287–1290.
56. HANSEN, A. & T. ZEUTHEN. 1981. Extracellular ion concentrations during spreading depression and ischemia in the rat brain cortex. Acta Physiol. Scand. 113: 437–445.
57. HARRIS, R. J., L. SYMON, N. M. BRANSTON & M. BAYHAN. 1981. Changes in extracellular calcium activity in cerebral ischaemia. J. Cereb. Blood Flow Metab. 1: 203–209.
58. HANSEN, A. J. 1985. Effects of anoxia on ion distribution in the brain. Physiol. Rev. 65: 101–148.
59. SIESJÖ, B. K. 1988. Acidosis and ischemic brain damage. Neurochem. Pathol. 9: 31–88.
60. HARRIS, R. J., T. WIELOCH, L. SYMON & B. K. SIESJÖ. 1984. Cerebral extracellular calcium activity in severe hypoglycemia: Relation to extracellular potassium and energy charge. J. Cereb. Flow Metab. 4: 187–193.
61. NICHOLSON, C. 1980. Dynamics of the brain cell microenvironment. Neurosci. Res. Program Bull. 18: 177–322.
62. NEEDERGAARD, M. & J. ASTRUP. 1986. Infarct rim: Effect of hyperglycemia on direct current potential and (^{14}C)2-deoxyglucose phosphorylation. J. Cereb. Blood Flow Metab. 6: 607–615.
63. KUDO, Y. & A. OGURO. 1986. Glutamate-induced increase in intracellular Ca^{2+} concentration in isolated hippocampal neurones. Br. J. Pharmacol. 89: 191–198.
64. THAYER, S. A., S. N. MURPHY & R. J. MILLER. 1986. Widespread distribution of dihydropyridine-sensitive calcium channels in the central nervous system. Mol. Pharmacol. 30: 505–509.
65. MURPHY, S. N., S. A. THAYER & R. J. MILLER. 1987. The effects of excitatory amino acids on intracellular calcium in single mouse striatal neurons *in vitro*. J. Neurosci. 7(12): 4145–4158.
66. HEINEMANN, U. & R. PUMAIN. 1980. Extracellular calcium activity changes in cat sensorimotor cortex induced by iontophoretic application of amino acids. Exp. Brain Res. 40: 247–250.
67. PUMAIN, R. & U. HEINEMANN. 1985. Stimulus- and amino acid-induced calcium and potassium changes in rat neocortex. J. Neurophysiol. 53 1–16.
68. PUMAIN, R., I. KURCEWICZ & J. LOUVEL. 1987. Ionic changes induced by excitatory amino acids in the rat cerebral cortex. Can. J. Physiol. Pharmacol. 65: 1067–1077.
69. SLADECZEK, F., J.-P. PIN, M. RÉCASENS, J. BOCKAERT & S. WEISS. 1985. Glutamate stimulates inositol phosphate formation in striatal neurones. Nature (London) 317: 717–719.
70. SLADECZEK, F., M. RÉCASENS & J. BOCKAERT. 1988. A new mechanism for glutamate receptor action: Phosphoinositide hydrolysis. Trends Neurosci. 11: 545–549.
71. NICOLETTI, F., J. L. MEEK, M. J. IADAROLA, D. M. CHUANG, B. L. ROTH & E. COSTA. 1986. Coupling of inositol phospholipid metabolism with excitatory amino acid recognition sites in rat hippocampus. J. Neurochem. 46: 40–46.
72. SUGIYAMA, H., I. ITO & C. HIRONO. 1987. A new type of glutamate receptor linked to inositol phospholipid metabolism. Nature (London) 325: 531–533.
73. VAN HARREVELD, A. 1970. A mechanism for fluid shifts specific for the central nervous system. *In* Current Research in Neurosciences. H. T. Wycis, Ed. Top Probl. Psychiatry Neurol. 10: 62–70.
74. ZHANG, E., M. LAURITZEN, T. WIELOCH & A. J. HANSEN. 1989. Calcium movements in

brain during failure of energy metabolism. *In* Cerebral Ischemia and Calcium. A. Hartmann & W. Kuschinsky, Eds. 162–168. Springer Verlag. Berlin.

75. BENVENISTE, H., M. B. JØRGENSEN, N. H. DIEMER & A. J. HANSEN. 1988. Calcium accumulation by glutamate receptor activation is involved in hippocampal cell damage after ischemia. Acta Neurol. Scand. **78:** 529–536.

76. PARK, C. K., D. G. NEHLS, D. I. GRAHAM, G. M. TEASDALE & J. MCCULLOCK. 1988. The glutamate antagonist MK-801 reduces focal ischemic brain damage in the rat. Ann. Neurol. **24:** 543–551.

77. COQUHOUN, D., E. NEHER, H. REUTER & C. F. STEVENS. 1981. Inward current channels activated by intracellular Ca in cultured cardiac cells. Nature (London) **294:** 752–754.

78. HOFMEIER, G. & D. LUX. 1981. The time courses of intracellular free calcium and related electrical effects after injection of $CaCl_2$ into neurons of the snail, Helix pomatia. Pflügers Arch. **391:** 242–251.

79. PARTRIDGE, L. D. & D. SWANDULLA. 1988. Calcium-activated non-specific cation channels. Trends Neurosci. **11**(2): 69–72.

80. MARTY, A. 1987. Control of ionic currents and fluid secretion by muscarinic agonists in exocrine glands. Trends Neurosci. **10:** 373–377.

81. KAILA, K. & J. VOIPIO. 1988. Postsynaptic fall in intracellular pH induced by GABA-activated bicarbonate conductance. Nature (London) **330:** 163–165.

82. PHILLIPS, J. M. & C. NICHOLSON. 1979. Anion permeability in spreading depression investigated with ion-sensitive microelectrodes. Brain Res. **173:** 567–571.

83. HÖLLER, M., H. DIERKING, K. DENGLER, F. TEGTMEIER & T. PETERS. 1986. Effect of flunarizine on extracellular ion concentration in the rat brain under hypoxia and ischemia. *In* Acute Brain Ischemia Medical and Surgical Therapy. N. Battistina, P. Fiorani, R. Courbier, F. Plum & C. Fieschi, Eds.: 229–236. Raven Press. New York, NY.

84. PETERS, T. 1986. Calcium in physiological and pathological cell function. Eur. Neurol. **25**(Suppl. 1): 27–44.

85. LANGER, G. A. 1982. Sodium-calcium exchange in the heart. Annu. Rev. Physiol. **44:** 435–449.

86. EISNER, D. A. & W. J. LEDERER. 1985. Na-Ca exchange: Stoichiometry and electrogenicity. Am. J. Physiol. **248:** C189.

87. SHEU, S.-S. & M. P. BLAUSTEIN. 1986. Sodium/calcium exchange and regulation of cell calcium and contractility in cardiac muscle, with a note about vascular smooth muscle. *In* The Heart and Cardiovascular System. H. A. Fozzard, E. Haber, R. B. Jennings, A. M. Katz & H. E. Morgan, Eds.: 509–535. Raven Press. New York, NY.

88. BARCENAS-RUIZ, L., D. J. BEUCKELMANN & W. G. WIER. 1987. Sodium-calcium exchange in heart: Membrane currents and changes in Ca^{2+}_i. Science **238:** 1720–1722.

89. NOMA, A. 1983. ATP-regulated K^+ channels in cardiac muscle. Nature (London) **305:** 147–148.

90. FOSSET, M., J. R. DE WEILLE, R. D. GREEN, H. SCHMID-ANTOMARCHI & M. LAZDUNSKI. 1988. Antidiabetic sulfonylureas control action potential properties in heart cells via high affinity receptors that are linked to ATP-dependent K^+ channels. J. Biol. Chem. **263:** 7933–7936.

91. NICHOLLS, D. G. 1989. The release of glutamate, aspartate and GABA from isolated nerve terminals. J. Neurochem. **52:** 337–341.

92. KACZMAREK, L. K. 1987. The role of protein kinase C in the regulation of ion channels and neurotransmitter release. Trends Neurosci. **10:** 30–34.

93. EBERHARD, D. A. & R. W. HOLZ. 1988. Intracellular Ca^{2+} activates phospholipase C. Trends Neurosci. **11:** 517–520.

94. MODY, I., J. D. LAMBERT & U. HEINEMANN. 1987. Low extracellular magnesium induces epileptiform activity and spreading depression in rat hippocampal slices. J. Neurophysiol. **57:** 869–888.

95. GOROLEVA, N. A., V. I. KOROLEVA, T. AMEMORI, V. PAVLIK & J. BURES. 1987. Ketamine blockade of cortical spreading depression in rat. Electroencephaloger. Clin. Neurophysiol. **66:** 440–447.

96. HANSEN, A. J., M. LAURITZEN & T. WIELOCH. 1988. NMDA antagonists inhibit cortical spreading depression, but not anoxic depolarization. *In* Frontiers in Excitatory Amino Acid

Research. E. A. Cavalheiro, J. Lehmann & L. Turski, Eds.: 661–666. Alan R. Liss, New York, NY.

97. NEEDERGAARD, M. & A. J. HANSEN. 1988. Spreading depression is not associated with neuronal injury in the normal brain. Brain Res. **449:** 395–398.

98. PELLIGRINO, D., L.-O. ALMQUIST & B. K. SIESJÖ. 1981. Effects of insulin-induced hypoglycemia on intracellular pH and impedance in the cerebral cortex of the rat. Brain Res. **221:** 129–147.

99. WIELOCH, T., R. J. HARRIS, L. SYMON & B. K. SIESJÖ. 1984. Influence of severe hypoglycemia on brain extracellular calcium and potassium activities, energy, and phospholipid metabolism. J. Neurochem. **43:** 160–168.

100. NORBERG, K. & B. K. SIESJÖ. 1976. Oxidative metabolism of the cerebral cortex of the rat in severe insulin-induced hypoglycaemia. J. Neurochem. **26:** 345–352.

101. NILSSON, B., C.-D. AGARDH, M. INGVAR & B. K. SIESJÖ. 1981. Cerebrovascular response during and following severe insulin-induced hypoglycemia: CO_2-sensitivity, autoregulation, and influence of prostaglandin synthesis inhibition. Acta Physiol. Scand. **111:** 455–463.

102. OZYURT, E., D. I. GRAHAM, G. N. WOODRUFF & J. McCULLOCK. 1988. Protective effect of the glutamate antagonist, MK-801 in focal cerebral ischemia in the cat. J. Cereb. Blood Flow Metab. **8:** 138–143.

103. STRONG, A. J., G. S. VENABLES & G. GIBSON. 1983. The cortical ischaemic penumbra associated with occlusion of the middle cerebral artery in the cat: 1. Topography of changes in blood flow, potassium ion activity, and EEG. J. Cereb. Blood Flow Metab. **3:** 86–96.

104. NISHIZUKA, Y. 1984. The role of protein kinase C in cell surface signal transduction and tumor promotion. Nature (London) **308:** 693–698.

Calcium-Induced Neuronal Degeneration: A Normal Growth Cone Regulating Signal Gone Awry (?)

S. B. KATER, M. P. MATTSON, AND P. B. GUTHRIE

Program in Neuronal Growth and Development
Department of Anatomy and Neurobiology
Colorado State University
Fort Collins, Colorado 80523

INTRODUCTION

Why do nerve cells degenerate? The answer to this question may lie in the answer to another question: How does the nervous system increase its integrative capabilities during development? These two questions are not frequently linked together. However, recent evidence compels us to consider the possibility that the same mechanisms which endow the developing brain with increased coding capability might actually malfunction later in life and produce highly deleterious effects. While the ability to change is the *sine qua non* of the higher nervous system, change beyond constrained limits could in fact be deleterious.[1] What could be more effective for increased integration than increasing the extent of dendritic arborizations, and similarly what could be more deleterious than the radical shrinking or even elimination of such dendritic surfaces? Taken together, these questions form the basis for the present discussion which centers around our recent work demonstrating that intracellular calcium levels can determine whether a neuron is in an expanding or degenerating state of development.

Hypothesis

The general focus of this paper is the interrelationship of those mechanisms normally used to form connections during development and modify connections in the mature nervous system, and whether these same mechanisms are also playing a role in the catastrophic processes of neuronal degeneration. Our working hypothesis is that neuronal architecture can be selectively modified by a set of underlying processes now well understood for their roles in neural coding.[2-6] Namely, **action potentials, neurotransmitters, and intracellular calcium levels play a role in governing the shape of the neuron's dendritic arbor.** These regulators can determine when dendritic arbors grow, when they stop, when they are pruned back and regress and, finally, when they are completely eliminated. The logical extension of these processes is the basis of our final hypothesis, namely, that the **same processes regulating the plasticity of neuroarchitecture are, in fact, one part of a continuum which ultimately can lead to neuronal death.**[5-7]

The Neuronal Growth Cone

The neuronal growth cone is well recognized as the primary neuronal structure responsible for the shaping of a dendritic tree. The behavior of this organelle is responsible for process outgrowth and branching, target recognition, and connectivity. It is now well established that the neuronal growth cone is highly responsive to a variety of agents which are more usually recognized for their roles in information coding.[4,8] These regulatory mechanisms can determine whether the growth cone advances, whether it turns, branches, stops, or even regresses. The neuronal growth cone must also be a key organelle of neuroplasticity. It seems likely that the growth cone represents a common (transient?) dynamic form that neurite tips must become in order to be capable of morphological change. The growth cone has been found in all stages of neuronal life and probably represents the form which neurite tips take that underlies change. In many ways, the neuronal growth cone can be regarded as the most prominent morphological substrate of at least one major class of neuronal plasticity.

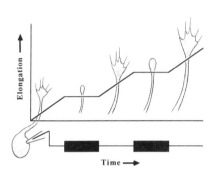

Time →

FIGURE 1. A summary of the results of Cohan, *et al.*[3] on the effect of electrical activity on neurite outgrowth. An extracellular patch electrode, on the cell body, was used to evoke action potentials. Simultaneously, neurite elongation and the status of growth cones were monitored photographically. The results of these studies demonstrated that, with the electrode in place, normal elongation and growth cone behavior were observed. During periods of evoked electrical activity, however, both growth cone behavior and neurite elongation were greatly changed. The firing of action potentials inhibited neurite elongation, filopodia withdrew, and the lamellipodial surface area was greatly decreased. Subsequent to cessation of action potentials, general motility resumed. This is a reversible, and highly reproducible, effect.

Activity Regulation of Growth Cone Behavior

One of the best established concepts in neurobiology is that the nervous system can be changed as a function of its activity. At the cellular level, the most direct test of this general statement comes from our observations on the neuronal growth cone. In experiments employing the giant growth cones of molluscs, we have been able to demonstrate conclusively that whether the growth cone advances or is retarded is directly dependent upon the activity state of the neuron: the generation of action potentials within a neuron inhibits growth cone behavior. This work[3] brought home in a very simple fashion the idea that the normal integrative events of processing in the mature nervous system also have important ramifications for shaping the structure of that very same system. The generation of activity in a given circuit would perforce stabilize that circuit. Alternatively, one must ask, when activity ceases, would such circuitry break down and reconfigure into a different (more usable?) network?

The experiments shown in FIGURE 1 demonstrate the behavior of the neuronal growth cone during neurite elongation in cell culture. Under these conditions, where the growth cone can be monitored most precisely, it can be seen that the rate of advance can proceed in a highly linear fashion. Upon generation of action potentials by electrical stimulation of the cell body, however, activity of the growth cone ceases. Filopodia and lamellipodia pull in and the neurite ceases its elongation. This arrested status is maintained for as long as activity is maintained in the neuron. When activity ceases, the growth cone resumes its elongation and its now classical behavior pattern of filopodial and lamellipodial exten-sion. These observations demonstrate the precise regulation of neuronal morphology through well understood mechanisms normally considered only with the context of the mature nervous system.

Neurotransmitter Regulation of Growth Cone Behavior

We can now identify a new role for classical neurotransmitters in addition to their fundamental role in interneuronal communication. The same neurotransmitters which are the mainstay of neuronal circuit function can also act as prominent sculptors of that same developing neuronal circuitry. This role for neurotransmitters positions them prominently as potential agents of neuronal dysfunction. Thus, inherent in their ability to selectively sculpt and shape dendritic arborizations, lies the potential for highly deleterious actions.

Neurotransmitter effects on neuronal growth cones seem to be quite general and prominent within neural circuits throughout the animal kingdom. The first observation of neurotransmitter effects on growth cones were made on the giant growth cones of the mollusc, *Helisoma*. A very significant observation was that neurotransmitters could se-lectively affect the growth cones of specific identified neurons. For example, serotonin has a highly stereotyped action on the growth cones of neuron B19; the growth cones of neuron B19 retract their filopodia and lamellipodia and cease elongation in the continued presence of serotonin.[2] The effect is highly selective in that other neurons (*e.g.*, neuron B5) are unaffected by this particular neurotransmitter. We now know that different neu-rotransmitters can effect different but overlapping subsets of neurons.[9] Additionally, there can be interactive effects between neurotransmitters. For instance, the outgrowth-ar-resting effects of serotonin on neuron B19 can be completely negated by the presence of acetylcholine[10] which, alone, has no effect on growth.

Many observations initially made on these invertebrate neurons are now known to occur in mammalian primary neuronal cell cultures. For instance, pyramidal neurons from the hippocampal cortex respond dramatically to the presence of neurotransmitters.[6] The neurotransmitter glutamate can selectively cause a pruning of dendrites of these hippoc-ampal neurons while having no visible effects on their axonal growth cones. In addition, it is possible to demonstrate that selective application to individual dendrites can cause a selective pruning of individual dendrites without affecting those around it. Taken to-gether, these data have engendered a view wherein selected afferents could, by releasing glutamate, locally affect the specific morphology of a hippocampal pyramidal neuron. Such effects could occur during development or even in the mature nervous system. It is easy to envision how the dendritic architecture of such neurons could actually be sculpted by afferent activity impinging on one of the dendrites and thus, through the release of glutamate, selectively cause a stabilization, or even a retraction of that dendrite. In fact, a recent study demonstrated that endogenous glutamate, released from entorhinal cortex axons, can inhibit dendrite outgrowth in target hippocampal neurons.[11] Furthermore, the results of the latter study implicate glutamate as a signal for synaptic differentiation in the hippocampus. In this way glutamate apparently acts as a sculptor of the very circuits in which it participates in the coding processes. Similar findings have been made in recent *in situ* studies of the vertebrate retino-tectal system where NMDA type glutamate recep-tors appear to be important in the formation of the mapping of synapses in the tectum.[12]

Within the hippocampal pyramidal neurons, as within invertebrate neurons, it is possible to demonstrate combinatorial effects of neurotransmitters. Glutamate is not the sole neurotransmitter impinging on the dendrites of pyramidal neurons. It is well known that while glutamate is the major excitatory transmitter, its effects in neural integration can in part, be negated by the inhibitory neurotransmitter, GABA. In fact, GABA (plus its potentiator, diazepam) can negate the glutamate pruning effects on dendritic architecture.[13] There emerges, from this data, a picture wherein the key concept is balance (FIG. 2). Under conditions of balanced excitation and inhibition, dendrites may be stable or outgrowth can occur. Under conditions, however, of excitatory imbalance, we note profound dendritic regression and accordingly would expect the loss of synaptic inputs along such routes.

Calcium as a Final Common Pathway Governing the Behavior of Growth Cones

Recent experiments have demonstrated that changes in intracellular calcium levels are both necessary and sufficient to account for most of the observed effects of neurotransmitters and electrical activity on neuroarchitecture. Analyses of growth cones in identified snail neurons have given us insight into roles for intracellular calcium in the regulation of

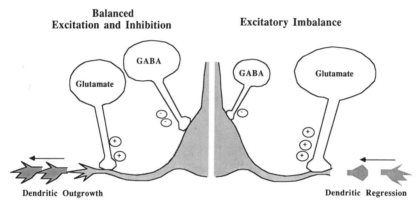

FIGURE 2. A schematic representation of the control of neurite outgrowth and regression by neurotransmitters. A pyramidal cell is diagramatically represented as having both excitatory (glutamate) and inhibitory (GABA) inputs. Periods of dendritic outgrowth occur during periods of balanced excitation and inibition. Dendritic regression and ultimately degeneration can occur during excitatory imbalance. Such an excitatory imbalance could occur through either increased excitation or decreased inhibition.

neuroarchitecture. A combination of pharmacological manipulations of intracellular calcium levels[14,15] and direct mapping of intracellular calcium concentrations using fura-2[16,17] demonstrated that neurite outgrowth can occur only when intracellular calcium levels fall within a specific range. Furthermore, different components of neurite outgrowth (*e.g.*, growth cone motility, elongation, branching) may be controlled by different levels of calcium within the outgrowth-permissive range.[4,14,18] Neurotransmitters such as serotonin, which inhibit neurite outgrowth, do so by causing calcium influx.[14,16] Similarly, elevations in intracellular calcium mediate the outgrowth-inhibiting effects of evoked action potentials.[16] Studies of developing hippocampal neurons have also linked the inhibiting effects of glutamate on dendritic outgrowth to rises in intracellular calcium levels.[6,19] The data suggest that glutamate activates receptors on dendrites[20] which results in a localized influx that selectively affects dendritic outgrowth. In addition, activation of GABA receptors apparently prevents glutamate inhibition of outgrowth by reducing calcium influx.[13]

Extension of a Common Mechanism: Inhibition of Dendritic Outgrowth, Dendritic Regression, and Ultimately, Neuronal Degeneration

Our work has demonstrated that neurotransmitters can have profound effects on neuronal architecture through such processes as the inhibition of dendritic outgrowth, or when glutamate is present in higher concentrations, dendritic regression. FIGURE 3 demonstrates that a clear line of continuity exists from the outgrowth-regulating actions of glutamate to neuronal death. An overabundance of the excitatory transmitter glutamate is now well known to produce excitatory amino acid toxicity.[21-23] We believe we are looking at a

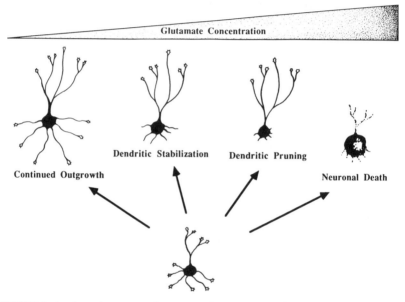

FIGURE 3. A schematic representation of the responses of hippocampal neurons to varying concentrations of glutamate. A pyramidal neuron that is exposed to a particular glutamate concentration may either continue normal outgrowth or have specific dendrites stabilized. Individual dendrites could be stabilized by localized application of glutamate. Increasing concentrations of glutamate would result in dendritic pruning and, ultimately, in the death of a neuron by excitatory amino acid toxicity. The important feature of this representation is that these processes represent a continuum along a set of possible responses to glutamate.

continuum of possible actions of these neurotransmitters; an extension of these important and sophisticated mechanisms for shaping neuronal architecture could well result in the pathological condition of neural degeneration. If this is an extension of such a mechanism, then the same kinds of rules which operated in regulation of dendritic outgrowth might well operate in regulation of neuronal death. For instance, the combinatorial actions of glutamate and GABA, should prove effective not only in regulating neurite outgrowth but also in regulating cell death. In fact, we have been able to demonstrate that the activation of GABA receptors can significantly inhibit neural degeneration evoked by glutamate. Thus, there seems to exist a continuum of actions that can be brought about by these common mechanisms.

Calcium as a Final Common Pathway Governing the State of Generation or Degeneration of Neuronal Arbors

What, then, of calcium and neurodegeneration? It is clear that large, sustained rises in intracellular calcium can lead to cell death.[23] In the mammalian brain, excitatory amino acids have been implicated in the neurodegeneration seen in ischemic brain damage,[21,22,24] epilepsy,[25] and Alzheimer's disease.[26] Recent cell culture studies have demonstrated that large, sustained rises in intracellular calcium are necessary and sufficient for glutamate neurotoxicity.[6,13,23,27] Glutamate-induced rises in intracellular calcium can be prevented by concurrent activation of GABA receptors.[13,28] Thus, we find that the same basic signalling mechanisms (*i.e.*, neurotransmitters and electrical activity) employed in normal brain functions, can be causally involved in neurodegeneration. There seems to be a fine line between the adaptive transmitter-induced calcium influxes involved in neuronal development and function on the one hand, and maladaptive degenerative processes that occur with excess calcium influx.

Calcium Homeostasis

Calcium signalling is obviously a very sophisticated process which can control a variety of cellular functions.[29] It is now clear that individual neurons display a high degree of calcium homeostasis. We know that individual neurons of different classes can display quite different intracellular calcium levels. We also know that even within neurons, the pattern of distribution of intracellular calcium can be highly specific.[30] These facts point to the degree of very fine tuning with which calcium levels are regulated within nerve cells. What one must now investigate are the precise mechanisms which hold this balance. *A priori*, there are several general ways in which specific levels of intracellular calcium could be regulated. The equation in FIGURE 4 describes the critical variables for setting intracellular calcium levels. Essentially, calcium levels are the sum of those sources of intracellular calcium which increase calcium in the cytosol and those sinks of intracellular calcium which decrease intracellular calcium in the cytosol. The sources of intracellular calcium are transmembrane fluxes and releases from intracellular stores. The sinks of intracellular calcium are calcium sequestration and pumping. The precise spatial distributions of each of these components are responsible for establishing the specific spatial patterns of calcium concentration we have seen, which can be so heterogeneous within a single neuron.

It is the components of calcium homeostasis described above which should be the targets of considerable investigation with respect to neuronal dysfunction and aging. Do calcium sequestration and/or pumping mechanisms go awry and, thus, calcium concentrations rise despite normal inputs into the system? Alternatively, are particular ion channels gated incorrectly, allowing more calcium into the cell? By a systematic investigation of each of the components of the calcium homeostasis equation we will gain insight into the mechanisms by which neuronal calcium can be regulated in normal function for processes such as the pruning of dendritic arborization. We can also examine how calcium-regulating processes malfunction, and how excessive calcium concentrations can result in the pathological conditions of neurodegeneration.

Finally, it should be recognized that neurodegeneration resulting from a rise in intracellular calcium may be an extension, not only of the processes for selectively pruning individual dendrites, but also of normal processes for selectively eliminating individual neurons from neural circuitry. Thus, the nervous system seems to be endowed with a set of properties which provide it with an incredible degree of flexibility and plasticity.[31]

$$[Ca^{++}]_i = [I + R] - [E + S]$$

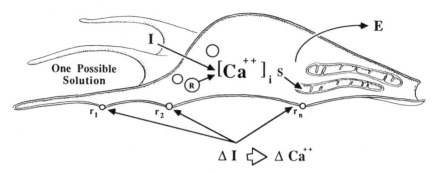

FIGURE 4. Factors determining calcium concentrations within growth cones. The *equation* at the top indicates the various factors determining intracellular calcium levels within the growth cone. Calcium levels are increased by Influx, and by Release from intracellular stores; levels are decreased by Efflux (pumping) and by Sequestration. There are several possible solutions to the above equation which would produce the known distribution of calcium within a growth cone. The *representation below* shows one example. If influx were localized primarily at the leading edge of the main body of the growth cone, and/or efflux or sequestration were localized primarily in the growth cone-neurite junction, then calcium concentrations would be higher in the leading edge than the neurite. One important implication of the existence of a gradient is the possibility that anything perturbing the gradient could have profound effects on the behavior of the growth cone. For example, if the neurotransmitter receptors were localized at the leading edge of the growth cone, activation of these receptors would directly increase the influx of calcium at that point, and increase the gradient. Activation of other receptors at the trailing edge of the growth cone (*e.g.*, receptors interacting with the intracellular matrix molecules) could reduce the gradient by increasing influx at that point.

When such processes malfunction, however, it is not surprising that highly deleterious and even pathological conditions can result.

CONCLUSIONS

Taken together, our work has now demonstrated that action potentials and neurotransmitters converge at the level of membrane potential and thus regulate growth cones.[17] Membrane potential, in large part, determines intracellular calcium levels.[4,13,14,27] Specific intracellular calcium levels can result in modified outgrowth. Increased calcium can yield decreased neurite outgrowth, pruning of dendrites and ultimately the death of individual neurons. Our final summary figure (FIG. 5) depicts this phenomenon. Taken together, the current **Calcium Hypothesis** provides a working model for a basic mechanism for the regulation of neuronal growth cones. There are now strong indications that this mechanism can potentially malfunction and ultimately result in the death of particular neurons. We offer this view of a continuum of regulatory events by a specific and singular regulatory agent, and propose calcium as a potential nodal point and causal agent for at least some neurodegenerative diseases.

SUMMARY

The neuronal growth cone is involved in neurite elongation, directional pathfinding, and target recognition. These activities are essential for proper assembly of functional circuits within the developing nervous system, for regeneration of functional circuitry following damage, and also, perhaps, for remodeling of the nervous system in response to environmental stimuli. Our studies of both molluscan and mammalian neurons in culture have shown that neurite outgrowth can only proceed when intracellular calcium levels lie within a specific outgrowth-permissive range. Cessation of outgrowth can be induced by a variety of signals normally used for communication within the adult nervous system, including neurotransmitters, and action potentials; all of these signals elevate levels of intracellular calcium above the outgrowth-permissive range. For example, glutamate, whether added to the medium or released from co-cultured entorhinal explants, can selectively inhibit dendritic outgrowth. Conversely, inhibitory neurotransmitters can block the outgrowth-inhibitory effects of glutamate and actually promote expansion of dendritic arbors. Dendritic outgrowth is therefore regulated by a balance between excitatory and inhibitory neurotransmitter activity. Extreme excitatory imbalance in neurotransmitter input to pyramidal neurons causes cell death. Each of these changes in neuroarchitecture is mediated by changes in levels of intracellular calcium. We therefore put forward the hypothesis that key mechanisms which normally control the development and

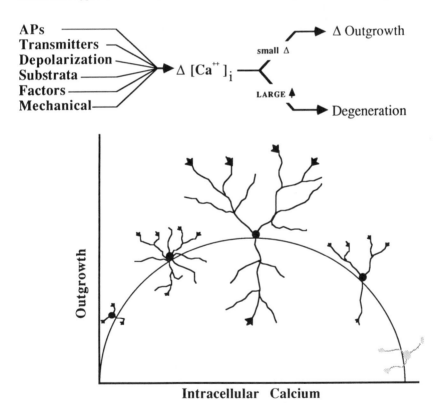

FIGURE 5. Calcium ultimately can determine the status of neurite outgrowth and neurodegeneration within neurons. A variety of agents including action potentials, neurotransmitters, membrane potential surfaces, substrate factors, and mechanical forces can change intracellular calcium levels. Changes in intracellular calcium levels can lead to either increased neurite outgrowth and growth cone behavior, or decreased growth cone behavior and cessation of neurite outgrowth. Very large increases in intracellular calcium can ultimately produce neuronal degeneration.

plasticity of neural circuitry, are also involved in neurodegeneration. Local, moderate elevations in calcium result in dendritic pruning. Higher, global elevations in calcium result in cell death. This cell death may serve an important function during normal development; aging may result in the same mechanism being employed pathologically. When intracellular calcium levels are not regulated within normal limits, as may occur in aging, neurodegeneration may occur.

REFERENCES

1. COTMAN, C. W. & L. L. IVERSEN. 1987. Excitatory amino acids in the brain focus on NMDA receptors. Trends Neurosci. **10:** 263–265.
2. HAYDON, P. G., D. P. MCCOBB & S. B. KATER. 1984. Serotonin selectively inhibits growth cone dynamics and synaptogenesis of specific identified neurons of Helisoma. Science **226:** 561–564.
3. COHEN, C. S. & S. B. KATER. 1986. Suppression of neurite elongation and growth cone motility by electrical activity. Science **232:** 1638–1640.
4. KATER, S. B., M. P. MATTSON, C. S. COHAN & J. A. CONNOR. 1988. Calcium regulation of the neuronal growth cone. Trends Neurosci. **11:** 315–321.
5. MATTSON, M. P. 1988. Neurotransmitters in the regulation of neuronal cytoarchitecture. Brain Res. Rev. **13:** 179–212.
6. MATTSON, M. P., P. DOU & S. B. KATER. 1988. Outgrowth-regulating actions of glutamate in isolated hippocamapal pyramidal neurons. J. Neurosci. **8:** 2087–2100.
7. LIPTON, S. A. & S. B. KATER. 1989. Neurotransmitter regulation of neuronal outgrowth, plasticity, and survival. Trends Neurosci. **12:** 265–270.
8. KATER, S. B. & M. P. MATTSON. 1988. Extrinsic and intrinsic regulators of neuronal outgrowth and synaptogenesis in identified Helisoma neurons in isolated cell culture. *In* Cell Culture Approaches to Invertebrate Neurosciences. D. J. Beadle, G. Lees & S. B. Kater, Eds.: 1–32. Academic Press. London.
9. MCCOBB, D. P., P. G. HAYDON & S. B. KATER. 1987. Dopamine and serotonin regulation of neurite outgrowth in different identified neurons. J. Neurosci. Res. **19:** 19–26.
10. MCCOBB, D. P., C. S. COHAN, J. A. CONNOR & S. B. KATER. 1988. Interactive effects of serotonin and acetylcholine on neurite elongation. Neuron **1:** 377–385.
11. MATTSON, M. P., R. E. LEE, M. E. ADAMS, P. B. GUTHRIE & S. B. KATER. 1988. Interactions between entorhinal axons and target hippocampal neurons: A role for glutamate in the development of hippocampal circuitry. Neuron **1:** 865–876.
12. CLINE, H. T., E. A. DEBSKI & M. CONSTANTINE-PATON. 1987. N-methyl-D-aspartate receptor antagonist desegregates eye-specific stripes. Proc. Natl. Acad. Sci. USA **84:** 4342–4345.
13. MATTSON, M. P. & S. B. KATER. 1989. Excitatory and inhibitory neurotransmitters in the generation and degeneration of hippocampal neuroarchitecture. Brain Res. **478:** 337–348.
14. MATTSON, M. P. & S. B. KATER. 1987. Calcium regulation of neurite elongation and growth cone motility. J. Neurosci. **7:** 4034–4043.
15. MATTSON, M. P., A. TAYLOR-HUNTER & S. B. KATER. 1988. Neurite outgrowth in individual neurons of a neuronal population is differentially regulated by calcium and cyclic AMP. J. Neurosci. **8:** 1704–1711.
16. COHAN, C. S., J. A. CONNOR & S. B. KATER. 1987. Electrically and chemically mediated increases in intracellular calcium in neuronal growth cones. J. Neurosci. **7:** 3588–3599.
17. MCCOBB, D. P. & S. B. KATER. 1988. Membrane voltage and neurotransmitter regulation of neuronal growth cone motility. Dev. Biol. **130:** 599–609.
18. MATTSON, M. P., P. B. GUTHRIE & S. B. KATER. 1988. Components of neurite outgrowth which determine neuroarchitecture: Influence of calcium and the growth substrate. J. Neurosci. Res. **20:** 331–347.
19. MATTSON, M. P., P. B. GUTHRIE & S. B. KATER. 1989. Intracellular messengers in the generation and degeneration of hippocampal neuroarchitecture. J. Neurosci. Res. **21:** 447–464.
20. CONNOR, J. A., W. J. WADMAN, P. E. HOCKBERGER & R. K. S. WONG. 1988. Sustained

dendritic gradients of Ca^{2+} induced by excitatory amino acids in CA1 hippocampal neurons. Science **240:** 649–653.

21. SCHWARCZ, R., A. C. FOSTER, E. D. FRENCH, W. O. WHETSELL & C. KOHLER. 1984. Excitotoxic models for neurodegenerative disorders. Life Sci. **35:** 19–32.

22. ROTHMAN, S. M. & J. W. OLNEY. 1986. Glutamate and the pathophysiology of hypoxic-ischemic brain damage. Ann. Neurol. **19:** 105–111.

23. CHOI, D. W. 1988. Calcium-mediated neurotoxicity: Relationship to specific channel types and role in ischemic damage. Trends Neurosci. **11:** 465–469.

24. SIMON, R. P., J. H. SWAN, T. GRIFFITHS & B. S. MELDRUM. 1984. Blockade of N-methyl-D-aspartate receptors may protect against ischemic damage in the brain. Science **226:**850–852.

25. SLOVITER, R. S. 1983. "Epileptic" brain damage in rats induced by sustained electrical stimulation of the perforant path. I. Acute electrophysiological and light microscopic studies. Brain Res. Bull. **10:** 675–697.

26. MARAGOS, W. F., J. T. GREENAMYRE, J. B. PENNEY & A. B. YOUNG. 1987. Glutamate dysfunction in Alzheimer's disease: An hypothesis. Trends Neurosci. **10:** 65–68.

27. MATTSON, M. P. & S. B. KATER. 1989. Development and selective neurodegeneration in cell cultures from different hippocampal regions. Brain Res. **490:** 110–125.

28. MATTSON, M. P., P. B. GUTHRIE, B. C. HAYES & S. B. KATER. 1989. Roles for mitotic history in the generation and degeneration of neuroarchitecture. J. Neurosci. **9:** 1223–1232.

29. BLAUSTEIN, M. P. 1988. Calcium transport and buffering in neurons. Trends Neurosci. **11:** 438–444.

30. GUTHRIE, P. B., M. P. MATTSON, L. R. MILLS & S. B. KATER. 1988. Calcium homeostasis in molluscan and mammalian neurons: Neuron-selective set-point of calcium rest concentration. Soc. Neurosci. Abstr. **14:** 582.

31. LYNCH, G. 1986. Synapses, circuits, and the beginnings of memory. MIT Press. Cambridge, MA.

Changes in Calcium Homeostasis during Aging and Alzheimer's Disease[a]

CHRISTINE PETERSON,[b] RATIV RATAN,[c] MICHAEL
SHELANSKI,[d] AND JAMES GOLDMAN[e]

[b]Department of Psychobiology
University of California, Irvine
Irvine, California 92717

[c]Department of Pharmacology
New York University
New York, New York 10016

[d]Department of Pathology
Columbia University
College of Physicians and Surgeons
New York, New York 10032

[e]Departments of Pathology and Psychiatry
Columbia University
College of Physicians and Surgeons
and
The New York State Psychiatric Institute
722 West 168th Street
New York, New York 10032

Several lines of evidence suggest that the regulation of calcium homeostasis may be altered by aging and Alzheimer's disease (for review see REF. 1). Aging decreases calcium uptake by isolated nerve endings from aged rodent brain,[2-4] but increases total calcium associated with the synaptosomal plasma membrane.[2] Fibroblasts from aged and Alzheimer's donors also have decreased calcium uptake and increased superficial binding when compared to cells from young donors (FIG. 1[5]). Aging also reduces the lectin-induced mitogenic response of T cells from aged mice[6] and humans.[7] Furthermore concanavalin-A-stimulated calcium uptake declines in lymphocytes from Alzheimer's patients[8] when compared to normal aged individuals. Some of these deficits can be partially ameliorated by treatment with drugs that promote calcium uptake, such as 3,4-diaminopyridine[2,3] or the calcium ionophore, ionomycin.[9]

The study of biochemical alterations in neurodegenerative disease through the use of peripheral tissues is an attractive approach. Cellular regulation can be more readily examined in vitro since many in vivo factors that can affect metabolism can be controlled. Skin fibroblasts and neurons share many common features. For example, both have a similar neuroectodermal origin[10] and similar ion channels, including calcium-regulated potassium channels,[11] and respond to a variety of polypeptide growth factors. Fibroblast

[a]Supported in part by the Allied-Signal Corp./Alzheimer's Association Faculty Scholar Award, the French Foundation, and National Institute on Aging Grants #AG07855 and AG05142.

growth factor, for example, was first purified from brain sources[12] and was recently shown to have neurotrophic activity in cortical and hippocampal neurons.[13,14] Fibroblasts produce[15,16] and have receptors for nerve growth factor.[16] Other similarities include but are not limited to processes that transport glutamate[17] and produce protease nexin.[18] Dynamic cell functions cannot be easily studied in the central nervous system, particularly in the human brain. Thus, whether neurons manifest similar alterations in calcium homeostasis during aging or neurodegenerative disease is unclear. Post-mortem tissues are appropriately used to measure certain enzyme activities but are inadequate to study dynamic processes, such as calcium homeostasis.

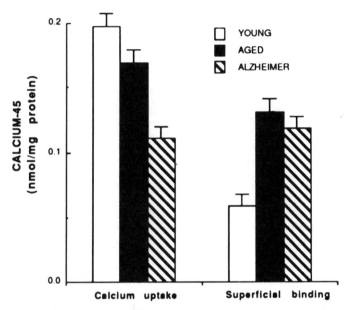

FIGURE 1. Altered calcium uptake and superficial binding in fibroblasts from aged and Alzheimer's donors. Fibroblasts were maintained in 16.6% fetal calf serum in Dulbecco's modified Eagle's medium. The cells were rinsed with calcium-free medium and then scraped and incubated with calcium-45 for 40 min. Calcium uptake was estimated after the cells were rinsed with EGTA. Superficial binding was determined by subtraction (calcium uptake after no EGTA rinse minus calcium uptake after EGTA rinse[5]).

There are several indications that one might find abnormalities in peripheral tissues from Alzheimer's patients. Both the appearance of a familial form of Alzheimer's disease and the increased risk that first degree relatives of Alzheimer's patients have of expressing the disorder support the possibility of a strong genetic component.[19,20] The search for a genetic cause(s) of Alzheimer's Disease has become the subject of intensive investigations. In early-onset (before age 60) familial forms of the disease there is evidence that a genetic defect lies on the proximal part of the long arm of chromosome 21.[21] In late onset the genetic focus does not appear to be at the same locus.[22] Thus there is some evidence that various genetic defects, *i.e.*, different genes, might lead to the brain lesions associated with Alzheimer's disease. Apart from the obvious pattern of inheritance and a somewhat younger age of onset cases with typical familial Alzheimer's disease are phenotypically indistinguishable from those with the sporadic form of the disorder.[23] Furthermore, elderly individuals with Down's syndrome, a disease with a known chromosomal abnormality, develop Alzheimer's-like neuropathological changes[24,25] (for review see REF. 26). Identifying the gene(s) involved in the pathogenesis of the Alzheimer's type brain lesions in Down's syndrome would be potentially very helpful to our understanding of Alzheimer's disease in general. A number of investigators report metabolic changes in peripheral tissues from Alzheimer's patients (for review see REFS. 27,28).

Due to the depressed calcium uptake in aged rodent brain and the detection of alterations in peripheral tissues from aged and Alzheimer's patients the present studies examined calcium homeostasis in cultured skin fibroblasts. In particular, both bound and cytosolic free calcium concentrations were determined.

The present studies used skin fibroblast cell lines from the National Institute on Aging Cell Repository. Eight lines from Alzheimer's patients from six different families were studied (for a detailed description of the cell lines see REF. 29). Cells were used at early passages to avoid complications due to *in vitro* aging. Furthermore, because confluence and the presence of serum alter various aspects of cell function the metabolic studies were performed with subconfluent cultures that were serum-starved for 24 hours. Biochemical characteristics of the cells such as protein and DNA content are unaltered by aging or Alzheimer's disease. The number of cells at confluence decreases with aging but is not further altered due to Alzheimer's disease. Thymidine incorporation into DNA and the percentage of labeled nuclei were similarly reduced due to aging and Alzheimer's disease when compared to cells from young individuals.

Aging and Alzheimer's disease lead to deficits in biosynthetic and oxidative processes. Glucose and glutamine decarboxylation decline 50% in cells from old donors and 75% in cells from patients with Alzheimer's disease when compared to young donors. There was a parallel decline in glucose and glutamine incorporation into protein and lipids. In contrast, leucine into proteins declined due to aging but was only slightly reduced in cells from Alzheimer's patients. These findings suggest that energy-producing processes that depend upon mitochondrial function may be more severely altered by Alzheimer's disease than cytosolic biosynthetic processes.

Calcium exists as bound and free (ionized) forms in the cell. Cells generally maintain the free calcium concentration around 10^{-7} M by processes that are complex and tightly regulated. Calcium is sequestered by intracellular organelles such as mitochondria and smooth endoplasmic reticulum and extruded across the plasma membrane by energy-dependent transport systems. The set point for calcium is balanced by an equilibrium between uptake and efflux mechanisms across the plasma and organelle membranes. Calcium can also be bound to specific calcium binding proteins, such as calmodulin, calbindin, troponin, and vitamin D-dependent proteins. Furthermore, sialic acid and other negatively charged residues within or on the surface of the cell membranes can serve as potential calcium binding sites. Alterations in any of these aspects of calcium homeostasis may appear as changes in calcium transport and subsequently be reflected in altered calcium-dependent processes.

Bound cell calcium increases during aging and Alzheimer's disease. Total cell calcium as measured by flameless atomic absorption spectroscopy was elevated in fibroblasts from aged and Alzheimer's disease donors (FIG. 2[29]). Bound calcium in fibroblasts from old donors increased 52% whereas calcium in Alzheimer's disease fibroblasts was elevated 197% when compared to cells from young donors. About 40% of this bound calcium could be removed by brief exposure to a calcium chelator EGTA, although the same relative differences between the groups were still apparent. *In vitro* aging of fibroblasts leads to an increase in calcium content.[30] All the cells in the described study, however, were matched for passage number (<10). In the brains of patients with Guam Parkinsonium dementia the calcium content of tangle bearing neurons is elevated.[31]

The free ionized intracellular calcium compartment was also altered by aging and Alzheimer's disease. The development of the calcium-sensitive fluorescent dyes, quin-2 and fura-2 has made the determination of cytosolic free calcium inside living cells possible.[32,33] With quin-2 the estimated concentration of ionized calcium was 54 nM in fibroblasts from young donors,[34] which resembles values published for other mammalian cells.[35] Cytosolic free calcium in cells from elderly donors averaged 37 nM, whereas Alzheimer's cells were even lower at 15 nM (FIG. 3[34]). These differences were not due to changes in intracellular pH or binding of the dyes to heavy metals. Similar results were obtained with the dye fura-2. The individual points were clustered and data from the three separate groups did not overlap (FIG. 4). The cytosolic free calcium concentration in the Alzheimer's cells are lower than that reported for mammalian cells during any phase of the cell cycle.[35]

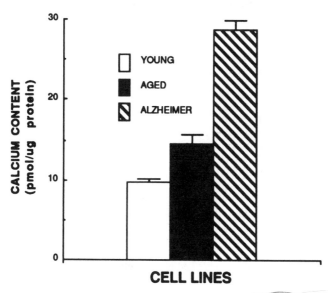

FIGURE 2. Increased calcium content during aging and Alzheimer's disease. Fibroblasts were maintained for 7 days in Dulbecco's modified Eagle's medium with 16.6% fetal bovine serum. The cells were rinsed with calcium-free medium and acidified with perchloric acid. Calcium content was estimated by atomic absorption spectroscopy.[29]

The ability of drugs to elevate intracellular calcium is altered by aging and Alzheimer's disease. A number of compounds transiently increase intracellular free calcium.[36-38] Whether cells from young, aged, and Alzheimer's disease donors respond similarly to such drugs was tested. After cells were deprived of serum for 24 hours, the addition of low concentrations of serum, bradykinin, 3,4-diaminopyridine or the peptide formyl-methionyl-leucyl-phenylalanine, a chemotactic agent for neutrophils produced rapid by transient elevations of free calcium.[39] Compared to cells from young donors, fibroblasts from older normal individuals showed lower peak calcium levels and slower rates in reaching the maximal free calcium concentrations. The differences were exaggerated further in the Alzheimer's disease cells where the responses where markedly blunted compared to normals. Six minutes after drug treatments only 1% serum and

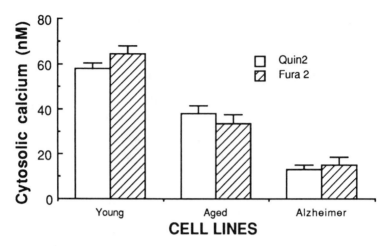

FIGURE 3. Decreased cytosolic free calcium during aging and Alzheimer's disease. Cells were deprived of serum for 24 hours, loaded with either quin-2 AM or fura-2 AM. Fluorescence was estimated with a microscope photometer.[34]

3,4-diaminopyridine maintained basal calcium concentrations above the initial resting values in the aged or Alzheimer's cells.

Do such low levels of cytosolic free calcium interfere significantly with cell processes? To answer this question processes that depend upon changes in calcium concentrations within the cell were studied.

Fibroblasts in the resting state do not manifest gross morphological differences. For example, the cell shape in all three groups of fibroblasts is similar. Cytoskeletal structure as examined by indirect immunofluorescence using antibodies to myosin, alpha-actinin, tubulin, microtubule-associated proteins, and vimentin as well as rhodamine-coupled phallodin, as a probe for filamentous actin revealed patterns of normal organization (Peterson, unpublished observations).

Differences between Alzheimer's and normal cells were seen during dynamic events but no obvious changes were observed in resting cells. All cells attached to plastic surfaces at similar rates. Cells from aged donors, however, did not go from a rounded three dimensional state to a more flattened two dimensional configuration as quickly as those from young donors. The Alzheimer's cells remained rounded far longer than even the aged cells. This defect in cell spreading can be reversed by treatment with the calcium ionophore, A23187.[39] Furthermore, fibroblasts from aged donors adhere better to a plastic substrate than those from Alzheimer's patients.[40] Amyloid beta-protein precursor mRNA is reduced in Alzheimer's cells and may play a role in reduced cell adhesion.

Fibroblasts from aged and Alzheimer's donors are unable to regulate intracellular calcium levels. These deficits appear in several separate and independent measures. (1) calcium-45 uptake decreases; (2) bound calcium increases; (3) cytosolic free calcium concentrations decline; and (4) dynamic calcium-dependent processes (*i.e.*, cell spreading), which require the generation of calcium transients are impaired. The precise mechanisms for these alterations are unknown; however, they may suggest changes in calcium extrusion processes. The Ca^{2+}/Mg^{2+} ATPase is altered in fibroblasts from Alzheimer's donors when compared to normal elderly individuals.[41]

The specificity of the changes in calcium homeostasis during Alzheimer's disease needs to be determined and extended to larger numbers of individuals with Alzheimer's disease as well as to patients with other neurodegenerative disease. First degree relatives of patients with Alzheimer's disease also need to be studied as well as individuals with Down's syndrome, given the neuropathological changes that accompanies older Down's patients. A recent study with transformed lymphocytes (*e.g.*, lymphoblast) from Alzheimer's patients demonstrates no alteration in basal or stimulated levels of cytosolic free calcium.[42] However, in the CD4[+] subpopulation of lymphocytes from Alzheimer's donors after PHA stimulation depressed levels of cytosolic free calcium are observed.[43]

Given the multitude of cellular processes that depend upon calcium homeostasis, it is easy to speculate on the effect disordered calcium regulation could have on cell function. Low free calcium levels might stabilize the cellular cytoskeleton which may make cells less responsive to growth factors and other exogenous agents. It may also underlie the alterations in protein kinase C activity.[44] If one could generalize these findings to other cell types, neurons may be especially sensitive. Processes such as synaptic transmission, dendritic conduction, axonal transport, calcium-activated proteolysis, and calcium/calmodulin-stimulated kinase activity are all critical for neuronal function (for review see REF. 45). Alterations in calcium homeostasis have been suggested in Alzheimer's disease brain in considering the mechanisms underlying the formation of neurofibrillary tangles[46,47] and altered protein kinase activity.[48] Low levels of calcium might lead to insufficient activation of calcium-dependent proteases and therefore lead to the abnormal accumulation of cytoskeletal elements.[49] Recent studies demonstrate that calbindin[50] or paravalbumin[51] containing neurons escape the degeneration associated with Alzheimer's disease.

The limitation of current *in vivo* neurochemical approaches and the lack of an animal model of Alzheimer's disease limit the study of metabolic abnormalities in Alzheimer's

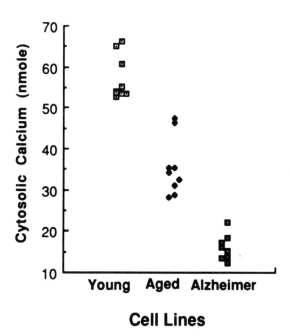

Cell Lines

FIGURE 4. Scatterplot of cytosolic free calcium in fura-2-loaded fibroblasts. Cells were grown in serum-free medium for 24 hours. The cells were loaded with fura-2 AM and fluorescence was estimated with a microscope photometer.[34] Each point represents a separate cell line in the indicated group.

disease. Post-mortem tissues are inadequate to study dynamic cell functions. The further examination of metabolic deficits in non-CNS cells may help to illuminate basic mechanisms during aging and age-related degenerative diseases.

SUMMARY

Several observations indirectly suggest that intracellular calcium regulation may be altered by aging and Alzheimer's disease. Thus, calcium homeostasis was examined directly in skin fibroblasts from Alzheimer's patients and compared to cells from normal young and elderly controls. Alterations in both bound and free calcium were noted; cells from Alzheimer's donors have higher levels of bound calcium but lower concentrations of free intracellular calcium when compared to cells from young and normal aged donors. These changes in calcium homeostasis may be physiologically significant, since processes that require transient elevations of intracellular free calcium, such as cell spreading, decline in the Alzheimer's cells. In summary, cultured skin fibroblasts from normal aged and Alzheimer's patients demonstrate deficits in calcium homeostasis and other metabolic processes when compared to cells from young donors.

REFERENCES

1. GIBSON, G. E. & C. PETERSON. 1987. Calcium and the aging nervous system. Neurobiol. Aging 9: 329–343.
2. PETERSON, C. & G. E. GIBSON. 1983. Aging and 3,4-diaminopyridine alter synaptosomal calcium uptake. J. Biol. Chem. 258: 11482–11486.
3. PETERSON, C., D. G. NICHOLLS & G. E. GIBSON. 1985. Subsynaptosomal calcium distribution during aging and 3,4-diaminopyridine treatment. Neurobiol. Aging 6: 297–304.
4. LESLIE, S. W., L. J. CHANDLER, E. BARR & R. P. FARRAR. 1985. Reduced calcium uptake by rat brain mitochondria and synaptosomes in response to aging. Brain Res. 329: 177–183.
5. PETERSON, C., G. GIBSON & J. P. BLASS. 1985. Altered calcium uptake in cultured skin fibroblasts from patients with Alzheimer's disease. N. Engl. J. Med. 312: 1063–1065.
6. MILLER, R. A. 1986. Immunodeficiency of aging: Restorative effects of phorbol ester combined with calcium ionophore. J. Immunol. 137: 805.
7. GROSSMANN, A., J. A. LEDBETTER & P. S. RABINOVITCH. 1989. Reduced proliferation in T lymphocytes in aged humans is predominantly in the CD8[+] subset and is unrelated to defects in transmembrane signaling which are predominantly in the CD4[+] subset. Exp. Cell Res. 180: 367–382.
8. GIBSON, G. E., P. NIELSEN, K. SHERMAN & J. P. BLASS. 1987. Diminished mitogen-induced calcium uptake by lymphocytes from Alzheimer patients. Biol. Psychiatry 22: 1079–1086.
9. MILLER, R. A. & E. R. SIMONS. 1986. T cell dysfunction in aged mice: Altered production and response of intracellular calcium transients (Abstract). Intl. Congr. Immunol. 6: 78.
10. JACOBSEN, . 1979. Developmental Neurobiology. Plenum Press. New York, NY.
11. GRAY, P. T. A., S. Y. CHIU, S. BEVAN & J. M. RITCHIE. 1986. Ion channels in rabbit cultured fibroblasts. Proc. R. Soc. Lond. 227: 1–16.
12. PETTMANN, B., G. LABOURDETTE, M. WEIBEL & M. SENSENBRANNER. 1986. The brain fibroblast growth factor (FGF) is localized in neurons. Neurosci. Lett. 68: 175–180.
13. MORRISON, R. S., A. SHARMA, J. DE VELLIS & R. A. BRADSHAW. 1986. Basic fibroblast growth factor supports the survival of cerebral cortical neurons in primary culture. Proc. Natl. Acad. Sci. USA 83: 7537–7541.
14. WALICKE, P., M. COWAN, N. NAMO, A. BAUD & R. GUILLEMIN. 1986. Fibroblast growth factor promotes survival of dissociated hippocampal neurons and enhances neurite extension. Proc. Natl. Acad. Sci. USA 83: 3012.

15. OGER, J., B. G. W. ARNASON, N. J. PANTAZIS, J. R. HEHRICH & M. YOUNG. 1974. Synthesis of nerve growth factor by L and 3T3 cells in culture. Proc. Natl. Acad. Sci. USA **71:** 1554.

16. SCHWARTZ, J. P. & X. O. BREAKEFIELD. 1980. Altered nerve growth factor in fibroblasts from patients with familial dysautonomia. Proc. Natl. Acad. Sci. USA **77:** 1154–1158.

17. DALL' ASTA, V., G. C. GAZZOLA, R. FRANCHI-GAZZOLA, O. BUSSOLATI, N. LONGO & G. G. GIODOTTI. 1983. Pathways of L-glutamic acid transport in cultured human fibroblasts. J. Biol. Chem. **258:** 6371–6379.

18. ROSENBLATT, D. E., C. W. COTMAN M. NIETO-SAMPEDRO, J. E. ROWE & D. J. KNAUER 1987. Identification of a protease inhibitor produced by astrocytes that is structurally and functionally homologous to human protease nexin-1. Brain Res. **415:** 40–48.

19. LARSSON, T., T. SJOGREN & G. JACOBSON. 1963. Senile dementia: A clinical, sociomedical, and genetic stdy. Acta Psychiatr. Scand. (Suppl). **167:** 39–150.

20. HESTON, U., A. R. MASTRI, V. E. MASTRI & J. WHITE. 1981. Dementia of the Alzheimer type: Clinical genetics, natural history and associated conditions. Arch. Gen. Psychiatry **38:** 1085–1090.

21. ST. GEORGE-HYSLOP, P. H., R. TANZI, R. POLINSKY, J. L. HAINES, L. NEE, P. C. WATKINS, R. MYERS, R. G. FELDMAN, D. POLLEN, D. DRACHMAN, J. GROWDON, A. BRUNI, J-F. FONCIN, D. SALMON, P. FROMMELT, L. AMADUCCI, S. SORBI, S. PIACENTINI, G. D. STEWART, W. J. HOBBS, P. M. CONNEALLY & J. F. GUSELLA. 1987. The genetic defect causing familial Alzheimer's disease maps on chromosome 21. Science **235:** 885–890.

22. PERICAK-VANCE, M. A., L. H. YAMAOKA, C. S. HAYNES, J. L. HAINES, P. C. GASKELL, W.-Y. HUNG, C. M. CLARK, A. L. HEYMAN, J. A. TROFATTER, J. P. EISENMENGER, J. R. GILBERT, J. E. LEE, M. J. ALBERTS, M. C. SPEER, D. V. DAWSON, R. J. BARTLETT, N. L. EARL, T. SIDDIQUE & A. D. ROSES. 1988. Genetic linkage studies in late-onset Alzheimer's disease families. *In* Genetics and Alzheimer's Disease. P. M. Sinet, Y. Lamour & Y. Christen, Eds.: 116–123. Springer Verlag. Paris.

23. DAVIES, P. 1986. The genetics of Alzheimer's disease: A review and discussion of the implications. Neurobiol. Aging **7:** 459–466.

24. ROPPER, A. H. & R. S. WILLIAMS. 1980. Relationship between plaques, tangles and dementia in Down syndrome. Neurology **30:** 639–644.

25. WISNIEWSKI, K. E., A. J. DALTON, D. R. CRAPPER MCLACHLAN, G. Y. WEN & H. M. WISNIEWSKI. 1985. Alzheimer's disease in Down's syndrome: Clinicopathologic studies. Neurology **35:** 957.

26. SINEX, E. M. & C. R. MERRIL, Eds. 1982. Alzheimer's disease, Down's syndrome and aging. Ann. N.Y. Acad. Sci., Vol. 396.

27. BLASS, J. P. & A. ZEMCOV. 1984. Alzheimer's disease: A metabolic systems degeneration? Neurochem. Pathol. **2:** 103–114.

28. HOLLANDER, E., R. C. MOHS & K. L. DAVIS. 1986. Antemortem markers of Alzheimer's disease. Neurobiol. Aging **7:** 367–387.

29. PETERSON, C. & J. E. GOLDMAN. 1986. Alterations in calcium content and biochemical processes in cultured skin fibroblasts from aged and Alzheimer donors. Proc. Natl. Acad. Sci. USA **83:** 2758–2762.

30. SHAPIRO, B. L. & L. F.-H. LAM. 1982. Calcium and age in fibroblasts from control subjects and patients with cystic fibrosis. Science **16:** 417–419.

31. YANGIHARA, R., R. M. GARRUTO, D. C. GAJDUSEK, A. TOMITA, T. UCHIDAWA, Y. KNOAGAYA, L.-M. CHEN, I. SOBUE, C. C. PLATO & C. J. GIBBS. 1984. Calcium and vitamin D metabolism in guamanian chamorrow with amylotropic lateral sclerosis and parkinsonism-dementia. Ann. Neurol. **15:** 42–48.

32. GRYNKIEWICZ, G., M. POENIE & R. Y. TSIEN. 1985. A new generation of Ca^{2+} indicators with greatly improved fluorescence properties. J. Biol. Chem. **260:** 3440–3450.

33. TSIEN, R. Y., T. J. RINK & M. POENIE. 1985. Measurements of cytosolic free Ca^{2+} in individuals cells using fluorescence microscopy with dual excitation wavelengths. Cell Calcium **6:** 145–157.

34. PETERSON, C., R. RATAN, M. L. SHELANSKI & J. E. GOLDMAN. 1986. Cytosolic free calcium and cell spreading decrease in fibroblasts from aged and Alzheimer donors. Proc. Natl. Acad. Sci. USA **83:** 7999–8001.

35. KEITH, C. H., F. R. MAXFIELD & M. L. SHELANSKI. 1985. Intracellular free calcium levels are reduced in mitotic PtK2 epithelial cells. Proc. Natl. Acad. Sci. USA **82:** 800–804.

36. TSIEN, R. Y., T. POZZAN & T. J. RINK. 1982. T-cell mitogens cause early changes in cytoplasmic free Ca^{2+} and membrane potential in lymphocytes. Nature **95:** 68–71.

37. MOOLENAAR, W. H., L. G. J. TERTOLEN & S. W. DE LAAT. 1984. Growth factors immediately raise cytoplasmic free Ca^{2+} in human fibroblasts. J. Biol. Chem. **259:** 8066–8069.

38. MORRIS, D. H., J. C. METCALFE, G. A. SMITH, T. R. HESKETH & M. V. TAYLOR. 1984. Some mitogens cause rapid increases in free calcium in fibroblasts. FEBS Let. **169:** 189–193.

39. PETERSON, C., R. R. RATAN, M. L. SHELANSKI & J. E. GOLDMAN. 1988. Altered response of fibroblasts from aged and Alzheimer donors to drugs that elevate cytosolic free calcium. Neurobiol. Aging **9:** 261–266.

40. UEDA, K., G. GOLE, M. SUNDSMO, R. KATZMAN & T. SAITOH. 1989. Decreased adhesiveness of Alzheimer fibroblast: Is beta-protein precursor involved? Ann. Neurol. **25:** 246–251.

41. RIZOPOULOS, E., J. P. CHAMBERS, A. O. MARTINEZ & M. J. WAYNER. 1988. Kinetic properties of the $[Ca^{2+} + Mg^{2+}]$ATPase in Alzheimer and normal fibroblasts at low free calcium. Brain Res. Bull. **21:** 825–828.

42. GIBSON, G. E. & L. TORAL-BARZA. 1988. Calcium in lymphoblasts from aged and Alzheimer donors. Calcium, Membranes, Aging and Alzheimer's Disease Meeting. Beckman National Academy of Sciences. Irvine, CA. Abstracts, p. 29.

43. GROSSMAN, A. & P. S. RABINOVITCH. 1988. Deficiency in intracellular calcium regulation in CD4+ cells of patients with sporadic Alzheimer's Disease (abstract). Cytometry Suppl. **2:** 65.

44. SAITOH, T., G. COLE, T. HUYNH, R. KATZMAN & M. SUNDSMO. 1988. Abnormal protein kinase C in Alzheimer fibroblasts (abstract). Soc. Neurosci. **18:** 154.

45. CARVAHLO, A. P. 1982. Calcium in the nerve cell. *In* Handbook of Neurochemistry. A. Lajtha, Ed., Vol. 2: 69. Plenum Press, New York, NY.

46. PERL, D. P., D. C. GAJDUSEK, R. M. GARRUTO, R. T. YANAGIHARA & C. J. GIBBS. 1982. Aluminum accumulation in amyotrophic lateral sclerosis in Parkinsonism dementia of Guam. Science **217:** 1053–1055.

47. GARRUTO, R. M., R. FUKATSU, R. YANAGIHARA, D. C. GAJDUSEK, G. HOOK & C. E. TIORI. 1984. Imaging of calcium and aluminum in neurofibrillary tanglebearing neurons in Parkinsonism-dementia of Guam. Proc. Natl. Acad. Sci. USA **81:** 1875–1879.

48. SAITOH, T. & K. R. DOBKINS. 1986. Increased in vitro phosphorylation of a M_r60,000 protein in brain from patients with Alzheimer disease. Proc. Natl. Acad. Sci. USA **83:** 9764–9767.

49. NIXON, R. 1989. This volume.

50. HOFFMAN, S. R., N. W. KOWALL & A. C. McKEE. 1988. Calbindin D28 neurons in the hippocampal formation are resistant to degeneration and do not develop regenerative features in Alzheimer's disease (abstract). Soc. Neurosci. **14:** 154.

51. MORRISON, J. H., K. COX, P. R. HOF & M. R. CELIO. 1988. Neocortical paravalbumin-containing neurons are resistant to degeneration in Alzheimer's disease (abstract). Soc. Neurosci. **14:** 1085.

Defective Calcium Signal Generation in a T Cell Subset That Accumulates in Old Mice[a]

RICHARD A. MILLER

Department of Pathology
Boston University School of Medicine
Boston, Massachusetts 02118

Nearly all circulating T cells are in a resting, or G_o, state, and have a cytoplasmic free calcium ion concentration ("[Ca]$_i$") of approximately 100–150 nM. The addition of a mitogen (*e.g.*, Con A, or antibody to the CD3 component of the T cell's antigen receptor) leads within 1–2 min to an increase in [Ca]$_i$ to approximately 200–1000 nM. The generation of this calcium signal seems to be dependent on IP$_3$-mediated release of Ca^{2+} from internal stores in the familiar Jurkat tumor model, but in nontransformed T cells reflects primarily influx of Ca^{2+} across the plasma membrane. Calcium signals are thought to play an important role in triggering the subsequent stages of the activation process, including the expression of activation-specific genes (*e.g.*, myc, fos, IL-2, and IL-2R); and indeed calcium ionophores, in combination with activators of protein kinase C (PK-C), are potent T cell mitogens.[1] This essay describes our current model for how aging leads to alterations in the ability of T lymphocytes to produce internal calcium signals in response to mitogens.

Not All T Cells in Old Mice Are Equally Defective

Limiting dilution assays, which count the percentage of T cells that can participate in an *in vitro* immune reaction, have shown[2] that aging leads to a decline in the proportion of responsive T cells in mouse spleen and blood, but little or no change in the amount of response generated by those cells that are triggered. Flow cytometric analyses of human blood T cells have also documented[3] a decline in the fraction of cells that can leave the G_o state. Studies of IL-2 receptor expression, a necessary prelude to the IL-2 dependent G_1-S transition, have shown[4] a decline in the proportion of murine T cells that can reach this intermediate step of the activation sequence, a defect also reflected at the level of IL-2 receptor mRNA expression (Macauley and Miller, unpublished). Even expression of c-myc mRNA, detectable within 1 hr of the addition of Con A, is clearly diminished in aging T cells.[5] Our current efforts have therefore focussed on age-sensitive lesions in the very earliest stages of the activation process, in particular calcium signal generation.

A Decline, with Age, in the Number of T Cells That Generate a Calcium Signal in Response to Mitogens

Using indo-1 to monitor [Ca]$_i$ in splenic T cells from old and young mice, we found[6] a decline, with age, in the average change in [Ca]$_i$ produced by Con A triggering, and thus

[a]Work discussed in this review was supported by Grants AG-07114 and AG-03978 from the National Institutes of Health. Richard Miller is supported by a Scholar Award from the Leukemia Society of America and by a Research Career Development Award from the National Institute on Aging.

in the peak $[Ca]_i$ levels attained, with little if any difference in the baseline $[Ca]_i$ values of resting T cells. A flow cytometric method established that this difference in average $[Ca]_i$ signal generation reflected a decline, with age, in the proportion of T cells that could increase their $[Ca]_i$ values above the resting baseline levels. Similar results have also been obtained, in mice, by Grossmann and Rabinovitch (unpublished). Proust et al., somewhat in contrast, have suggested[7] that age-associated decline in Con A-induced changes of $[Ca]_i$ reflect higher baseline values in unstimulated T cells from old mice, rather than changes in peak levels apparent in our data and that of Grossmann and Rabinovitch. Our own work also demonstrates a decline, with age, in the number of T cells that generate a Ca^{2+} signal in response to anti-CD3 antibody, but we see no age-dependent change in the (smaller) fraction of T cells responding to the plant lectin phytohemagglutinin. The defects in Con A and anti-CD3 responses affect cells within both the CD4 "helper" and CD8 "cytotoxic" T cell subsets (Philosophe and Miller, submitted).

Altered Transmembrane Influx, and Altered Calcium "Buffering," May Both Contribute to Defective Signal Generation

In principle, and as documented[8] for other cell types also in practice, agonist-triggered changes in $[Ca]_i$ can reflect changes in the rate at which Ca^{2+} ions enter the cytoplasm from internal or external pools, or in the rate at which Ca^{2+} ions are removed from the cytoplasm, e.g., by the plasma membrane's ATP-dependent extrusion pump or through sequestration into internal organelles. Speculating that the age-dependent loss in Ca^{2+} signal generation in Con A-stimulated T cells might reflect a defect in receptor-operated calcium channels, we decided to look at $[Ca]_i$ levels in T cells exposed to very small doses of the calcium ionophore ionomycin. Since ionomycin would transport extracellular Ca^{2+} ions into the cytoplasm through a receptor-independent mechanism, we anticipated that there would be change, with age, in the $[Ca]_i$ levels of ionomycin-challenged T cells. In contrast, we found[9] a consistent age-associated decline in the steady-state levels of $[Ca]_i$ reached by T cells exposed to ionomycin doses large enough to perturb the cells from the resting state, but small enough not to overwhelm entirely the cells' internal Ca^{2+} buffering mechanisms. At any ionomycin dose tested (5–80 ng/ml), splenic T cells from old mice attained a level of $[Ca]_i$ only about 50% as high as that attained by T cells from a young mouse tested in parallel. Studies of ^{45}Ca uptake showed that the resistance to ionomycin challenge did not reflect any age-associated loss in the ability of ionomycin/Ca^{2+} complexes to cross the plasma membrane. This result cannot be attributed to alterations in receptor-operated Ca^{2+} channels, and suggests strongly that there are increases, with age, in the ability of T cells to buffer changes in their internal Ca^{2+} levels. We speculate that increases in the number, velocity, or calcium sensitivity of the plasma membrane calcium extrusion pumps might provide an explanation for the ionomycin results.

Increased resistance to changes in $[Ca]_i$ is not, however, the only mechanism that could contribute to poor Ca^{2+} signal generation in Con A-triggered T cells from old mice. Direct measures of ^{45}Ca uptake showed[10] a 2-fold decline, with age, in the rate of ^{45}Ca influx in T cells exposed to mitogenic concentrations of Con A. Since unstimulated T cells from old mice had a significantly higher "leak" rate for ^{45}Ca influx, the Con A-related change in the rate of ^{45}Ca influx declined about 4-fold in the older mice. At this stage it is hard to know whether the decline in ^{45}Ca influx, or the increased activity of the Ca^{2+} buffering system, contributes more to the deficit in Ca^{2+} signal generation. Since mitogen-induced changes in ^{45}Ca influx are, however, transient,[11] while changes in $[Ca]_i$ are relatively long-lasting, we suspect that altered Ca^{2+} extrusion plays a key role in the age-related lesion.

It is important to note that some Con A-dependent signals seem to be entirely unimpaired in T cells from old mice. Thus we find[5] no change in the production of c-myc primary transcripts after Con A activation of old T cells, and also no change[10] in the generation of inositol phosphates, including IP_3 and IP_4, after Con A stimulation of T cells from old donors. (The latter finding has been disputed.[7]) IP_3 generation, like increased ^{45}Ca influx, is tightly coupled to the initial receptor-dependent signalling event. The contrast between altered Ca^{2+} influx and unaltered IP_3 generation suggests that the T cell's receptors may be coupled to at least two semiindependent internal activation pathways, one impaired by aging, and one relatively spared. We speculate that these pathways may relate to differences in protein kinase function, with semiindependent regulation of tyrosine-specific and serine/threonine-specific kinase pathways.

Responsive Virgin and Hyporesponsive Memory T Cell Subsets

Although studies using CD4 and CD8 antibodies have failed to reveal any dramatic age-related alteration in T cell subset distribution that could account for the defective T cell function in aging mice, we thought it likely that there might exist changes in subset

TABLE 1. Responder Frequencies in Sorted PGP-1^{hi} and PGP-1^{lo} T Cells from Old and Young Mice[a]

Assay	Responder	Young	Old
IL-2 secretion	PGP-1^{hi}	48 ± 7	16 ± 2
	PGP-1^{lo}	438 ± 41	430 ± 52
	both	333 ± 32	188 ± 16
Cytotoxicity	PGP-1^{hi}	88 ± 18	31 ± 3
	PGP-1^{lo}	283 ± 57	203 ± 35
	both	121 ± 21	56 ± 7
Proliferation	PGP-1^{hi}	82 ± 10	32 ± 6
	PGP-1^{lo}	216 ± 36	166 ± 32
	both	121 ± 17	64 ± 11

[a]Values are mean frequencies (± SEM) for N = 9 experiments, each of which compared PGP-1^{hi} and PGP-1^{lo} T cells from one old and one young mouse. Frequencies are reported as the number of Con A-responsive cells per 1000 cells of the indicated responder cell type. "Both" refers to cell populations sorted without respect to the PGP-1 signal. (Data from Lerner et al.[13])

distribution not detectable with the CD4 and CD8 reagents alone. PGP-1 ("phagocyte glycoprotein 1") has recently been identified as a useful marker for discriminating virgin T cells (fresh from the thymus) from long-lived memory cells that have undergone at least one round of antigen-triggered proliferation; memory cells are PGP-1^{hi}, in contrast to PGP-1^{lo} virgin cells.[12] Since thymic export of new virgin T cells is thought to decline with age, while the conversion of virgin to memory cells is likely to continue unabated, we postulated that aging might lead to an accumulation of T cells within the PGP-1^{hi} subset. Indeed, we found[13] that PGP-1^{hi} T cells do increase with age, from about 20% of the T cell pool in young adult mice to about 60% of the T cells in 18–24-month-old animals. Similar increases were found within spleen, lymph node and blood T cell populations, and applied to cells within both the CD4 helper and CD8 cytotoxic subpopulations.

To see if this shift from the PGP-1^{lo} to the PGP-1^{hi} pool had functional consequences, we used a limiting dilution culture method to estimate the number of functional T cells within electronically sorted PGP-1^{hi} and PGP-1^{lo} T cell populations. We found (TABLE 1) that the PGP-1^{lo} virgin cells were substantially more likely to respond to Con A than were the PGP-1^{hi} T cells, whether the responses were measured by lymphokine production, generation of cytotoxic T cell effectors, or IL-2-dependent proliferation. Studies of [Ca]$_i$ changes also established (Philosophe and Miller, submitted) that PGP-1^{hi} T cells were substantially less likely than PGP-1^{lo} T cells to increase their [Ca]$_i$ levels when exposed to Con A, anti-CD3 antibodies, or even the receptor-independent stimulus, ionomycin. We concluded, therefore, that the age-dependent accumulation of PGP-1^{hi} T cells could account for much of the decline, with age, in the proportion of functionally competent T cells, and that lower Ca^{2+} signal generation by PGP-1^{hi} T cells might well contribute to the diminished functional abilities of these cells.

Developmental Origins of Defective T Cells in Old Mice

Where do the dysfunctional T cells come from? Are they the most recent thymic emigrants, the defective products of an involuted thymus gland? Or are they the antiquated progeny of the earliest cells to enter the peripheral immune system in embryonic and early adult life, made stale by a lifetime of residence in the periphery? Our data documenting severe defects in a class of T cells expressing high levels of PGP-1, a marker for memory T cells, suggest that long-lived T cells, rather than recent thymic emigrants, are likely to be the primary culprits. Indeed virgin T cells in old mice, at least as typified by the PGP-1^{lo} marker phenotype, seem almost as functional (by limiting dilution tests) as their counterparts in young control animals (see TABLE 1). It now seems more likely that the age-related decline in responses to polyclonal activators reflects an accumulation of hyporesponsive cells in the PGP-1^{hi} memory T cell class.

Why then are PGP-1^{hi} T cells so difficult to stimulate? It is clear from studies[12] of antigen-specific memory T cells in recently primed mice that these respond perfectly well when re-exposed to the original priming antigen. We can propose two models to explain the relative lack of responsiveness of PGP-1^{hi} T cells to Con A and anti-CD3 antibodies. It may be, for example, that all cells within the PGP-1^{hi} subclass have subtly altered activation requirements, changed in such a way that they respond well to cognate antigen as presented by specialized antigen-presenting cells, and yet poorly to mitogenic lectins and anti-receptor antibodies. An alternate idea would suggest that T cells freshly recruited to the PGP-1^{hi} memory pool might be as responsive as PGP-1^{lo} T cells, but that they then gradually lose responsiveness as time goes on. At any age, the PGP-1^{hi} population would contain a mixture of fresh, responsive T cells and stale, less responsive cells, with the latter type becoming progressively more prominent with age. Both limiting dilution (TABLE 1) and Ca^{2+} data show a decline with age in responsiveness even of cells within the PGP-1^{hi} subset, consistent with the latter idea. It is also hard to reconcile the notion of differential activation requirements with the finding (Philosophe and Miller, submitted) that PGP-1^{hi} T cells are relatively resistant to Ca^{2+} changes even when these are induced by ionomycin. A more definitive understanding of the relative responsiveness of "fresh" and "stale" PGP-1^{hi} cells will require experimentation with mice in which PGP-1^{hi} T cells of known age have been synchronously produced.

SUMMARY

TABLE 2 outlines our current understanding of the bases for activation defects in T cells from old mice, with speculative ideas indicated by a question mark. Many, though not all, T cells from old mice show defects in the generation of Ca^{2+} signals within the first few minutes of exposure to a mitogenic lectin or to antibody to components of the T cell receptor. These defects seem to involve an increased resistance to changes in cytoplasmic free calcium ion concentration, perhaps though changes in the calcium extrusion pump. Diminished rates of uptake of calcium from extracellular sources also contribute to defective calcium signal generation. Some aspects of the activation process, including production of inositol phosphates, seem in contrast not to be altered by aging in mice. T cells with defects in Ca^{2+} responses seem to be contained preferentially within the PGP-1^{hi} subsets of both the helper and cytotoxic lineages, subsets that accumulate dramatically in old mice. Since the PGP-1^{hi} phenotype is thought to distinguish memory T cells from virgin thymic emigrants, we propose a model in which aging leads to the progressive conversion of virgin to memory T cells, which in turn gradually lose the ability to respond to activating stimuli.

TABLE 2. Age-Associated Defects in T Cell Activation

Defects in calcium signal generation
Low influx rate (^{45}Ca after Con A)
Better "buffering" (resistance to ionomycin)
Origin of defective cells
More PGP-1^{hi} "memory" cells
PGP-1^{hi} cells less reactive than PGP-1^{lo} cells
Old PGP-1^{hi} cells less reactive than young PGP-1^{hi} cells
Control of c-Myc expression
Poor mRNA accumulation
Normal production of primary transcripts
(?) Abnormal nuclear processing
"Split" defect in signal transduction
Calcium influx impaired
Inositol phosphate production intact
(?) Control by different protein kinases

REFERENCES

1. MASTRO, A. M. & M. C. SMITH. 1983. Calcium-dependent activation of lymphocytes by ionophore, A23187, and a phorbol ester tumor promoter. J. Cell. Physiol. 116: 51–56.
2. MILLER, R. A. 1984. Age-associated decline in precursor frequency for different T cell-mediated reactions, with preservation of helper or cytotoxic effect per precursor cell. J. Immunol. 132: 63–68.
3. STAIANO-COICO, L., Z. DARZYNKIEWICZ, J. M. HEFTON, R. DUTKOWSKI, G. J. DARLINGTON & M. E. WEKSLER. 1983. Increased sensitivity of lymphocytes from people over 65 to cell cycle arrest and chromosomal damage. Science 219: 1335–1337.
4. VIE, H., & R. A. MILLER. 1986. Decline, with age, in the proportion of mouse T cells that express IL-2 receptors after mitogen stimulation. Mech. Ageing Dev. 33: 313–322.
5. BUCKLER, A., H. VIE, G. SONENSHEIN, & R. A. MILLER. 1987. Defective T lymphocytes in old mice: Diminished production of mature c-myc RNA after mitogen exposure not attributable to alterations in transcription or RNA stability. J. Immunol. 140: 2442–2446.
6. MILLER, R. A., B. JACOBSON, G. WEIL, & E. R. SIMONS 1987. Diminished calcium influx in lectin-stimulated T cells from old mice. J. Cell. Physiol. 132: 337–342.
7. PROUST, J. J., C. R. FILBURN, S. A. HARRISON M. A. BUCHHOLZ & A. A. NORDIN. 1987. Age-related defect in signal transduction during lectin activation of murine T lymphocytes. J. Immunol. 139: 1472–1478.
8. ALKON, D. L. & H. RASMUSSEN. 1988. A spatial-temporal model of cell activation. Science 239: 998–1005.
9. MILLER, R. A., B. PHILOSOPHE, I. GINIS, G. WEIL & B. JACOBSON. 1989. Defective control

of cytoplasmic calcium concentration in T lymphocytes from old mice. J. Cell. Physiol. **138:**2 175–182.

10. LERNER, A., B. PHILOSOPHE & R. A. MILLER. 1988. Defective calcium influx and preserved inositol phosphate generation in T cells from old mice. Aging Immunol. Infect. Dis. **1:** 149–158.

11. FREEDMAN, M. H., M. C. RAFF & B. GOMPERTS. 1975. Induction of increased calcium uptake in mouse T lymphocytes by concanavalin A and its modulation by cyclic nucleotides. Nature **255:** 378–382.

12. BUDD, R. C., J. C. CEROTTINI & H. R. MACDONALD. 1987. Phenotypic identification of memory cytolytic T lymphocytes in a subset of Lyt-2$^+$ cells. J. Immunol. **138:** 1009–1013.

13. LERNER A., T. YAMADA & R. A. MILLER. 1989. PGP-1^{hi} T lymphocytes accumulate with age in mice and respond poorly to concanavalin A. Eur. J. Immunol. **19:** 977–982.

Membrane Signal Transduction in T Cells in Aging Humans[a]

SUDHIR GUPTA

Division of Basic and Clinical Immunology
Medical Sciences I, C-264A
University of California, Irvine
Irvine, California 92717

INTRODUCTION

The nature of the process of aging is poorly understood, but it has been suggested that the key to its understanding lies within the immune system. It is widely accepted that with the advancing age there is a waning of immune system functions. It is striking that not all immune functions are affected to the same extent; some even appear essentially intact until a very advanced age. The biochemical basis of signal transduction events elicited as a consequence of interactions between specific T cell receptor (TcR) and its ligands which act sequentially to drive T lymphocytes from a resting stage, through activation and proliferation stage to final differentiation stage is presently an area of intense interest. These signal transduction events can be arbitrarily divided into *early* and *late* events. The *early* events include, changes in plasma membrane potentials, intracellular mobilization and fluxes of calcium, and the activation and redistribution of protein kinase C (PKC). The *late* events include DNA, RNA and protein synthesis, production of lymphokines, expression of lymphokine receptors, and clonal expansion. Recently two monoclonal antibodies, anti-CD3 that reacts with CD3 complex[1] and WT31 monoclonal antibody that reacts with epitope-defining α/β TcR[2,3] have been shown to activate human T cells by inducing changes in plasma membrane potentials, intracellular free calcium ($[Ca^{++}]_i$), and translocation and activation of protein kinase C.[4-7] In human aging, changes in late steps of T cell activation have been extensively documented;[8-13] however, no study of the early events of T cell activation in human aging has been published. In this investigation, we have examined the early steps of activation in aging humans using anti-CD3 and WT31 monoclonal antibodies and phytohemagglutinin (PHA) as activating signals and compared them with those in young subjects. The data show that in aging humans there are defects in early events of T cell activation and these changes could be the earliest and only changes in a subgroup of aging subjects who do not show any evidence of T cell defects by classical methods (late steps of activation).

MATERIALS AND METHODS

Subjects

Aging subjects (65–85 years) were drawn from the Leisure World Community and the young subjects (22–40 years) were the volunteers from the staff and the faculty of the University of California, Irvine.

[a]This work was supported in part by United States Public Health Service Grants AI-26465 and GM-14514.

Reagents

Anti-CD3 (T3) monoclonal antibody was purchased from Coultor, FL, and WT31 monoclonal antibody was a gift from Dr. Noel Warner, Becton-Dickinson, Mountain-view, CA. DIOC5 (3'-3'-dipentyloxacarbocyanine iodide) dye and quin-2 AM was purchased from Molecular Probes, Junction City, OR. For PKC assay, phosphatidylserine, 1,2 diolein, and ATP were purchased from SIGMA, St. Louis, MO. ^{32}ATP was purchased from Amersham, Orange, NJ.

Membrane Potentials

Plasma membrane potentials were measured with DIOC5 dye using FACS by a technique previously described.[14] Lymphocytes were stimulated with 50 ng/ml each of anti-CD3 and WT31 monoclonal antibodies and 10 μg/ml of PHA and plasma membrane potentials measured. Data are expressed as mean log channel numbers. Carbocyanine dyes have an inherent problem of partioning between plasma and mitochondrial membranes. However, in lymphocytes there is a very small component of mitochondrial potentials and a very brief exposure to the dye makes it unlikely that we are measuring any significant contribution of mitochondrial membrane potentials.

Free Intracellular Calcium

T cells were loaded with quin-2 AM dye and stimulated with anti-CD3 (100 ng/ml) and PHA (10 μg/ml) and changes in $[Ca^{++}]_i$ were measured using spectrofluorimeter by a technique described by Tsien et al.[15] Data are expressed in nM.

Protein Kinase C

Purified T cells from young and aging subjects were stimulated with 50 ng/ml of anti-CD3 and WT31, and PKC activity was measured in both the cytosol and the membrane fractions by enzyme assay.[16] Data are expressed as pmol/min/mg.

RESULTS

Membrane Potentials

Results of 3 separate experiments are shown in TABLE 1. Plasma membrane potentials were decreased (depolarized) in aging T cells as compared to young subject. Furthermore, the activation with anti-CD3 and WT31 monoclonal antibodies produced depolarization of plasma membrane of young T cells; in contrast, either no response or hyperpolarization of aged T cells was observed upon stimulation with anti-CD3 and WT31 monoclonal antibodies.

Free Intracellular Calcium

A representative experiment is shown in FIGURE 1. The basal levels of $[Ca^{++}]_i$ in aging T cells were lower as compared to young subjects. Anti-CD3 and PHA induced less increase in $[Ca^{++}]_i$ in aging T cells as compared to young T cells.

TABLE 1. Plasma Membrane Potentials in T Cells of Aging and Young Humans

Exp. #	Mean Fluorescence Channel Numbers (Log)		
	Resting T Cells	WT31	Anti-CD3
1. Aging	18.4	17.4	20.4
Young	34.2	15.4	19.3
2. Aging	20.2	26.2	22.4
Young	36.3	18.6	21.5
3. Aging	26.4	21.4	24.2
Young	31.6	24.5	21.2

Protein Kinase C

Purified T cells were stimulated with anti-CD3 and WT31 monoclonal antibodies, and PKC activity was measured after 10 minutes in both cytosol and membrane fractions. Data in TABLE 2 show that in young subjects 99% of cytosolic PKC was lost and translocated to the plasma membrane following activation with anti-CD3 and WT31 monoclonal antibodies, whereas only 60–70% of cytosolic PKC was translocated to the plasma membrane in aging T cells.

DISCUSSION

In this study we have shown defects in early events of signal transduction in T cells from aged humans. These defects could be the earliest changes during senescence of the immune system.

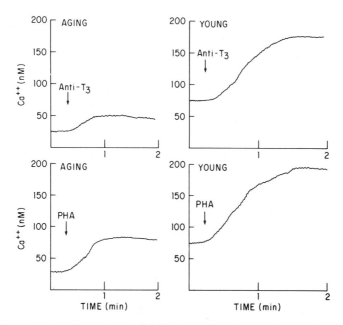

FIGURE 1. Intracellular free calcium in T cells from aging and young human subjects following activation with PHA and anti-CD3 (T3) monoclonal antibody using quin-2 AM dye.

The changes in the late steps of T cell activation in aging are well documented. The most prominent changes are the involution of thymus and decline in the levels of thymic hormone and cell-mediated immune functions[8–13] leading to increased incidence of autoimmunity, infection and cancer.[8,17] Schwab et al.[13] reported decreased T cell proliferative response to anti-CD3 monoclonal antibody in aging humans. Since anti-CD3 induces changes in "second messengers," the $[Ca^{++}]_i$ and PKC, our observations of the defects in both PKC and $[Ca^{++}]_i$ would explain poor T cell responses in aging observed by Schwab et al.[13] Miller[18] has demonstrated a reconstitution of aging T cell response to concanavalin A (Con A) with phorbol ester combined with calcium ionophore. Kennes et al.[19] have shown a defect of PHA-induced Ca^{++}-dependent pathway in aging. Proust et al.[20] have shown defects in Con A-induced translocation of PKC, and in the net increase of inositol triphosphate and $[Ca^{++}]_i$ in T cells from aging mice. Witkowski et al.[21] have linked the abnormality of cation transport across the T lymphocyte membrane to the immune dysfunctions of aged humans and mice. They demonstrated rapid decrease in transport activity of the lymphocyte membrane Na^+-K^+-ATPase in human T cells and decrease in the number of Na^+-K^+-ATPase molecules in T cells from aging mice. During activation of normal human and murine T cells with mitogens and antigens the plasma membrane is depolarized.[22–24] Witowski et al.[25] showed decreased resting membrane potentials in aged animals and hyperpolarization of membrane in aged mice as compared to depolarization in young mice following activation with Con A. In the present study we

TABLE 2. Protein Kinase C Activity in T Cells from Aging and Young Humans

| | PKC Activity (pmol $\times$ 10³/min/mg) | | | |
| | Young | | Aging | |
Stimulus	Cytosol	Membrane	Cytosol	Membrane
None	16.0 ± 4.0	1.0 ± 0.3	13.0 ± 3.1	0.8 ± 0.2
Anti-CD3 (50 ng/ml)	0.6 ± 0.2	14.4 ± 3.8	3.2 ± 0.5	9.4 ± 3.2
WT31 (50 ng/ml)	0.7 ± 0.3	14.1 ± 4.8	4.2 ± 0.3	10.1 ± 2.9

have also observed similar abnormalities of T lymphocyte plasma membrane in aging humans following activation with anti-CD3 and WT31 monoclonal antibodies and PHA. What is responsible for the changes in plasma membrane potentials in aged T cells remains to be elucidated. Membrane ion channels play an important role in the homeostasis of membrane potentials. It is then possible that there are abnormalities of membrane ion channels in aged T cells. Aging in many respects resembles autoimmune diseases and we have reported ion channel abnormality in autoimmune prone mice.[26] Rabinovitch et al.[27] have shown a differential rise in $[Ca^{++}]_i$ in CD^{4+} and CD^{8+} T cells following activation with anti-CD3 monoclonal antibody. It would be of interest to examine whether the defect in $[Ca^{++}]_i$ in aging is differentially partitioned between CD^{4+} and CD^{8+} cells. The role of serum factors (including anti-lymphocyte antibodies present in aging) in signal transductional defects remains to be explored.

SUMMARY

The interaction of the T cell receptor complex with the ligand is associated with early molecular events involved in the process of signal transduction. These events include, changes in membrane potentials, intracellular free calcium $[(Ca^{++})_i]$, and the activation and translocation of protein kinase C (PKC). The aim of this study was to elucidate the abnormalities in membrane signal transduction pathway as the basis of T cell-mediated immune deficiency associated with aging. Peripheral blood mononuclear cells from aging humans and sex-matched young subjects were stimulated with anti-CD3 monoclonal antibody, WT31 monoclonal antibody (defines epitopes for alpha/beta T cell receptor genes), and phytohemagglutinin (PHA), and examined for the changes in plasma membrane potentials, changes in $(Ca^{++})_i$, and the activation of PKC. The membrane potentials were measured with DiOC5 dye using FACS. Intracellular free calcium was measured with quin-2 dye and spectrofluorimeter, and the membrane and cytosolic PKC were measured by an enzyme assay. The resting membrane potentials in aging T cells were decreased (plasma membrane depolarized) when compared to T cells from young subjects. In T cells from aging humans, there was a lack of a change in the membrane potentials or membrane potentials were increased (plasma membrane hyperpolarized) following activation with anti-CD3 and WT31 monoclonal antibodies, whereas in young T cells membrane potentials were decreased (depolarized). The basal $[Ca^{++}]_i$ levels in aging T cells were less than that in young T cells, and a much smaller rise in $[Ca^{++}]_i$ was observed following activation with PHA and anti-CD3 monoclonal antibody in aging T cells than in the young T cells. In aging cells there was less translocation of PKC from cytosol to the membrane following activation with anti-CD3 and WT31. This study demonstrates that the abnormal membrane signal transduction pathway plays a role in T cell dysfunctions associated with human aging.

REFERENCES

1. MEUER, S., O. ACUTO, T. HERCEND, S. SCHLOSSMAN & E. REINHERZ. 1984. The human T cell receptor. Annu. Rev. Immunol. **2:** 23–50.
2. SPITS, H., J. BORST, W. TAX, P. J. A. CAPEL, C. TERHORST & J. E. DEVRIES. 1985. Characterization of a monoclonal antibody (WT31) that recognizes a common epitope on the human T cell receptor for antigen. J. Immunol. **135:** 1922–1928.
3. LANIER, L. L. & A. WEISS. 1986. Presence of Ti (WT31) negative T lymphocytes in normal blood and thymus. Nature **324:** 268–270.
4. IMBODEN, J. B. & J. D. STOBO. 1985. Transmembrane signalling by the T cell antigen receptor. J. Exp. Med. **161:** 446–456.
5. NEL, A. E., P. BOWIC, G. R. LATTANZE, H. C. STEVENSON, P. MILLER, W. DIRIENGO, G. F. STEFANINI & R. M. GALBRAITH. 1987. Reaction of T lymphocytes with anti-CD3 induces translocation of C-kinase activity in the membrane and specific substrate phosphorylation. J. Immunol. **138:** 3519–3524.
6. ALCOVER, A., M. J. WEISS, J. F. DALEY & E. L. REINHERZ. 1986. The T 11 glycoprotein is functionally linked to a calcium channel in precursor and mature T lineage cells. Proc. Natl. Acad. Sci. USA **83:** 2614–2618.
7. GUPTA, S., M. SHIMIZU, R. BATRA & B. VAYUVEGULA. 1989. Early and late changes in human T cell activation by WT31 MoAb. *In* Mechanisms of Lymphocyte Activation and Immune Regulation II. S. Gupta, & W. E. Paul, Eds. 35–43. Plenum Press. New York, NY.
8. MAKINODAN, T. & M. M. B. KAY. 1980. Age influence on the immune system. Adv. Immunol. **29:** 287–330.
9. ZATZ, M. M. & A. L. GOLDSTEIN. 1985. Thymosins, lymphokines, and the immunology of aging. Gerontology **31:** 263–277.
10. GUPTA, S. & R. A. GOOD. 1979. Subpopulations of human T lymphocytes. X. Alterations in T,B, third population cells and T cells with receptors for IgM (T_M) or G (T_G) in aging humans. J. Immunol. **122:** 1214–1219.
11. GUPTA, S. 1984. Autologous mixed lymphocyte reaction in man. IX. Autologous mixed lymphocyte reaction and lymphocyte subsets in aging humans. Scand. J. Immunol. **19:** 187–191.
12. VAYUVEGULA, B., M. SHIMIZU & S. Gupta. 1987. Autologous mixed lymphocyte reaction in

man. XX. The cellular and molecular basis of deficient T-T AMLR in aging humans. Clin. Immunol. Immunopathol. **44:** 364–370.

13. SCHWAB, R., P. B. HAUSMAN, E. RINOOY-KAN & M. E. WEKSLER. 1985. Immunologic studies of aging. X. Impaired T lymphocytes and normal monocyte response from elderly humans to the mitogenic antibodies OKT3 and Leu 4. Immunology **55:** 677–684.

14. VAYUVEGULA, B., L. SLATER, J. MEADER & S. GUPTA. 1988. Correction of altered plasma membrane potentials. A possible mechanism of cyclosporin A and verapamil reversal of pleiotropic drug resistance in neoplasia. Cancer Chemother. Pharmacol. **22:** 163–168.

15. TSIEN, R. Y., T. POZZAN & R. T. RINK. 1982. Calcium homeostasis in intact lymphocytes: Cytoplasmic free calcium monitored with a new intracellularly trapped fluorescent dye. J. Cell Biol. **94:** 325–334.

16. FARRAR, W. L. & F. W. RUSCETTI. 1986. Association of protein kinase C activation with IL-2 receptor expression. J. Immunol. **136:** 1266–1273.

17. MAKINODAN, T., S. J. JAMES, T. INAMIZU & M-P. CHANG. 1984. Immunologic basis for susceptibility to infection in the aged. Gerontology **30:** 279–289.

18. MILLER, R. A. 1986. Immunodeficiency of aging: Restorative effects of phorbol ester combined with calcium ionophore. J. Immunol. **137:** 805–808.

19. KENNES, B., C. HUBERT, D. BROHEE & P. NEVE. 1981. Early biochemical events associated with lymphocytes activation in aging. I. Evidence that Ca^{2+}-dependent process induced by PHA are impaired. Immunology **42:** 119–126.

20. PROUST, J. J., C. R. FILBURN & S. A. HARRISON. 1987. Age-related defect in signal transduction during lectin activation of murine T lymphocytes. J. Immunol. **139:** 1472–1478.

21. WITKOWSKI, J. M., A. MYSLIWSKI & J. MYSLIWSKA. 1985. Decrease of lymphocyte (Na^+, K^+) ATP-ase activity in aged people. Mech. Dev. **33:** 11–17.

22. FELBER, S. M. & M. D. BRAND. 1983. Early plasma membrane potential changes during stimulation of lymphocytes by concanavalin A. Biochem. J. **210:** 885–891.

23. KIEFER, H., A. J. BLUME & H. R. KARBACK. 1980. Membrane potential changes during mitogenic stimulation of mouse spleen lymphocytes. Proc. Natl. Acad. Sci. USA **77:** 2200–2204.

24. TATHAM, P. E. R. & P. J. DELVES. 1984. Flow cytometric detection of membrane potential changes in murine lymphocytes induced by concanavalin A. Biochem. J. **221:** 137–146.

25. WITKOWSKI, J. M. & H. S. MICKLEM. 1985. Decreased membrane potential of T lymphocytes in aging mice: Flow cytometric studies with a carbocyanine dye. Immunology **56:** 307–313.

26. CHANDY, K. G., T. E. DeCOURSEY, M. FISCHBACH, N. TALAL, M. D. CAHALAN & S. GUPTA. 1986. Altered K^+ channel expression in abnormal T lymphocytes from mice with lpr gene mutation. Science **233:** 1197–1200.

27. RABINOVITCH, P. S., C. H. JUNE, A. GROSSMANN & J. A. LEDBETTER. 1986. Heterogeneity among T cells in intracellular calcium responses after mitogen stimulation with PHA or anti-CD3. Simultaneous use of Indo-1 and immunoflourescence with flow cytometry. J. Immunol. **137:** 952–961.

Calcium and Proto-Oncogene Involvement in the Immediate-Early Response in the Nervous System

JAMES I. MORGAN AND TOM CURRAN[a]

Departments of Neuroscience and [a]Molecular Oncology
Roche Institute of Molecular Biology
Roche Research Center
Nutley, New Jersey 07110

BACKGROUND

Stimulation of a neuron results in two general types of responses. First, there are rapid, transcription-independent, events that are elicited by second messenger-mediated alterations in the posttranslational state of existing substrates. These rapid responses would include such mechanisms as cyclic nucleotide modulation of the phosphorylation of ion channels, receptors, or contractile and cytoskeletal proteins. The second type of response occurs over a longer time-frame and has a requirement for new protein synthesis. In brain, such transcription-dependent processes may be involved in adaptation, plasticity, or long-term modification of the output of a neuronal circuit as might occur, for instance, in learning and memory.[1,2] The basic question in this latter type of mechanism is how stimulation of the neuron results in alterations in gene transcription. Recent evidence indicates that at least part of the answer to this question lies in the coupling of stimulants, via second messengers, to a group of inducible genes that encode proteins that can participate in transcription factor AP-1 complexes; namely c-jun, c-fos and several genes encoding Fos-related antigens.[3,4] It is these proteins that can then either regulate the levels of ongoing transcription of some genes or perhaps participate in the induction of others. In this context the rapidly inducible genes, often referred to as immediate-early response genes,[2,5] should be viewed as third messengers coupling cell stimulation to long-term alterations in transcription rates in an as yet poorly defined stimulus-transcription coupling cascade. The basic outline of this intracellular coupling mechanism is depicted in FIGURE 1.

The Role of Calcium in the Immediate-Early Response in Neurons

In the PC12 pheochromocytoma, membrane depolarization elicited either by elevation of the extracellular potassium concentration or treatment with veratridine results in a calcium-dependent induction of c-fos.[3] Since this induction is blocked by dihydropyridine antagonists of the so-called L-type voltage-sensitive calcium channel it is believed that this class of channel gates the calcium that activates fos transcription. This conclusion is supported by the finding that an agonist of the L-channel, BAY K 8644, induces fos in a calcium-dependent manner.[3] Further pharmacological analysis indicates that calmodulin antagonists such as trifluoperazine, chlorpromazine, and W7 all block or attenuate depolarization-induced fos expression;[3] indicating that the calmodulin/calmodulin kinase system may represent an element of the intracellular pathway linking the calcium influx to transcriptional activation.

283

A major question yet to be answered is how calcium or calmodulin bring about alterations in gene transcription. Obviously one possibility is that calmodulin kinase phosphorylates one or more proteins that directly interact with transcription regulatory elements of c-fos resulting in an induction of the gene. Recently a DNA sequence located 60 nucleotides upstream of the transcription initiation site of c-fos was shown to be essential for induction by calcium-dependent agents.[6,7] In contrast, a number of calcium-independent inducers of fos, such as polypeptide growth factors and phorbol esters, do not require this element but rather utilize a distinct DNA sequence (often referred to as the serum response element) located some 300 nucleotides upstream of the start site.[8] The

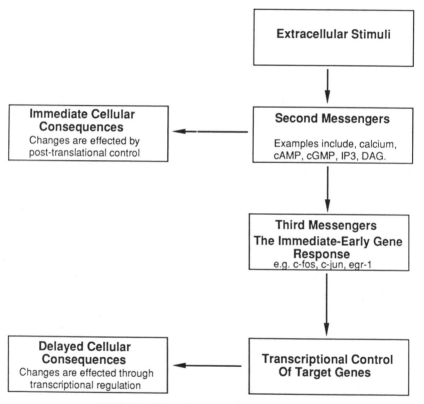

FIGURE 1. The cellular immediate-early response.

serum response element is known to have proteins associated with it that are candidates for transcriptional regulators;[9] however, no such proteins have yet been identified that interact with the calcium-dependent site.

Induction of fos and jun in the Central Nervous System

The demonstration of fos induction in a cultured cell line was extrapolated to neurons *in vivo* using the pentylenetetrazole (PTZ) seizure paradigm. Analogous to the induction of fos by membrane depolarization in PC12 cells, increased neuronal activity *in vivo, as* elicited by the convulsant PTZ, also results in the rapid but transient expression of fos.[4] The c-fos protein (Fos) is localized exclusively to the nuclei of neurons and shows a stereotyped pattern of expression in brain.[4] Thus certain brain areas show a rapid and intense staining for Fos (*e.g.*, pyriform cortex and dentate gyrus of the hippocampus), other regions appear Fos-positive after a slight delay (*e.g.*, cerebral cortex and the remainder of hippocampus), while some areas never contain detectable Fos (*e.g.*, cerebellum and brain stem). Subsequent studies have established that Fos expression can be triggered in brain by both electrical and physiological stimulation,[10,11] which, along with the presence of Fos-positive neurons in untreated mouse brain,[4] would suggest that the induction of this proto-oncogene is not in itself an indication of an insult to the neuron. Rather it would suggest that Fos fulfills a role in the normal economy of the neuron *in vivo* but that hyperexcitation results in higher levels of Fos expression and the recruitment of a greater number of neurons to express the proto-oncogene.

In a parallel study we have investigated the levels of expression of c-jun mRNA in mouse brain following PTZ-induced seizures, since this proto-oncogene encodes a protein (Jun) that also participates in transcription factor AP-1 complexes.[12] Like c-fos, the mRNA for c-jun increases soon after seizure (see FIG. 2). There are, however, qualitative differences between the expression patterns of the two transcripts. First, c-jun has, relative to c-fos, a higher level of expression in unstimulated brain. Second, while c-jun mRNA levels rise rapidly following seizure and then begin to decline they subsequently level off at values markedly above basal. These elevated values have been observed to persist throughout the period of study (about 8 hours). These results demonstrate that while c-fos and c-jun may be induced by identical stimuli the subsequent regulation of their transcription is different. We also infer that this situation results in the presence of elevated levels of Jun for at least 8 hours post seizure.

The results obtained with c-fos and c-jun point to there being a stringent and complex regulation of the immediate-early genes at the transcriptional level. Indeed, we have established that c-fos becomes refractory to reinduction by PTZ seizure.[4] This observation was stimulated by the finding that, after their characteristic increase, c-fos mRNA levels actually fall to below original basal values following seizure. Furthermore, these low levels persist for at least 4 hours. It is during this period of subbasal expression of c-fos that the refractoriness to reinduction is most pronounced.[4] We conclude, therefore, that the level of Fos permitted in a neuron is strictly proscribed and that it is subject to feedback regulation.

Transcription Factor AP-1 Levels and Composition Following Seizure

While the PTZ-seizure model is nonphysiological it does serve to amplify the responses seen with normal stimulation to an extent that it becomes technically feasible to perform a detailed analysis of the biochemical and molecular sequelae of neuronal excitation. Thus using this approach it is possible to perform immunoblot and gel shift assays on brain nuclear extracts to determine the levels and composition of the transcription factor AP-1 complex before and after seizure.

Transcription factor AP-1 (AP-1) was first described as an activity in Hela cell extracts that could support transcription from certain genes using *in vitro* assays.[13] Subsequently, AP-1 was shown to interact and bind to a consensus DNA sequence found in many genes that is essential for transcription.[14,15] Recently it has been recognized that AP-1 is a nucleoprotein complex comprised of the product of the jun proto-oncogene (Jun), Fos, and several Fos-related proteins;[16] (see FIG. 2). Since Fos and Jun are seizure-inducible components of AP-1 we determined whether total AP-1 levels also increased following seizure using a gel shift assay. In parallel with the increase in Fos, total AP-1 levels rose as determined by gel shift against an oligonucleotide containing the AP-1 consensus sequence. In whole brain this increase in AP-1 persisted for 6 to 8 hours, although c-fos mRNA had long since returned to basal values. To determine the relative contribution of Fos to these AP-1 complexes, two assays were performed. First, prior treatment of the nuclear extract with an anti-Fos antibody eliminated essentially all of the gel shift activity

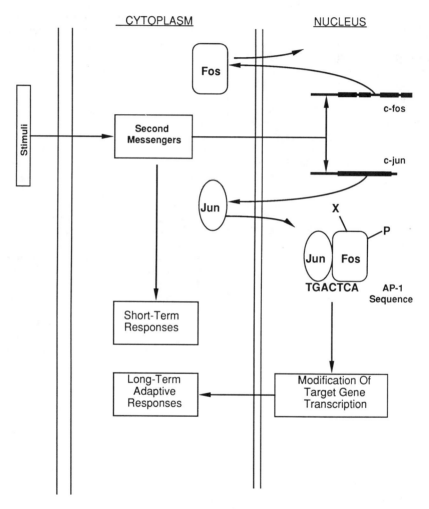

FIGURE 2.

against the AP-1 sequence, indicating that Fos or proteins reactive with the antiserum participated in the complex in some manner. Second, an immunoblot analysis was performed on the nuclear extracts using the same anti-Fos antiserum. In untreated mouse brain there was little or no Fos detectable, commensurate with the low level of resident AP-1 activity. However, two additional proteins were evident that reacted with the antiserum and were designated Fos-related antigens 46kDa (Fra-46K), and 35kDa (Fra-35K) respectively. Given the results of the antibody competition it is inferred that these proteins must participate in AP-1 complexes. This notion is solidified by the situation obtained following seizure. In the first 1 to 2 hours post-seizure Fos levels increase dramatically but then subside to near basal values by 4 hours. The two Fra, however, only begin to increase 2 hours after stimulation and their elevated levels persist for 6 to 8 hours (dependent upon the brain region studied). Indeed, the level of the Fra-46K increases before that of the Fra-35K. This clearly demonstrates that both Fra are components of AP-1, and that they may participate in the complex even in the complete absence of Fos. Furthermore, the Fra-containing (Fos-free) AP-1 complex is perfectly competent to bind to its target sequence (as determined by the gel shift assay).

We have determined that these relatively long-term changes in AP-1 levels and composition are all triggered by a brief period of neuronal stimulation. This can be shown by blocking convulsions with benzodiazepines a few minutes after their onset in the PTZ model or by giving a single one second electrical stimulation in the maximal electroshock paradigm. That is, once set in train the genetic sequelae proceed in the absence of further stimulation. One other point that should be made pertains to the refractoriness of c-fos to reinduction by PTZ at later times when fos mRNA has fallen to sub-basal levels. As noted above, refractoriness is greatest at four hours post-treatment. At this time, there is no fos mRNA and little or no Fos present in brain. It has been suggested that Fos is a negative autoregulator of its own transcription.[17,18] While this may be so, it does not account for the refractoriness observed in brain at the later times since no Fos is present. Rather, the appearance of the Fra correlate with the unresponsiveness of fos. Thus the Fra-Jun complexes may be negative regulators of c-fos.

Of key significance is the notion that the composition (as well as the level) of AP-1 undergoes dynamic changes following neuronal excitation because of the staggered appearance and disappearance of Fos and the Fra. This raises the issue of the precise targets and functions of these various AP-1 complexes. It is known that the AP-1 consensus sequence has a number of variations;[13,14,19] thus it is possible that the different AP-1 complexes could differentially interact with these. As these variant sequences are present in different genes it would provide a mechanism whereby a stimulus could induce specific groups of genes in a staggered manner. Another possibility is that the biological activity of the AP-1 complex is dependent upon its composition. That is, while the presence of the Fra in the complex does not seem to perturb DNA binding they may alter the transcriptional activity of AP-1. *In vitro* assays have shown that Fos may variously transactivate marker genes in a positive, negative, or neutral manner.[13,14,19] Its precise action may be dependent upon the specific AP-1 regulatory element under investigation and its position relative to other promoter and enhancer sequences. However, we would speculate that the net effect of AP-1 is also a function of its composition. The precise biochemical details of how this interaction between Jun and either Fos or one of the Fra brings about changes in transcription rates await to be elucidated. However, we have now provided the first biochemical framework upon which these pathways can be constructed.

PROSPECTS

The above findings leave us with a number of considerations regarding calcium, aging, and Alzheimer's disease. We have shown that there exists an intracellular cascade, involving calcium, that links neuronal stimulation to rapid alterations in gene expression. This coupling also includes the participation of specific DNA sequences within defined genes. Although there is absolutely no evidence to point to an involvement of the immediate-early genes in aging and neurodegenerative conditions we will take this opportunity to briefly outline some possible levels of interactions. While this represents some egregious speculation on our part and some outlandish suggestions, it may serve to focus some attention on potentially interesting and novel areas of research pertaining to the possible molecular deficits or correlates of Alzheimer's disease.

(1) The formal possibility exists that somatic mutation may occur in regulatory elements of genes that cause them to alter their responsiveness to calcium (or indeed other inducers). It is unlikely that a mutation in one or at most a few neurons would be detrimental to an organism. However, during development, neurons of the central nervous system are generated on a clonal basis. Thus a somatic mutation arising early in development in a neuronal precursor would be amplified by clonal expansion. Such a defective clone might form the focus of a degenerative process in later life.

(2) Some viruses are known to have usurped eukaryotic growth regulatory genes (such as fos, src, ras, etc.), certain of which (*e.g.*, fos and jun) may bind to specific transcriptional regulatory sequences. Thus there is always the possibility that other viruses may encode proteins that interact with, inter alia, DNA elements conferring calcium sensitivity. Such pathogens need not be retroviruses but rather any virus that can establish a persistent infection of a neuron. At the molecular level, we would speculate that in Alzheimer's disease a virus exists that produces as part of its repertoire a protein that interacts with one of the activating sequences present in fos or jun (or any number of other immediate-early genes). This would lead to the inappropriate expression of the cascade which we would argue over the long-term is detrimental to the neuron. This scenario has a number of variations. For instance, the viral product could repress the immediate-early cascade rather than induce it. Alternatively, the protein might act directly upon the target AP-1 sequences thus mimicking the presence of Fos-Jun.

(3) An analogous argument can be constructed for environmental factors impinging upon the immediate-early cascade. Thus it is conceivable that agents ingested during the life of the individual could influence any of the multitude of targets in the immediate-early cascade. Since this pathway may be involved in bringing about changes in the long-term behavior of neurons via alterations in gene transcription it is plausible that perturbing gene expression over many years could lead eventually to irreversible neuronal damage.

(4) Aging in neurons may be accompanied by alterations in levels or responsiveness of molecules involved in the coupling process. For instance, an old neuron, or one in the early phase of Alzheimer's degeneration, may not regulate its intracellular calcium so rigorously. This would result in changes in the responsiveness of the immediate-early genes. For instance, the reinduction of c-fos in brain is refractory and apparently under stringent regulation, possibly by several fos-related genes. Thus we would conclude that the absolute amount of Fos permitted in a neuron is sharply proscribed. By disrupting the stringent regulation of fos in the brain it may be that neuropathological changes could be set in train. The analog of this situation would be the disruption of correct bone formation by the overexpression of Fos in neonates. Clearly this argument is not confined to fos. Indeed, one could argue that a shift in the entire immediate-early cascade could result from limited alterations in the ability of a neuron to manage its intracellular calcium. The fundamental problem in addressing these possibilities is our lack of knowledge of the number of genes represented in the immediate-early category, the repertoire of these that can be expressed in neurons, and most important their precise function(s).

(5) We have demonstrated that enhanced neuronal activity produces a stereotyped induction of fos in specific brain regions. Particularly responsive areas include the hippocampus, amygdala, and pyriform cortex in the mouse. Subsequently, there is a recruitment of other neuronal populations for fos expression. In contrast, it is our impression that neurons with constitutively high firing rates do not induce for c-fos, suggesting that they have mechanisms that either repress the proto-oncogene or alternatively that maintain

cytoplasmic calcium (or other signaling molecules) at relatively low levels even during times of enhanced firing. Therefore, it could be argued that the depletion of some sets of particularly susceptible neurons in the initial phases of Alzheimer's disease may alter basic connections and thereby change the physiological properties and susceptibility of further sets of neurons to induction of c-fos. While this would not represent the cause of the disease it is a mechanism that could lead to its progressive nature.

SUMMARY

Depolarization of neurons either in culture or *in vivo* results in the rapid, calcium-dependent induction of several, so-called, immediate-early genes; the prototypes being c-fos and c-jun. The proteins encoded by c-jun, c-fos, and several fos-related genes all participate in a complex that interacts with the AP-1 consensus DNA sequence, previously shown to be important for the "transcriptional activation" of certain genes. Thus it is proposed that neuronal stimulation, via elevated intracellular calcium, leads to the induction of a series of genes, some of which encode proteins involved in transcriptional regulation, that contribute to long-term adaptive and plastic responses. Surprisingly, the molecular composition of the brain AP-1 binding complex varies with time after stimulation. This is because some of the inducible Fos-related proteins accumulate with much slower kinetics than Fos itself and only appear in significant amounts when Fos has disappeared. Of some considerable interest is the result these compositional alterations have upon the transcriptional activity of the AP-1 complex. Given the foregoing findings we consider some of the possible implications this might have for aging and neurodegenerative disorders particularly with regard to alterations in cellular calcium homeostasis.

REFERENCES

1. GOELET, P., V. F. CASTELLUCCI, S. SCHACHER & E. R. KANDEL. 1986. Nature **322:** 419–422.
2. CURRAN, T. & J. I. MORGAN. 1987. Memories of fos. BioEssays **7:** 255–258.
3. MORGAN, J. I. & T. CURRAN. 1986. Role of ion flux in the control of c-fos expression. Nature **322:** 552–555.
4. MORGAN, J. I., D. R. COHEN, J. L. HEMPSTEAD & T. CURRAN. 1987. Mapping patterns of c-fos expression in the central nervous system after seizure. Science **237:** 192–197.
5. LAU, L. F. & D. NATHANS. 1987. Expression of a set of growth-related immediate early genes in BALB/c 3T3 cells: Coordinate regulation with c-fos or c-myc. Proc. Natl. Acad. Sci. USA **84:** 1182–1186.
6. GILMAN, M. Z. 1988. The c-fos serum response element responds to protein kinase C-dependent and independent signals but not to cyclic AMP. Genes Dev. **2:** 394–402.
7. SHENG, M., S. T. DOUGAN, G. MCFADDEN & M. E. GREENBERG. 1988. Calcium and growth factor pathways of c-fos transcriptional activation require distinct upstream regulatory sequences. Mol. Cell Biol. **8:** 2787–2796.
8. TREISMAN, R. 1985. Transient accumulation of c-fos RNA following serum stimulation required a conserved 5' element and c-fos 3' sequences. Cell **42:** 889–902.
9. TRIESMAN, R. 1987. Identification and purification of a polypeptide that binds to the c-fos serum response element. EMBO J. **6:** 2711–2717.
10. HUNT, S., A. PINI & G. EVAN. 1987. Induction of c-fos-like protein in spinal cord neurons following sensory stimulation. Nature **328:** 632–634.
11. SAGAR, S. M., F. R. SHARP & T. CURRAN. 1988. Expression of c-fos protein in brain: Metabolic mapping at the cellular level. Science **240:** 1328–1331.
12. BOHMANN, D., T. J. BOS, A. ADMON, T. NISHIMURA, P. K. VOGT & R. TJIAN. 1987. Human

proto-oncogene c-jun encodes a DNA binding protein with structural and functional properties of transcription factor AP-1. Science **238:** 1386–1392.

13. LEE, W., A. HASLINGER, M. KARIN & R. TJIAN. 1987. Activation of transcription by two factors that bind promoter and enhancer sequences of the human metallothionein gene and SV40. Nature **325:** 368–372.

14. ANGEL, P., M. IMAGAWA, R. CHUI, B. STAIN, R. J. IMBRA, H. J. RAHMSDORF, C. JONAT, P. HERRLICH & M. KARIN. 1987. Phorbol ester-inducible genes contain a common cis element recognized by a TPA-modulated trans-acting factor. Cell **49:** 729–739.

15. LEE, W., P. MITCHELL & R. TJIAN. 1987. Phorbol ester-inducible genes contain a common cis element recognized by a TPA-modulated trans-acting factor. Cell **49:** 741–752.

16. RAUSCHER, F. J., III., D. R. COHEN, T. CURRAN, T. J. BOS, P. K. VOGT, D. BOHMANN, R. TJIAN & B. R. FRANZA, JR. 1988. Fos-associated protein p39 is the product of the jun proto-oncogene. Science **240:** 1010–1016.

17. SCHONTHAL, A., P. HERRLICH, H. J. RAHMSDORF & H. PONTA. 1988. Requirement for fos gene expression in the transcriptional activation of collagenase by other oncogenes and phorbol esters. Cell **54:** 325–334.

18. SASSONE-CORSI, P., J. SISSON & I. M. VERMA. 1988. Transcriptional autoregulation of the proto-oncogene fos. Nature **334:** 314–319.

19. FRANZA, B. R., JR., F. J. RAUSCHER III, S. F. JOSEPHS & T. CURRAN. 1988. The Fos complex and Fos-related antigens recognize sequence elements that contain AP-1 binding sites. Science **239:** 1150–1153.

Index of Contributors